Commission of the European Communities

environment and quality of life

Climate and global change

Proceedings of the European School
of Climatology and Natural Hazards course,
held in Arles/Rhône, France, from 4 to 12 April 1990

Edited by :
J. C. Duplessy, A. Pons, R. Fantechi
Commission of the European Communities
200, rue de la Loi
B-1049 Brussels

Directorate-General
Science, Research and Development

1991

EUR 13149 EN

Published by the
COMMISSION OF THE EUROPEAN COMMUNITIES
Directorate-General
Telecommunications, Information Industries and Innovation

L-2920 Luxembourg

LEGAL NOTICE

Cataloguing data can be found at the end of this publication

Luxembourg: Office for Official Publications of the European Communities, 1991

ISBN 92-826-2779-9   Catalogue number: CD-NA-13149-EN-C

Printed in France

Table of Contents

IV

EUROPEAN SCHOOL OF CLIMATOLOGY AND NATURAL HAZARDS

Foreword

The training of new scientific and technical staff and the development of highly qualified scientific specialists are and always have been among the prime concerns of the Commission of the European Communities, which regards this task as a major requirement for the implementation of a common policy in science and technology.

The European School of Climatology and Natural Hazards is a part of the training and education activities of EPOCH (European Programme on Climatology and Natural Hazards).

It annually organizes courses open to graduating, graduate or post graduate students in appropriate fields of Climatology, Natural Hazards and closely related fields.

The courses are organized in cooperation with European institutions involved in the carrying out of the Community's R&D programmes on Climatology and Natural Hazards, and are aimed at allowing students to attend formal lectures and to participate in informal discussions with leading research workers. The opportunity for demonstrations, case studies or presentation of posters is given to the students attending the courses.

The teachers are selected among European Scientists who are leading authorities in their respective fields.

The present volume contains the lessons delivered at the course held in Arles/Rhône, France, from 4th to 12th April 1990 on the subject **"Climate and Global Change"**, together with short presentations from the students of their own research activities and interests.

It is hoped to constitute a valuable permanent record of the course, and to stimulate further interest on the part of teachers and students alike.

R. FANTECHI
Head of Service

List of Teachers

Prof. A. Berger	Université Catholique de Louvain, Institut d'Astronomie et de Géophysique Georges Lemaître, Louvain-la-Neuve, B
Dr. J.C. Duplessy	C.N.R.S., Centre des Faibles Radioactivités, Laboratoire Mixte CNRS-CEA, Gif/Yvette, F
Dipl. Met. M. Eckardt	Institut für Meteorologie, Freie Universität Berlin, Berlin, D
Prof. R. Frassetto	CNR — ISDGM, National Research Council Venice, Venice, I
Dr. A.C. Imeson	Fysisch Geografisch Bodemkunding Laboratorium, Landscape and Environmental Research Group, University of Amsterdam, Amsterdam, NL
Dr. P.S. Liss	University of East Anglia, School of Environmental Sciences, Norwich, UK
Dr. E. Maier-Reimer	Max-Planck-Institut für Meteorologie, Hamburg, D
Dr. J.F.B. Mitchell	Meteorological Office (Met. 020), Bracknell, Berkshire, UK
Prof. H. Oeschger	Physics Institute, University of Bern, Bern, CH
Prof. Dr. J. Pinto Peixoto	Instituto Geofísico do Infante D. Luís, Faculdade de Ciências, Universidade de Lisboa, Lisboa, P
Prof. A. Pons	Faculté des Sciences et Techniques de St. Jérôme, Université de Droit, d'Economie et des Sciences d'Aix-Marseille, Laboratoire de Botanique Historique et Palynologie, Marseille, F
Prof. S. Rivas-Martinez	Universidad Complutense, Facultad de Farmacia, Departamento de Biologia de las Plantas, II, Madrid, E
Dr. J. Rudolph	Institut für Atmosphärische Chemie, Jülich, D
Dr. C.J.E. Schuurmans	IMOU, Department of Physics, University of Utrecht, Utrecht, NL

List of Students

to which the Commission of the European Communities
has a grant for participation

ADAMO Paola	I	HOVINE Stéphane	B
ALONSO SARRIA Francisco	E	JOLLY Dominique	F
AMORUSO Antonella	I	KAKALIAGOU Olga	GR
ANDRE Jean—Frédéric	B	KING Stephen	UK
BARTZOKAS Aristides	GR	KWADIJK Jaap	NL
BOER Matthias	NL	LEAN Jennifer	UK
CONNOLLEY William	UK	LORENZETTI Maria Chiara	I
ESTEBAN PARRA Maria Jesus	E	MELETI Chariklea	GR
FLOCAS Helena	GR	MICHEL Elisabeth	F
FRIEDLINGSTEIN Pierre	B	MOTA Ana Maria	P
GARVEY Liam	IRE	PEÑALBA Maria Cristina	E
GAVILAN Rosario Gloria	E	PLASS Christian	D
HEALY Michael	IRE	SCHULTZ Hartmut	D
HEINZE Christoph	D	TOURPALI Kleareti	GR
HEWER Fiona	UK		

COMMISSION OF THE EUROPEAN COMMUNITIES

Climatology and Natural Hazards Research Programme

EUROPEAN SCHOOL OF CLIMATOLOGY AND NATURAL HAZARDS

Course on

"Climate and Global Change"

Natural Variability of the Geosphere – Biosphere System

Variability of the Earth's climate;
Variability of the continental biosphere;
Variability of the chemical composition of the atmosphere and ocean;
Modelling.

Biogeochemical Cycles and their Perturbation by Human Activities

The CO_2 cycle;
The sulphur cycle;
The nitrogen cycle;
The methane and hydrocarbon cycles;
Global ozone variability.

Monitoring and Forecasting Global Changes

Climatic evolution during the last century;
Albedo changes and satellite observation of the Earth;
Modelling global change;
The hydrological cycle;
Impact on ecosystems;
Land degradation;
Human response to global change.

EUROPEAN SCHOOL OF CLIMATOLOGY AND NATURAL HAZARDS

Course on

"Climate and Global Change"

1. NATURAL VARIABILITY OF THE GEOSPHERE — BIOSPHERE SYSTEM

VARIABILITY OF THE EARTH'S CLIMATE

J.C. DUPLESSY

Centre des Faibles Radioactivités
Laboratoire mixte CNRS-CEA
F-91198 Gif Sur Yvette Cédex

ABSTRACT

The climate of our planet has remained within a range compatible with life since several billion years. Theories of star evolution suggest that the amount of heat emitted by the sun increased by about 10% since $4 \ 10^9$ years, so that several factors of the climatic system, such as the chemical composition of the atmosphere, must have changed in order to maintain the air temperature within a range permitting the occurrence of abundant liquid water at the earth's surface.

On the 10^7 -10^8 year time scale, climatic variations are important and strongly linked to the plate tectonics and to the atmospheric CO_2 concentration changes. In particular warm conditions prevailing during the Cretaceous are best explained by the absence of continental masses in polar position and by higher amount of carbon dioxide in the atmosphere. On the 10^4 -10^5 year time scale, seasonal insolation changes which are linked to the variations of the earth's orbit around the sun, exhibit periodicities which are dominant within the spectrum of climatic variations. This so-called Milankovitch mechanism seems to be responsible for most of the Quaternary climatic variations, including the major glaciations. Although climatic variations on the 10^3 year time scale may be explained as the non-linear response of the climatic system to the Milankovitch forcing, most of the variability on the 10 -10^3 year range is still poorly understood. The variability of the cryosphere and of the ocean circulation must be taken into account and both are still poorly documented.

The last glacial-interglacial cycle (the last 150,000 years) provides the best reference frame to which mechanisms of climatic change and their impact on the atmosphere, ocean, and biosphere may be analyzed. Data can be obtained from the analysis of deep sea and continental sediments as well as from ice cores drilled through the major ice caps. Temperature estimates can be derived from micropaleontological data (pollen, foraminifera, diatoms, coccoliths) and isotopic measurements in ice and fossil marine foraminifera. Precipitation can be estimated from pollen and diatom fossils and sedimentological analysis. The chemical composition of the atmosphere can be reconstructed from the analysis of air bubbles included in ice cores and from geochemical indices. results show that during the peak of the last glaciation, the mean temperature of the Earth was about 4°C lower than today, but that the cooling was much stronger at high latitudes than in the tropics. The water cycle experienced dramatic changes, with a strong reduction of the amount of water available at the surface of the continents. Changes in ocean circulation and chemistry have been important in triggering both important changes in the atmospheric CO_2 content and abrupt climatic variations during the glacial to modern interglacial transition.

INTRODUCTION

Geologists have for long been interested in past climate, and there are many known climatic indicators in the geological record. Evidence such as moraines and other features of glacial conditions have been observed in northern Europe and northern America. However, such isolated pieces of evidence are not sufficient for understanding climatic change and its causes. For this one requires measurements of a particular climatic variable, either as a function of time at one location or as a function of geographical location at one time. For the last million years or so, this is most easily achieved by using radioactive isotopes for dating, and stable isotope techniques and transfer functions calibrating micropaleontological data in climatic terms.

In the last 25 years, a great deal of information about long-term climatic change in the last million years or more have been obtained from the study of marine and continental sediments. A reasonably consistent picture emerges, showing alternations between glacial and interglacial conditions, manifested by waning and waxing of large continental ice sheets coinciding with temperature, salinity and circulation changes in the ocean, temperature and humidity changes over the continents. In this paper, we shall describe the global evolution of the Earth's climate since the Precambrian and concentrate on the reconstruction of the last major oscillations generally referred to as the last climatic cycles, which occurred during the Quaternary. A general review on climatic variations and variability has been given by Berger (1981) and a detailed review of the last climatic cycle by

Duplessy (19 8).

CLIMATIC EVOLUTION BEFORE 65 MYR

The geographical distribution of ancient climates in a given geological period is the result of the utilisation of the solar radiation by the climatic system (atmosphere, ocean, cryosphere, surface of the continents). The atmospheric content of gas absorbing infra-red thermal radiation (CO_2, CH_4, H_2O, O_3, N_2O,....) and the transparency of the atmosphere first govern the heat budget of the Earth's surface. Then, the surface interactions depend on the geographical distribution of continents and oceans as well as phytogeographic and topographic factors characterizing the continental regions (Bernard, 1964).

The climate of our planet has remained within a range compatible with life since more than 2 000 Myr. Theories of star evolution suggest that the amount of heat emitted by the sun increased by about 10% since 4 000 Myr, so that several factors of the climatic system, such as the chemical composition of the atmosphere, must have changed in order to maintain the air temperature within a range permitting the occurrence of abundant liquid water at the earth's surface. Climatic variations can be recognized but the data available are inadequate to shed significant light on the factors associated these very ancient paleoclimatic variations.

Ancient glaciations

At least four major ice ages have been recognized within Precambrian times, about 2 300, 960, 750 and 680 Myr ago. Paleomagnetic studies suggest that glaciated areas,now recognized by striated pavements and glacial deposits (clays), were in high latitudes when their climate was cold.

The most well-known glacial deposits are in Ordovician rocks (450 Myr) in the Sahara. Evidences for striated rocks overlain by glacial deposits (tills), for permafrost have been widely found. The presence of marine sediments within some of the deposits shows that the ice was close to the sea. Glacial deposits of similar age have been recognized in South America, Newfoundland and possibly Britain.

During the Permo-carboniferous period (about 280 Myr), glacial deposits are present in all the Gondwana continents (South America, Africa, India, Australia, and Antarctica).

Younger geological sediments do not show any more evidence for glacial conditions, until the upper Cenozoic. This rapid review of past glaciations indicates that these cold climatic conditions have not been rare along the Earth's history. The presence of continents in quasi-polar position seems to be a prerequisite for glaciation, but other conditions are necessary as indicated by the occurrence of very long periods of global warm conditions, as those of the Mesozoic (245-65 Myr ago). Simulations performed with climate models suggest that high atmospheric CO2 concentrations were necessary to explain the small temperature gradient between equator and poles.

CLIMATIC EVOLUTION AFTER 65 MYR

This more recent period can be studied with much more details, thanks to cores from the ocean floor or from the bottom of lakes or peat bogs. These sediments contain material deposited at a fairly constant rate, generally in the order of a few centimeters per thousand years in the ocean and of a few centimeters per century in the continent. In order to derive a useful information on past climates, sediment must first be dated (See Roth and Potty, 1989). Then its fossil content (foraminifera, coccoliths or radiolaria in deep sea sediments, pollen or diatoms in continental deposits) are translated into quantitative estimates of those climatic variables that physical model use and simulate. This is achieved thanks to either the isotope geochemistry or micropaleontological transfer functions

Isotope geochemistry

Most elements occur in nature as a mixture of stable isotopes. These isotopes of an element differ slightly in their chemical and physical properties, as a consequence of the differences in their thermodynamic properties. Therefore, many natural processes entail slight separation of stable isotopes. For example, the evaporation and precipitation of water give rise to variations in the $^{18}O/^{16}O$ ratio, which reach an extreme in the difference between the isotopic composition of ice stored in the Antarctic ice sheet, and that of water in the ocean. Chemical processes also fractionate the isotopes, so that the $^{18}O/^{16}O$ ratio in calcite shells laid down by organisms is slightly greater than that of the water they inhabit. In this case, the fractionation factor is particularly interesting, because it is temperature dependent. This means that it may be possible to estimate past ocean temperatures by measuring the $^{18}O/^{16}O$ ratio of calcitic fossils (Emiliani, 1955; Duplessy, 1978; 1981; Shackleton, 1981).

As these isotopic variations are small, they are generally expressed as:

$$\delta^{18}O = \frac{(^{18}O/^{16}O)_{sample}}{(^{18}O/^{16}O)_{PDB}} - 1 \qquad \text{which is expressed in per mil.}$$

PDB is a reference standard.

A precise interpretation of the variations of the $^{18}O/^{16}O$ ratio of the calcitic shells of foraminifera in deep sea sediments is however difficult, because both the sea water temperature and $^{18}O/^{16}O$ ratio vary with the climate: since continental ice sheets are depleted in ^{18}O, as the amount of ice present on the continent increases during a glaciation, the water remaining in the ocean becomes richer in ^{18}O and its $^{18}O/^{16}O$ ratio increases. This isotopic enrichment is entirely reflected in the calcitic test of foraminifera.

As the temperature at which the shell formed decreases, the $^{18}O/^{16}O$ ratio of the calcite increases. Thus heavy isotopic ratios indicate cold climate. Conversely, temperature increases and ice cap melting will be reflected by light $^{18}O/^{16}O$ ratio. It has thus been necessary to estimate separately the variations of the two variables, sea water temperature and $^{18}O/^{16}O$ ratio (Labeyrie et al. 1987). Their results show that the variations of the isotopic composition of the sea water constitute 60 to 100% of the quaternary benthic records (depending on their location and depth). The $\delta^{18}O$ variations in a deep sea core can thus be used to determine unambiguously the sediment deposited during the major events of the last climatic cycle. This is the basis of the isotope stratigraphy (Shackleton and Opdyke, 1973).

Another useful isotope is ^{13}C, which is a heavy stable isotope of carbon. The $^{13}C/^{12}C$ ratio varies in nature, because photosynthesis discriminates carbon isotopes aginst the heavy one. As a consequence, the carbon isotopic composition of the CO_2 dissolved in sea water exhibits noticeable variations. Some species of benthic foraminifera (Genus *Cibicides*) closely record the modern $\delta^{13}C$ distribution in the world ocean, providing a good proxy for the reconstruction of the past $\delta^{13}C$ gradients in the ocean (Duplessy et al., 1984). Kroopnick (1985) has demonstrated that the distribution of $^{13}C/^{12}C$ ratio of the total CO_2 (ΣCO_2) dissolved in sea water delineates the general distribution of water masses and that modern $\delta^{13}C$ gradients in the deep ocean follow the net flow of the deep waters. The distribution of the $\delta^{13}C$ value of the ΣCO_2 dissolved in intermediate, deep and bottom waters of the ocean can be used as a tracer of the abyssal circulation under the following conditions: (i) within a deep water mass, the $\delta^{13}C$ value decreases with increasing oxidation of organic matter, such that the $\delta^{13}C$ decrease is related both to the surface productivity and the time elapsed since the water mass was isolated from the atmosphere; (ii) when two water masses are compared, the $\delta^{13}C$ value in each depends not only on their residence time at depth, but also on the ratio between the CO_2 produced from organic matter oxidation and the bicarbonate and carbonate ions derived from carbonate dissolution (note that this ratio is poorly controlled under past conditions); (iii) mixing between two water masses results in $\partial^{13}C$ values intermediate between the two original components (see a detailed discussion in Duplessy and Shackleton, 1985).

Transfer functions

Transfer functions are empirically derived equations for calculating quantitative estimates of past atmospheric and oceanic conditions from paleontological data. Ideally, this problem requires the formulation of a model that explains the biological responses in terms of the multitude of factors that comprise the observed ecological system. The basic assumptions are:
1. No significant changes occur to the interactions among organisms over the time scale studied and within the ecosystems sampled.
2. The modern observations provide sufficient information for analyzing the fossil observations, and further that a snapshot of modern spatial patterns is a sufficient basis for interpreting changes through time.
3. The biological responses are systematically related to the physical attributes of the biotic environment. This assumption implies that the relationships can be represented by mathematical expressions.

Transfer functions techniques, which use multivariate statistical procedures, were first applied to marine plankton data by Imbrie and Kipp (1971), to terrestrial pollen data by Webb and Bryson (1972), to tree ring data by Fritts et al. (1971). They provide typically temperature estimates either for the atmosphere (pollen) or for the sea water (foraminifer), with an accuracy of ± 1.5°C. Estimates for continental humidity and surface water salinity are more difficult to derive from the same fossil populations. Changes in diatom flora have been interpreted in terms of variations of the chemical composition of lake water.

TERTIARY AND QUATERNARY CLIMATES

At the beginning of the Tertiary, warm conditions apparently prevailed everywhere over the Earth's. Broad-leaved vegetation extended to latitude 70-75° in both hemispheres. As no ice was present over the continents, the interpretation of the $^{18}O/^{16}O$ ratio of planktonic and benthic foraminifera analyzed in the drilling performed by drilling vessels is straghtforward. It demonstrates that surface and deep waters were warm from 65 to 36 Myr. However, during this time interval, climatic conditions deteriorated very slowly, at a rate estimated close to 0.1°C/Myr and this change is still unexplained. Deep water temperature rapidly dropped at the Eocene/Oligocene boundary, about 36 Myr ago. This dramatic oceanic cooling is explained by the formation of deep waters according to a processs similar to the present one, i.e. by sinking of surface waters in cold areas during winter conditions, because the density of surface water may reached high values by cooling ans sea ice formation. This suggest that mountain glaciers were present in Antarctica and extended until the sea level at that time.

The next major step in the oxygen isotopic record of deep sea foraminifera is the strong $\delta^{18}O$ increase which occurred during the middle Miocene. It is well-documented in all the oceans and is best explained by a $\delta^{18}O$ change of the whole oceanic water mass due to the development of a huge ice cap over Antarctica. This interpretation is supported by several geological evidences, notideably the increase of ice rafted deposits in the Southern Ocean. The Antarctic ice cap experienced important volume variations and exhibited a marked extension about 6 Myr ago.

Until about 3 to 2.4 Myr, a marked dissymetry occurred between high latitude areas of both hemispheres, with a glaciated southern hemisphere and an unglaciated northern hemisphere. At that time, glaciers expanded also over Canada and Iceland. This cooling is not yet well-explained, but the relatively rapid uplift of mountains both in North America and in Himalaya may have been responsible for an inflow of cold air bringing snowy precipitations over continental areas of the northern hemisphere.

Since that time, the earth's climate has been characterized by an alternation of glacial and interglacial episodes, marked in the northern hemisphere by the waxing and waning of continental ice sheets and in both hemispheres by petriods of rising and falling temperatures and marked precipitation changes. These temporal variations generally exhibit most frequencies observed in the insolation spectra.

THE LAST CLIMATIC CYCLE

Ice volume and sea level

Figure 1 shows the variations of the global mean $\delta^{18}O$ of the ocean during the last 135,000 years. These variations reflect those of the volume of ice frozen over the high latitude continents. After a rapid deglaciation which occurred about 128,000 years ago, the Earth's climate was characterized by a volume of continental ice smaller than the present one. As a consequence, the sea level was higher than today by about 6 meters. This interglacial called Eemian lasted for about 10,000 years and was followed by a first phase of glaciation, which had culminated by about 107,000 years B.P. (Before Present). This ice advance was followed by several phases of ice disintegration, but the volume of ice was always larger than today: The sea level was estimated to be about -18 meters during the two high levels around 80,000 and

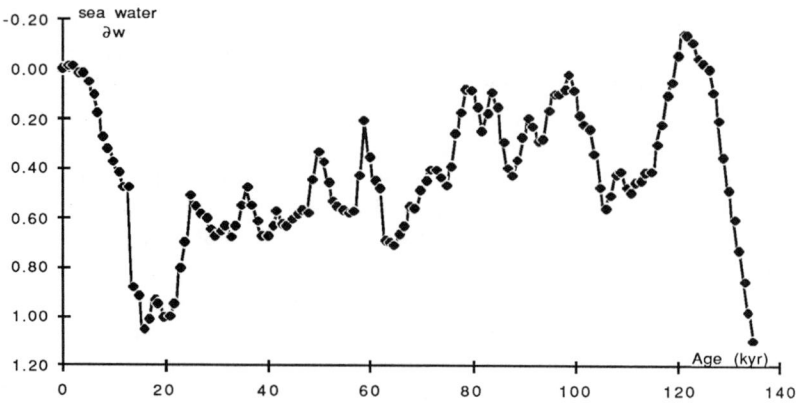

Figure 1: Variations of the $^{18}O/^{16}O$ ratio of the global mean oceanic water during the last 135,000 years. The value 0 indicates that the isotopic composition of the ocean is identical to that of the standard, which is the mean modern sea water isotopic composition (SMOW). Positive δw values (reported downward and expressed in per mil) indicate that the $^{18}O/^{16}O$ ratio of the global mean oceanic water was higher than the modern one, in relation with the waxing of continental ice sheets strongly depleted in ^{18}O.

99,000 years B.P.. Several phases of continental ice accumulation developed until the peak of the glaciation 18,000 years ago, when about 50,000,000 km^3 of ice were frozen, mainly over Canada and northern Europe. The final deglaciation began about 15,000 years ago and the ice melted in two separate phases, each lasting about 2,000 - 3,000 years. During these phases, the sea level increased at a rate of about 2 meters per century.

Temperatures

During the last interglacial, most low and mid-latitude oceanic areas experienced climatic conditions very similar to the present ones. Departures from the modern conditions have been recorded only at high latitudes, and only as barely significant deviations, taking into account of the standard error on the transfer function estimates (CLIMAP, 1984). Over the continents, quaternary deposits (i.e. formed during the last 1.8 M. years) are rare and, on a world-wide scale, only the broad features of the Eemian climate can be reconstructed. All the pollen data indicate that the climate was warmer and moister than today in northern Eurasia and northern America. A similar warming is suggested by the southern limit of permafrost in Eurasia (Flohn and Fantechi, 1984).

By the end of the Eemian, sea surface temperatures generally decreased. Close to the american continent in the Atlantic, the temperature drop was in phase with the ice volume increase. On the european side of the Atlantic, warm temperatures persisted for several thousand years, so that the glaciation of northern Europe began after that of northern America. However, pollen data indicate glacial conditions over northern Europe by about 107,000 B.P..

The pattern of temperature change during the last climatic cycle is only well known in the ocean. It strongly depends on the location and varies with latitude as expected from variations in solar radiation. For example, the high latitudes exhibit a strong response in the 41,000 year band, an indication that the insolation budget primarily depends on the obliquity of the Earth's axis. At lower latitude, the temperature show sharp variations with a frequency close to 23,000 years, characteristics of the precession.

The surface of the ice-age Earth at the last glacial maximum (LGM) has been reconstructed by CLIMAP (1976). The northern hemisphere differed markedly from today by the presence of huge land-based ice sheets, which were as much as 3 km in thickness, and by a significant increase in the extent of sea ice and marine-based ice sheets. In the southern hemisphere, the most striking contrast was the greater extent of sea ice during LGM. The sea level was about 100 m below the modern level. On the continents, grasslands, steppes, sandy outwash plains, and deserts spread at the expense of forests and the extent of snow-covered land was significantly greater. The global average sea surface temperature change associated with ice-age cooling was close to 2°C, but the magnitude of the cooling depended strongly on the geographic location: it was as much as 10°C in the north Atlantic, but almost zero in the central gyres of the Atlantic, Indian and Pacific oceans. This pattern resulted in a marked steepening of thermal gradients along frontal systems. As the cooling was much stronger over the continents than over the oceans, the estimate of the global average surface air temperature cooling is about 4-5°C, as calculated by General Circulation models of the atmosphere.

The deglaciation is a climatic event, abrupt on the geological time scale. The chronological resolution of the dating of paleoclimatic signals appears to be the most critical factor in quantifying the rate of temperature change of sea surface temperatures. Detailed analysis of deep sea cores with high sedimentation rates enables one to minimize the effects of bioturbation (the activity of benthic organisms, which permanently mix the upper centimeters of sediment). The dating of these cores by Accelerator Mass spectrometry demonstrated that, in the North Atlantic, the polar front retreated at a mean rate higher than 2 km/year. Off Portugal, this retreat resulted in a 10°C sea surface temperature increase in less than 400 years (Figure 2). This first warming, which led to temperatures similar to those of today, was followed by a sharp an abrupt cooling called the Younger Dryas which occurred in the whole high latitude northern hemisphere. The occurrence of a similar cooling in the southern hemisphere is still a matter of debate and the Younger Dryas is still poorly understood. The Younger Dryas lasted less than a millenia and was followed by a warming, less abrupt than the preceding one, leading to the Holocene conditions.

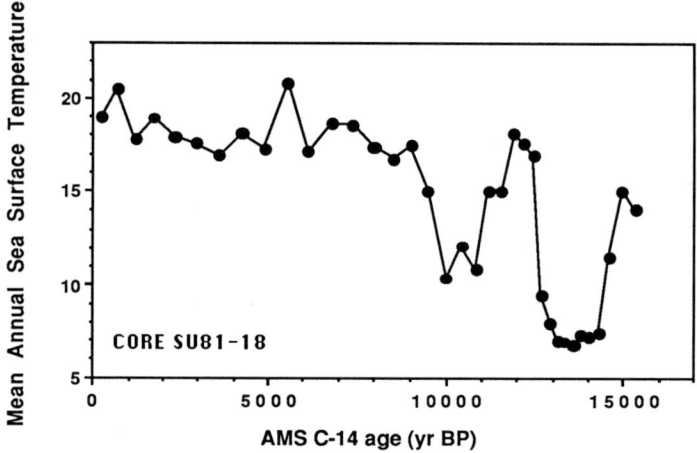

Figure 2: Paleoclimatic record of the last deglaciation in core SU 81-14 off Portugal. C-14 ages have been obtained by measuring the C-14 content of the shells of the planktonic foraminifera *Globigerina bulloides* at different levels in the core by Accelerator Mass Spectrometry. The upper curve shows the variations of the $^{18}O/^{16}O$ ratio of the planktonic foraminifera *Globigerina bulloides* . The lower curve displays the sea surface temperature estimates deduced from the variations of the foraminiferal fauna in the core

Oxygen isotope analysis of benthic foraminifera from the various basins of the ocean indicates that the deep waters had cooled by several degrees by the end of the Eemian, probably in response to the high latitude surface water cooling (Oppo and Fairbanks, 1987). By contrast, the temperature of intermediate waters did not change significantly (Kallel et al., 1988).

The $\delta^{13}C$ distribution of the total CO_2 dissolved in the deep ocean during the last interglaciation shows that the deep water sources were active in both hemispheres, but that the production of Antarctic Bottom Water (AABW) was stronger than that of North Atlantic deep Water (NADW) (Duplessy et al., 1984).

The global deep water circulation is very sensitive to the climate of high latitudes. For example, the production of NADW ceased during the peak of the melting of the ice sheets during the penultimate deglaciation. By contrast, this production increased during the inception of the glaciation (Duplessy and Shackleton, 1985).

During the last glacial maximum, the northern hemisphere source of deep water was weak in the North Atlantic and most bottom and deep water in the whole ocean originated from poorly-oxygenated surface water, sinking in the high latitudes of the Southern Ocean (Shackleton et al., 1983; Duplessy et al, 1988).

By contrast, Intermediate waters extended somewhat deeper than under modern conditions and were much more oxygenated than today in the Atlantic, Indian and Pacific oceans. We do not have enough data to interpret in detail the hydrology of the Intermediate waters in the Atlantic Ocean. In the Indian and Pacific oceans, intermediate waters were separated from the deep water by a well-developed deep thermocline, resulting in a strong stratification of the deep water realm. This resulted in strong changes in ocean chemistry which caused the decrease of the atmospheric CO_2 content (Broecker, 1982; Boyle and Keigwin, 1985).

The global circulation of deep and intermediate waters is strongly dependent on the earth's climate. Accelerator Mass Spectrometry (AMS) radiocarbon determination has made it possible to measure changes in the radiocarbon age of the deep ocean. Initial attempts to achieve this were thwarted by the difficulties imposed by bioturbation, which permanently mixes the upper ten or more centimeters of sediment. Shackleton et al (1988) have overcome this difficulty by making measurements in a core with a sufficiently high accumulation rate ($\geq$ 10 cm/kyr) to make the effects of bioturbation negligible. These authors demonstrated that the uncorrected age difference between the planktonic foraminifera N. *dutertrei* and the benthic foraminifera *Uvigerina* has changed during the last 30,000 years. During the last glaciation, it was close to 2,100 years, indicating that the ventilation age of the deep Pacific was about 500 years greater than it is today (Fig. 3).

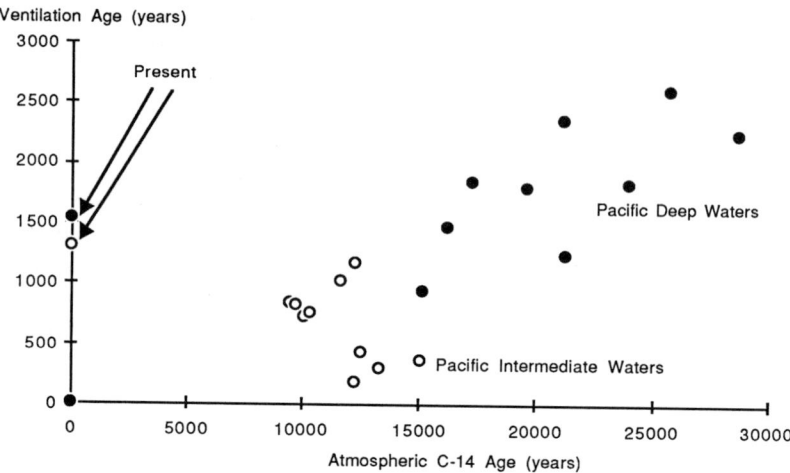

Figure 3: Variations of the ventilation rate of the Pacific deep and intermediate waters oceanic water during the last 30,000 years, deduced from the C-14 age difference between planktonic and benthic foraminifera found at the same sediment level in two deep sea cores.

Duplessy et al. (1990) followed the same strategy in order to determine the ventilation age of the Pacific Intermediate waters and compared the uncorrected ^{14}C ages of planktonic and benthic foraminifera in core CH 84-14 (41°44' N, 142°33' E, 978 m). They dated only the peaks of abundance of planktonic and benthic foraminifera, in order to minimize the effects of bioturbation, because the abundance of foraminifera vary by one to two orders of magnitude in response to local climatic variations. The development of benthic organisms is directly related to the surface productivity because food is brought to the abyss by sinking of particulate matter formed in the surface water. As a consequence, the peaks of abundance of planktonic and benthic foraminifera are generally found in the same sediment levels.

Carbon-14 determinations indicate that the mean difference is always lower than 1,000 years in this core before 10,000 year B.P., indicating that the ventilation of the intermediate water was much more important during the deglaciation than today (1250 years). During the Holocene, dissolution is high in this core, as in most North Pacific sediment cores, and planktonic foraminifera are not abundant enough to pursue this study.

In order to compare the variations of the ventilation time of the intermediate and deep waters of the Pacific oceans during the deglaciation, we show in Figure 3 the age difference between benthic and planktonic foraminifera in core CH 84-14 and in core TR 163-31B (3°37' S, 83°58' W, 3210 m). The ventilation rate of intermediate and deep pacific waters exhibits a similar pattern of variation and was smaller than under the modern conditions. An active deep and intermediate water circulation was an efficient factor diluting the water resulting from the melting of the ice sheets within the ocean water, which was salty and therefore denser than the meltwater.

This is C.F.R. Contribution N° xxx

BIBLIOGRAPHY

Bard, E., Arnold, M. , Duprat, J., Moyes, J., and Duplessy, J.C., Reconstruction of the last deglaciation: deconvolved records of $\partial^{18}O$ profiles, micropaleontological variations and accelerator mass spectrometry ^{14}C dating, Clim. Dyn., 1, 101-112, 1987.

Bard, E., Arnold, M. , Maurice, P., Duprat, J., Moyes, J., and Duplessy, J.C., Retreat velocity of theNorth Atlantic Polar Front during the last deglaciation determined by [14]C Accelerator Mass Spectrometry, Nature, 328, 791-794, 1987.

Berger, A. Ed., Climatic Variations and Variability: Facts and Theories, NATO A.S.I. Series, D. Reidel, 1981.

Bernard, E.A., The laws of physical paleoclimatology and the logical significance of paleoclimatic data., in "Problems of Paleoclimatology, ED. by A.E.M. Nairn, J. Wiley and Sons, New York, 1964.

Boyle, E. A., and Keigwin L. D., Comparison of Atlantic and Pacific paleo-chemical records for the last 215,000 years: Changes in deep ocean circulation and chemical inventories, Earth Planet. Sci. Lett., 76, 135–150, 1985.

Broecker, W. S., Glacial to interglacial changes in ocean chemistry, Prog. Oceanogr., 11, 151–197, 1982.

Broecker, W.S. and Peng, T.H., Tracers in the Sea, Eldigio Press, 1982.

CLIMAP Project Members, The surface of the Ice Age Earth, Science, 191, 1131–1137, 1976.

CLIMAP Project Members, The last interglacial ocean, Quat. Res., 21, 123-224, 1984.

Duplessy, J. C., Isotope Studies. In Climatic Change, Ed. by J. Gribbin, Cambridge University Press, 46-67, 1978.

Duplessy, J. C., and Shackleton N. J., Response of global deep-water circulation to the Earth's climatic change 135,000–107,000 years ago, Nature, 316, 500–507, 1985.

Duplessy, J. C., Shackleton, N. J., Matthews, R. K., Prell, W., Ruddiman, W. F., Caralp, M., and Hendy, C. H., ^{13}C record of benthic foraminifera in the last interglacial ocean: Implications for the carbon cycle and the global deep water circulation, Quat. Res., 21, 225–243, 1984.

Duplessy, J. C., Shackleton, N. J., Fairbanks, R. G., Labeyrie, L. D., Oppo, D. and Kallel, N., Deep water source variation during the last climatic cycle and their impact on the global deep water circulation, Paleoceanography, 1988.

J.C. Duplessy, Arnold, M., Bard, E., Juillet-Leclerc, A., Kallel, N., and Labeyrie, L. D., AMS ^{14}C study of transient events and of the ventilation rate of the Pacific Intermediate Water during the last deglaciation, Radiocarbon, 1990.

Emiliani, C., Pleistocene temperatures, J. Geol., 63, 538-578, 1955.

Flohn, H. and Fantechi, R. Eds, The climate of Europe: Past, Present and Future, D. Reidel, 1984.

Hedges, R. E. M. and Gowlett, J. A. J., Radiocarbon dating by Accelerator Mass Spectrometry, Scientific American, 82-89, 1986.

Imbrie, J. and Kipp, N. G., A new micropaleontological method for quantitative paleoclimatology: Application to a late Pleistocene Caribbean core. In "The Late Cenozoic Glacial Ages", Ed. by K.K. Turekian, 77-181, Yale Univ. Press., 1971.

Kallel, N., Labeyrie, L. D., Juillet-Leclerc, A. and Duplessy, J. C., A deep hydrological front between intermediate and deep-water masses in the Glacial Indian Ocean, Nature, 651–655, 1988.

Kroopnick, P., The distribution of ^{13}C of ΣCO_2 in the world oceans, Deep Sea Res., 32, 57–84, 1985.

Labeyrie, L. D., Duplessy, J. C., and Blanc, P. L., Variations in mode of formation and temperature of oceanic deep waters over the past 125,000 years, Nature, 327, 477–482, 1987.

Mix, A., and Fairbanks, R. G., North Atlantic surface-ocean control of Pleistocene deep ocean circulation, Earth Planet. Sci. Lett., 73, 231–243, 1985.

Oppo, D. W., and Fairbanks, R. G., Variability in the deep and intermediate water circulation of the Atlantic during the past 25,000 years: Northern hemisphere modulation of the Southern Ocean, Earth Planet. Sci. Lett., 86, 1–15, 1987.

Roth, E. and Potty, B., Eds., Nuclear Methods of Dating, CEA Paris, 1989.

Shackleton, N. J., Imbrie, J., and Hall, M., Oxygen and Carbon isotope record of east Pacific core V 19-30: Implications for the formation of deep water in the Late Pleistocene North Atlantic, Earth Planet. Sci. Lett., 65, 233–244, 1983.

Shackleton, N.J. and Opdyke, N. D., Oxygen isotope and paleomagnetic stratigraphy of equatorial Pacific core V28-238: oxygen isotope temperatures and ice volumes on a 10^5 year and 10^6 year scale, Quat. Res., 3, 39-55, 1973.

Shackleton, N. J., Duplessy, J. C., Arnold, M., Maurice, P., Hall, M., and Cartlidge, J., Radiocarbon age of Last Glacial Deep water, Nature, in press.

Webb, T., and Bryson, R. A., Late and post glacial climatic change in the northern Midwest, USA: quantitative estimates derived from fossil pollen spectra by multivariate statistical analysis, Quat. Res., 2, 70-115, 1972.

VARIABILITY OF THE CONTINENTAL BIOSPHERE

A. PONS

Faculté des Sciences St-Jérôme et CNRS, MARSEILLE

Summary

The different vegetation types of the last climatic cycle, well recorded in pollen sequences, may give a good idea of the variability of the continental biosphere. The characteristics of interglacial, glacial, interstadial and interphasic vegetations in the European subcontinent are described, together with some features relating to the Global vegetation. The two possible approaches to palaeoclimates from pollenanalytical data are presented. Principles, drawbacks and use of the nearest analogue approach are detailed. In particular, the "loading factor" technique, which enables to minimize the disturbing effect of man's impact, is described. Finally climate reconstructions (annual and seasonal parameters) from European sequences are briefly commented.

1. INTRODUCTION

On the continents the major part of the biosphere is vegetation, and the latter constitutes the best evidence of climate changes in space.

The same is true about climate changes that have occurred in the course of time.

In fact, all living beings are more or less dependent on climates but :
- vegetation/climate relationships are stronger than any other,
- a great part of the relations existing between climate and animals are more or less governed by vegetation itself.
- fairly good records of past vegetations are available, which have no equivalent in animals.

For all these reasons the main interest here will be vegetation (with some considerations about animals).

First it is necessary to explain briefly why and how past vegetation is more easily recorded than other elements of the continental biosphere.

This is due to the fact that vegetation is permanently represented by pollen owing to its three marvellous properties : high specificity, large dispersal and good preservation capacity.

Now, the first concern is "what range of ecosystems do these records of past vegetations suggest in correlation with the climate variability described in this same course by DUPLESSY.

Before answering this question, two digressions are necessary :

2. THE PRESENT VEGETATION

In the present world there are two kinds of vegetation :

First, there is a vegetation for which relationships with climate are obvious : this is the zonal vegetation. It has been recently well described (1) : "Vegetation is first of all sensitive to climate and to the general atmospheric circulation because the growth of plants closely depends on climatic factors such as solar radiation, temperature and rainfall. The tropical rainforest can develop only if temperature and moisture are favourable during the twelve months of the year. It is replaced by a dry forest or a savanna if the water content in soils decreases under a certain threshold value during several months. The savanna itself becomes clearer and clearer when aridity increases and finally it turns into desert. Outsides the Tropics, seasonal variations in temperature and moisture are the main climatic features. Vegetation develops into a deciduous forest in humid temperate regions, into more xerophytic formations in Mediterranean regions or into a grassland or a steppe in dry areas with a marked seasonal contrast. Farther off, there is the boreal forest (taiga) composed of birches and conifers, or the meager tundra, when trees cannot develop" (p.78). Thus, the distribution of the present vegetation follows the major climatic zones.

But there is another kind of vegetation, for which relationships with climate are secundary, if not negligible : it is the <u>stationary</u> vegetation. It is represented by all particular ecosystems in which a dominant and strictly local ecological factor (for example salts, permanent running water, moving and very poor soils..) excludes a zonal vegetation.

This kind of vegetation sets a problem in climate reconstruction.

3. THE LAST CYCLE

The best and so far sufficient way to obtain a fairly complete idea of the variability of vegetation as a whole is to analyse the different vegetation types that have existed during the last climatic cycle, because

- this cycle is the only one which is more or less correctly known,
- as yet, there is no clear evidence for a significant peculiarity distinguishing this cycle from previous cycles (even the latter are poorly known).

Moreover the last cycle has been more or less entirely recorded in Europe in a set of good pollen sequences (Fig.1).

In the world as a whole, a very small number of good records are available from Japan, Africa, Australia, Central and North America.

4. THE VARIABILITY OF CONTINENTAL HEMISPHERE DURING THE LAST CYCLE

A. European outline

Pollenanalysis enables one to distinguish four types of vegetation : interglacial, glacial, interstadial and interphasical.

<u>Interglacial vegetation</u> : The interglacial vegetation was fundamentally a forested vegetation type (Fig.2). But its most valuable characteristic consists in very clear, well-ordered and complete dynamics.

In contrast, <u>glacial vegetation types</u> (or stadial, or pleniglacial) are almost completely treeless (Fig.3): birch are dwarf trees, pine is rare and sparse (but with an efficient pollen production and dispersal). Vegetation is composed of shrubs and steppic plants (Artemisia, orach, meadow-rue, helianthemum), which are to-day typical of Thibet and North African high plateaus.

<u>Interstadial</u> vegetation types constitute a third category (Fig.4). They were forested like interglacial vegetation types, but with a striking difference : i.e. the almost total absence of several elements of the interglacial dynamics. Ecological evidences suggest temperate conditions more or less similar to interglacial conditions, but of very short duration, this forbidding a complete geological expression of climate. Of course, when dynamics are oversimplified, it is difficult to determine whether there really is an interstadial.

Finally, there is a fourth type of vegetation. It is characterized by a small number of arboreal taxa with low percentages, but evidencing brief and small peaks of Arboreal Pollen (essentially, if not exclusively, of Pine) in spite of a clear dominance of Non Arboreal Pollen : it is difficult to imagine the corresponding vegetation (Fig.5). Moreover, it is impossible to assert that various peaks in the arboreal percentages coïncide in time from on site to another. To this state of the European vegetation, Max WELTEN (2) gave the name of "interphase".

B. Three remarks about Europe

- In the infilling of Les Echets, owing to particularly propitious conditions and on the basis of comparative analyses of pollen percentages and absolute frequencies (3), it has been demonstrated that brief peaks in pollen curves corresponding to the Last Pleniglacial can be accounted for by interannual fluctuations in the flowering of herbaceous plants, as a result of local meteorological events, as often happens nowadays in desertic regions : this may be a key to the difficulty related to the interphasic vegetation.

- during the two great interstadials which followed the last interglacial, there appeared a strong N-S gradient (boreal vegetation with conifers in Northern Germany or Denmark and deciduous forest in northern and central France which has no equivalent to-day (4).

- at the latitude of La Grande Pile, Les Echets and Le Bouchet, the major characteristic of stadial vegetations is related to low temperature, but at lower latitudes (at Padul, near Granada or in Valle de Castiglione, near Rome), the most striking feature is not cold but dryness. (So that the Lower Pleniglacial - between 70 000 and 60 000 years ago - affected most the composition of subsequent ecosystems) (5).

C. At the global scale

Two general observations can be made :

- at middle and high latitudes : the European outline is also valid for other continents. There is only one difference : owing to the modest surface of our subcontinent, and also because of the orientation of the Alps and the barrier formed by the Mediterranean Sea, European floras have early suffered from ancient Pleistocene climatic changes, and they are comparatively very poor. But, above all the survival of steppes and tundra during interglacial episodes - just as the survival of forest during glacial episodes - has been very hard, so that it is difficult to find now in Europe all the equivalent ecosystems that prevailed in the past (the "analogues") : we will deal with this problem later;

- at low latitudes, there are very few data available. Besides, they almost strictly originate from African mountains and only cover the last 35-40 000 years (6).

It is difficult to make a significant synthesis of these data. Only four pieces of information are of general interest :

1) During the last Pleniglacial, there occurred an altitudinal lowering of the vegetation belts, which amounted to some one thousand meters (this is attested by the development of non arboreal vegetations);

2) The altitudinal lowering of mountain vegetations was associated with a breaking up of these vegetations.

3) At low elevation, scanty available data indicate :

- an important expansion of open ecosystems during the late Pleniglacial;

- the existence of a decidedly humid vegetation (perhaps more stationary than zonal) during the time-span between about 9 000 and 4500 B.P. (which, it seems, could be assimilated to an anathermic interglacial phase).

4) For the last 2000 years, the structure and distribution of the ecosystems have been affected by human impact.

D. Molluscs and insects

The contribution to the knowledge of past ecosystems, that is offered by the study of the remains of two groups of continental animals : insects (especially beetles) and molluscs, is worth emphasizing. As a matter of fact:

- insects and snails have a fairly good preservation capacity in a large range of places where sediments accumulate;

- most of them have strict ecological and climatic requirements;

- some of them depend on the vegetal cover (these can only provide additional information in relation to pollenanalytical results), however, a lot of them are carnivorous or adapted to a great variety of plants with a large climatic distribution (such as mosses, for example). Consequently, for these insects, some climatic parameters (especially summer temperature) are clearly limiting factors; thus, valuable palaeoclimatic indications can be derived from the study of corresponding fossil remains assemblages.

This is particularly true for beetles (Scarabeidae), which, moreover, have the capacity to expand so quickly that they indicate every abrupt climatic change more faithfully than plants (especially trees).

4. FROM THE CONTINENTAL BIOSPHERE TO PALAEOCLIMATES

There are two possible kinds of approaches to palaeoclimates from pollenanalytical data :

* The first is to consider the present-day taxa distribution in relation to two or more physiologically important climatic variables. This approach was pioneered by IVERSEN (7) (Fig. 6a).

The drawback of this method is that it only provides parameters in a discontinuous manner, because over a long period there are either taxa linked to a given climate parameter or taxa linked to other parameters.

It should be mentioned here that a similar but more elaborate approach, using "natural climatic range" of beetle remains has recently provided a fairly good reconstruction of seasonal temperatures in Britain since 22 000 B.P. (8).

* The other approaches take into account all the ancient pollen spectra thanks to more or less elaborate statistical procedures.

The first consists to establish a linear relation between the pollen frequencies ratio of two taxa and a given geographic gradient (altitude, for instance). Then past variations in the pollen ratio are converted into variations of parameters corresponding to the geographic gradient (9) (Fig.6b).

We chose to use another procedure : the nearest analogue approach.

In this approach, the climatic reconstruction is based upon the closest analogues in existing vegetation for fossil pollen spectra.

We assume that in any particular study area for which sufficient data are available over a given span of time, present-day analogues can be found for past climates, contemporaneous vegetation types and corresponding pollen spectra.

However, this methodology has four drawbacks :

1) Climate requirements of modern plants may be different from those in the past. In fact, the rate of evolution of tree-species and of herbs genera (these alone being studied through pollen analysis) has been unsufficient to introduce actual differences.

2) The range of past climatic conditions is perhaps not fully represented to-day and some "good analogues" for a lot of ancient spectra may be absent from our modern referential.

This problem cannot be overcome in an *a priori* fashion. Only the use of the greatest possible variety of modern spectra offers the best way to compensate for it.

3) Now, the third drawback is that past and present human impact on nature has disturbed the vegetation/climate (and hence the pollen/climate) relationships in the modern pollen data referential.

This is the major drawback.

Palaeoecology is particularly well informed about the consequence of human impact. For instance, REILLE (10) obtained a striking schema of man-induced perturbations in the altitudinal zoning of vegetation in Corsica, mainly since the fifteenth century (Fig.7).

But our first concern in elaborating our method was precisely to find an acceptable way to minimize the effects of this drawback. It consists to attribute, in the calculation of similarities between modern and ancient spectra, a "loading factor" to all pollen percentages. For each taxon, it is not the direct correlation with present climatic parameters which is expressed, but the loading factor takes into account the overall fluctuations of that taxon throughout the period covered by the reconstruction.

An other important mean to lessen the human impact effects is the use of logarithms of percentages in order to reduce the role of the to-day most common plants which are also those that expanded most under human action (for example Pine and Gramineae).

4) The relationships between vegetation and climate may be influenced by stationary peculiarities, but the "loading factor", which expresses the role locally played in the past by each taxon at each site, is a fairly good means of taking these peculiarities into account.

Finally our work comprises four tasks :

a) The first task is to build up the to-day referential set : for this, 667 spectra are now used, originating from the whole Europe, North Africa and Siberia (a total amount of 28 taxa was considered). The corresponding climate parameters were interpolated from 2000 meteorological stations.

b) The second task consists to select "good" analogues : for this, an Euclidian distance is calculated between each ancient spectrum and all the present spectra (Fig.8).

In this distance, for each taxon, the loading factor results from a Principal Component Analysis of the correlation of the frequence of this taxon in each spectrum of the sequence, with its own frequence in the underlying spectrum, on the one hand, and with the frequence of all the other taxa in the underlying spectrum, on the other hand.

Thus, the loading of a taxon whose frequencies never or constantly change, or change without any correlation with other taxa frequencies, has no climatic value and a minor loading (near zero). Taxa showing an opposite behaviour have a solid loading tending towards +1 for the majority of trees, and reversely tending towards -1 for non arboreal plant and some rare trees.

There is another approach to the loading, which is entirely based upon the present vegetation/climate relationships. It consists in the calculation of non linear functions (called "responses surfaces"), which describe the way a taxon depends on the joint effects of two or more environment variables (11). So far, this approach has only been experimented for the Holocene time (12).

c) Our third task is to calculate palaeoclimate from the present-day parameter of the best analogues.

This is the most technical point of the procedure. We use the best analogues among p sets of modern spectra (here p is chosen equal to 40 and each set comprises 10 spectra) drawn by lots (with replacement of the spectra that have been drawn). The mean of the climatic characteristics of these 40 "best analogues" is "the reconstruction". A "bootstrap" procedure allows one to simulate the probability of these results according to the homogeneity of the analogues. At 0.66 this probability is called "confidence interval". It is an expression of the lack of "perfect analogues".

d) Finally, the method as a whole is validated by "reconstructing" climatic parameters for modern pollen spectra for which corresponding climatic parameters have been obtained independently. The correlation obtained between estimated and actual modern climatic parameters is always close to 0.8 and argues in favour of the soundness of our method.

This method was used in the reconstruction of the last cycle from la Grande Pile, Les Echets and Le Bouchet sequences.

First it provided the reconstruction published by GUIOT et al. (13) (Fig.9).

But since this paper, we have achieved new results :

- first, reconstructions from a new original pollen sequence of La Grande Pile (14) and from the Lac du Bouchet sequence were in very good agreement with the two first ones;

- second, curves have been obtained for July and January temperatures and July and January precipitation (Fig. 10 and 11).

Two remarks should be made about these curves :

1) The two Pleniglacials (the first one between about 72 and 60 thousand years years ago, the second between about 25 to about 14 thousand years ago) have closely similar annual parameters; nevertheless : the first is characterized by very low January temperatures and July precipitation, the second is marked by even lower January temperature but also by much lower July precipitation.

2) The Eemian and the two following Interstadials have nearly the same annual climate, but the Interglacial (as well as our Postglacial) is marked by a rather oceanic seasonality contrasting with the continentality of the Interstadials, this making clear the steepness of these Interstadials vegetational gradient.

Seasonal reconstructions appear as an efficient means to understand past climates.

REFERENCES

(1) DUPLESSY, J. Cl. and MOREL, P. (1990). Gros temps sur la Planète. Ed. O. Jacob, Paris, 296 p.

(2) WELTEN, M. (1982). Stande der palynologishen Quartärforschung am schweizerischen Nordalpenrand. Geogr. Helv., 2, 75-83.

(3) BEAULIEU, J.L de and REILLE, M. (1986). The problem of abrupt climatic changes in the Würm Pleniglacial. The example of Les Echets (France). In Book of abstracts and reports from the Conference on Abrupt Climatic Change. SIO Reference series, 86-8, 102-103.

(4) BEHRE, K.E. (1989). Biostratigraphy of the last glacial period in Europe. Quaternary Science Reviews, 8, 25-44.

(5) PONS, A. and REILLE, M. (1988). The Holocene- and Upper Pleistocene pollen record from Padul (Granada, Spain) : a new study. Palaeogeography, Palaeoclimatology, Palaeoecology, 66, 243-263.

(6) BONNEFILLE, R. (1987). Evolution des milieux tropicaux africains depuis le début du Cénozoïque. Palynologie et milieux torpicaux. G. Cambon, P. Richard et J.P. Suc edit., EPHE Montpellier, 101-110.

(7) IVERSEN, J. (1944). Viscum, Hedera and Ilex as Climate Indicators. A contribution to the study of the Post-Glacial Temperature Climate. Geol. Fören.Dörhandl., 66 (3), 463-483.

(8) ATKINSON, T.C., BRIFFA, K.R. and COOPE G.R. (1987). Seasonal temperatures in Britain during the past 22,000 years, reconstructed using beetle remains. Nature, 325 (6105), 587-592.

(9) ADAM, D.P. and WEST, G.J. ((1983). Temperature and precipitation estimates through the last Glacial cycle from Clear Lake, California, Pollen data. Science, 219, 168-170.

(10) REILLE, M. (1975). Contribution pollenanalytique à l'histoire tardiglaciaire et holocène de la végétation de la montagne corse. Thèse ès Sciences, Université Aix-Marseille III, 306 p., 44 diag., 5 pl. h.t.

(11) BARTLEIN, P.J., PRENTICE, I.C. and WEBB, T. (1986). Climatic response surfaces from pollen data for some eastern North American taxa. Journal of Biogeography, 13, 35-57.

(12) HUNTLEY, B. and PRENTICE I.C. (1988). July temperatures in Europe from pollen data, 6000 years before Present. Science, 241, 687-689.

(13) GUIOT, J., PONS, A., BEAULIEU, J.L. de, . and REILLE, M. (1989). A 140,000yr climatic reconstruction from two European pollen records. Nature, 338, 309-313.

(14) BEAULIEU, J.L. de and REILLE, M. (1989). A new pollen sequence at La Grande Pile (Vosges, France). Terra, abstracts, 1 (1), 65-66.

(15) BEAULIEU, J.L. de and REILLE, M. (1984). The pollen sequence of Les Echets (France) : a new element for the chronology of the upper Pleistocene. Géogr. Phys. Quaternaire, 38 (1), 3-9.

(16) WOILLARD, G. (1978). Grande Pile peat bog : a continuous pollen record for the last 140,000 years. Quaternary Research, 9, 1-21.

(17) BEAULIEU, J.L. de and REILLE, M. (1984). A long Upper Pleistocene pollen record from Les Echets, near Lyon, France. Boreas, 13, 111-132.

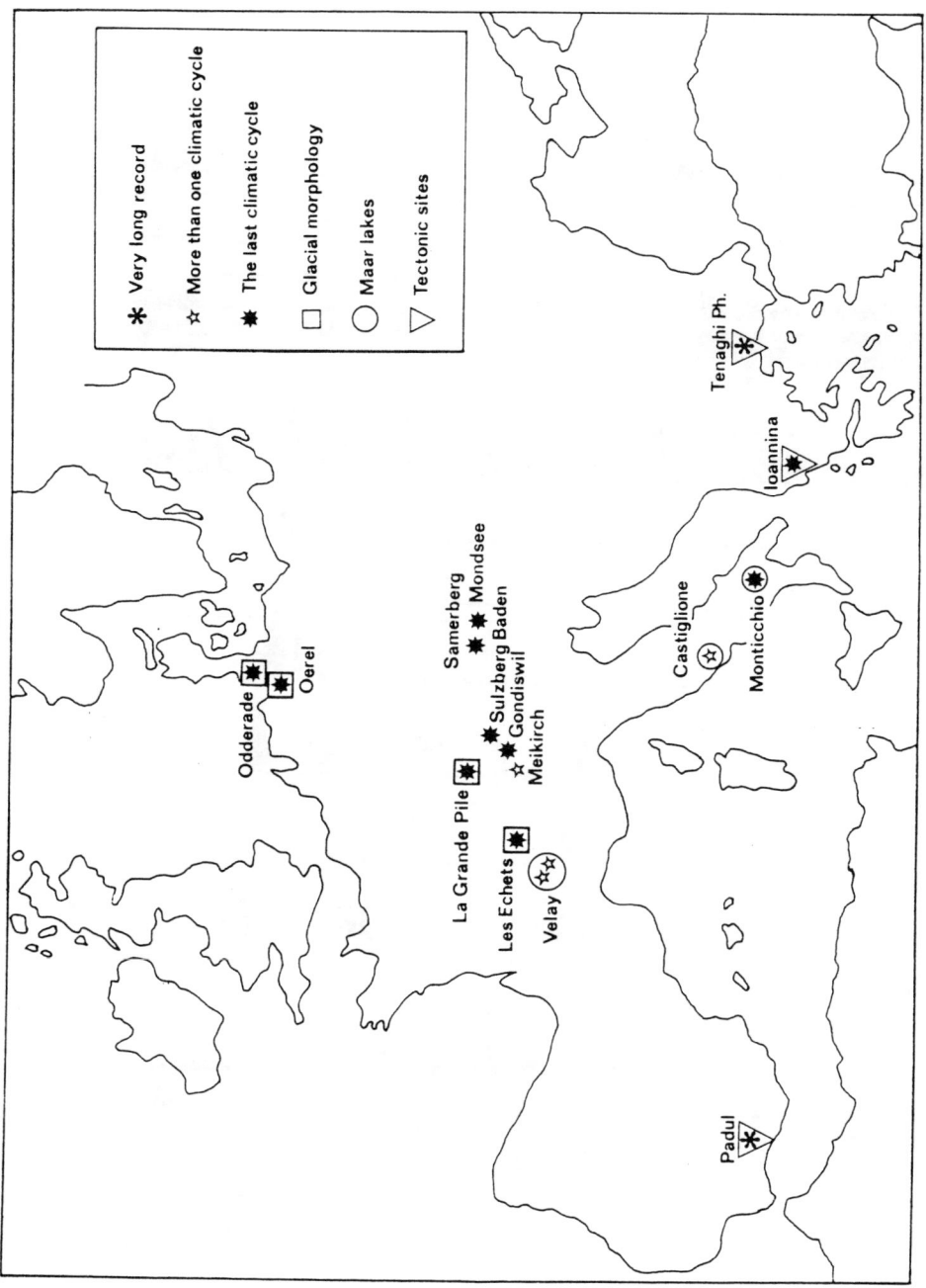

Figure 1 : Location map of the European sites with a pollen sequence from the last cycle.

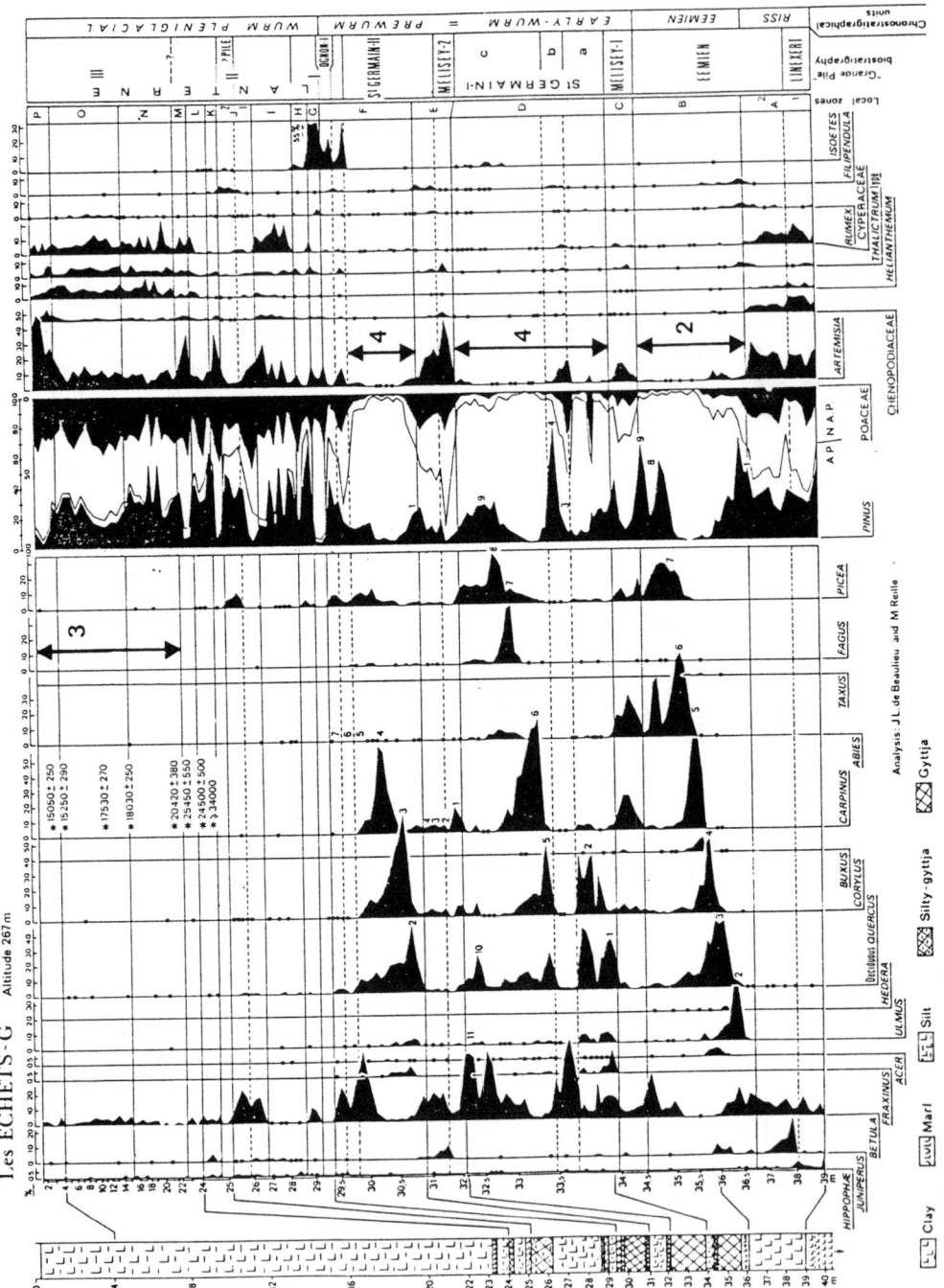

Les ÉCHETS-G   Altitude 267 m

Analysis: J.L. de Beaulieu and M. Reille

□ Clay   □ Marl   □ Silt   ⊠ Silty-gytija   ⊠ Gytija

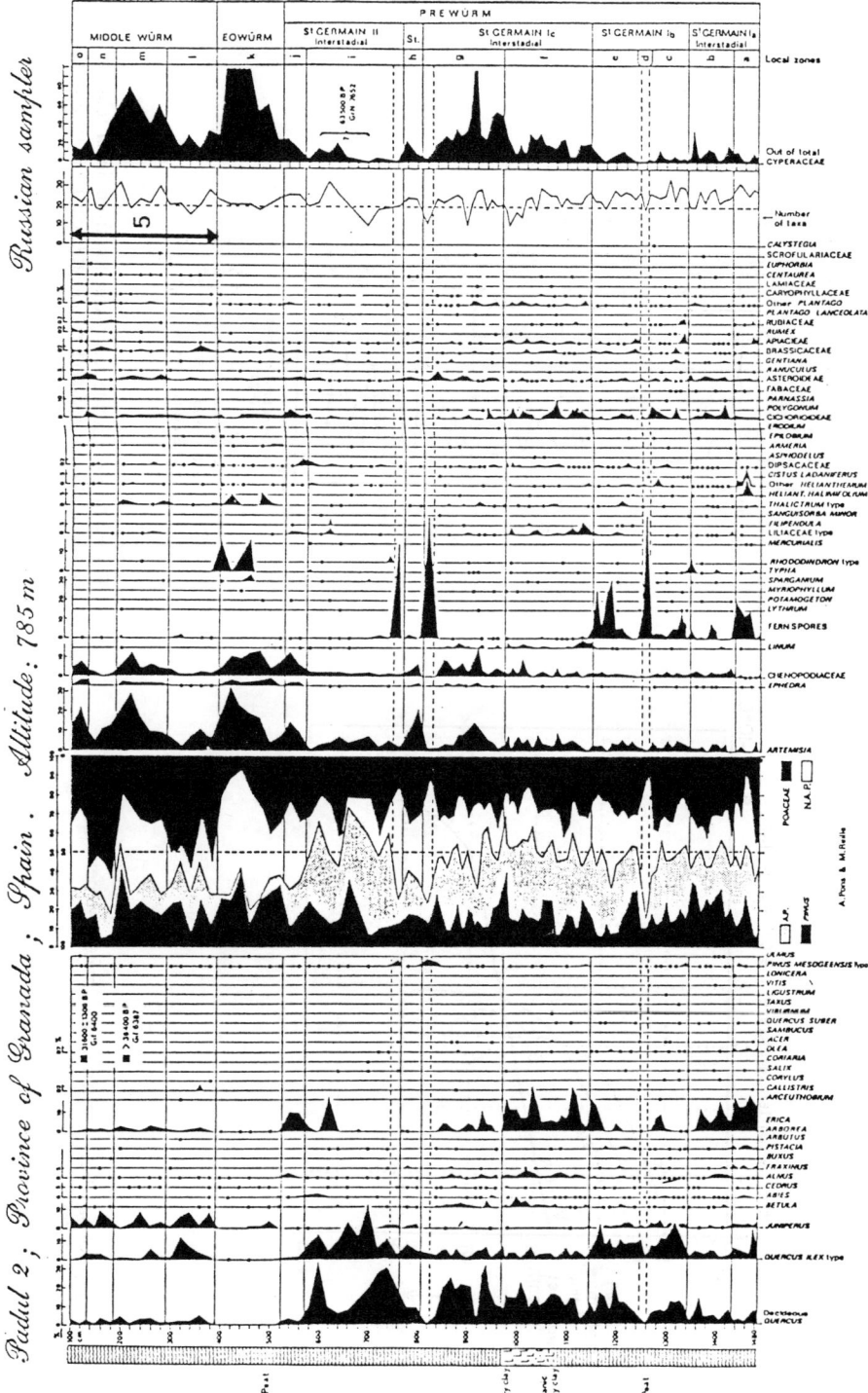

Figure 5 : Interphase at Padul (5).

27

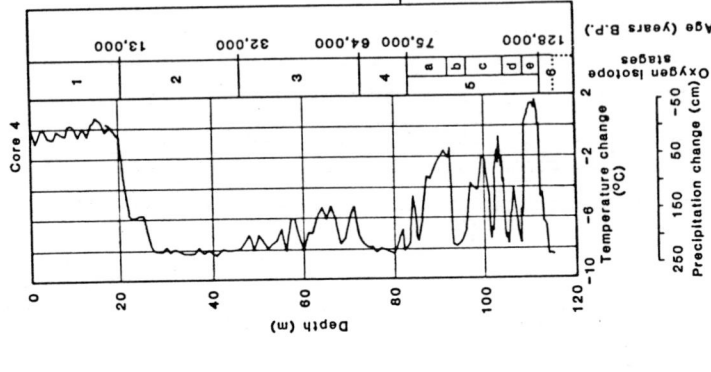

Figure 6b : A climate reconstruction from linear relation between altitude and pollen ratio at Clear Lake (California) (9).

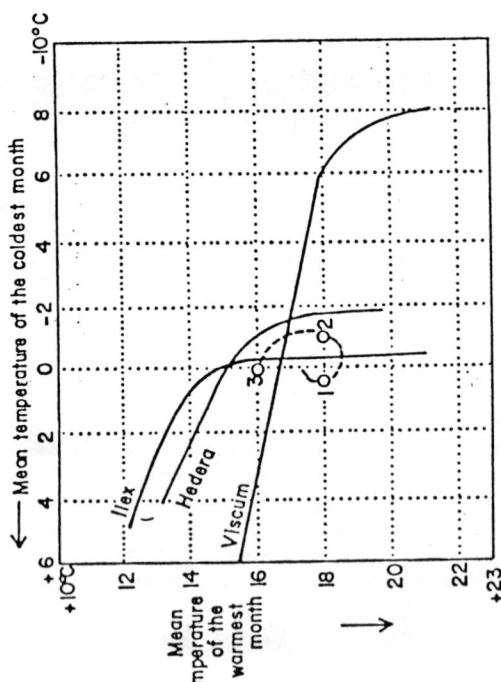

Figure 6a : Reconstruction of climate parameters from taxa distribution (7).

28

Before human action	Present state
Asylvatic vegetation	Asylvatic vegetation
Betula verrucosa	Betula verrucosa
Pinus laricio forest	*Pinus laricio forest*
Deciduous *Quercus* forest	
Quercus ilex forest	*Quercus ilex forest*
Olea and *Pistacia* lentiscus formations	*Olea* and *Pistacia* lentiscus formations

Figure 7 : The result of human impact in the Corsica mountain vegetation.

$$d_{ap}^2 = \sum_{j=1}^{m} \left[w_j \log \left(f_{pj} - f_{aj} \right) \right]^2$$

d = distance
a = ancient spectrum
p = present spectrum
j = each observed taxon
m = total number of considered taxa
w = loading factor
f = frequencies
log = natural logarithm

Figure 8 : The distance used in climate reconstruction from pollen data.

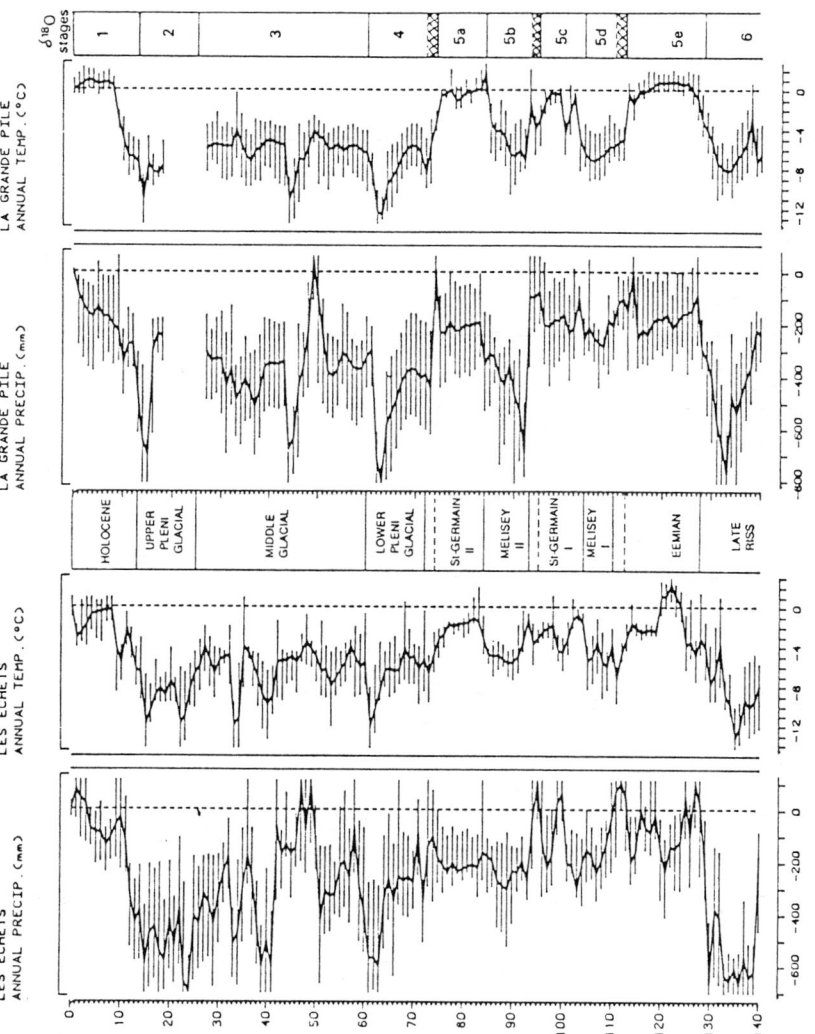

Figure 9 : A 140 000 year Climate reconstruction from two European pollen records (13 : Nature, 38, 309-313). Results are expressed as deviation from modern values. The error bars result from the simulation method. The chronological axis beyond the 14C time scale is based upon a linear interpolation of the dates provided by a general agreement on some precise events (end of the Eemian, beginning of the Lower Pleniglacial...). Fossil data from (16) and (17).

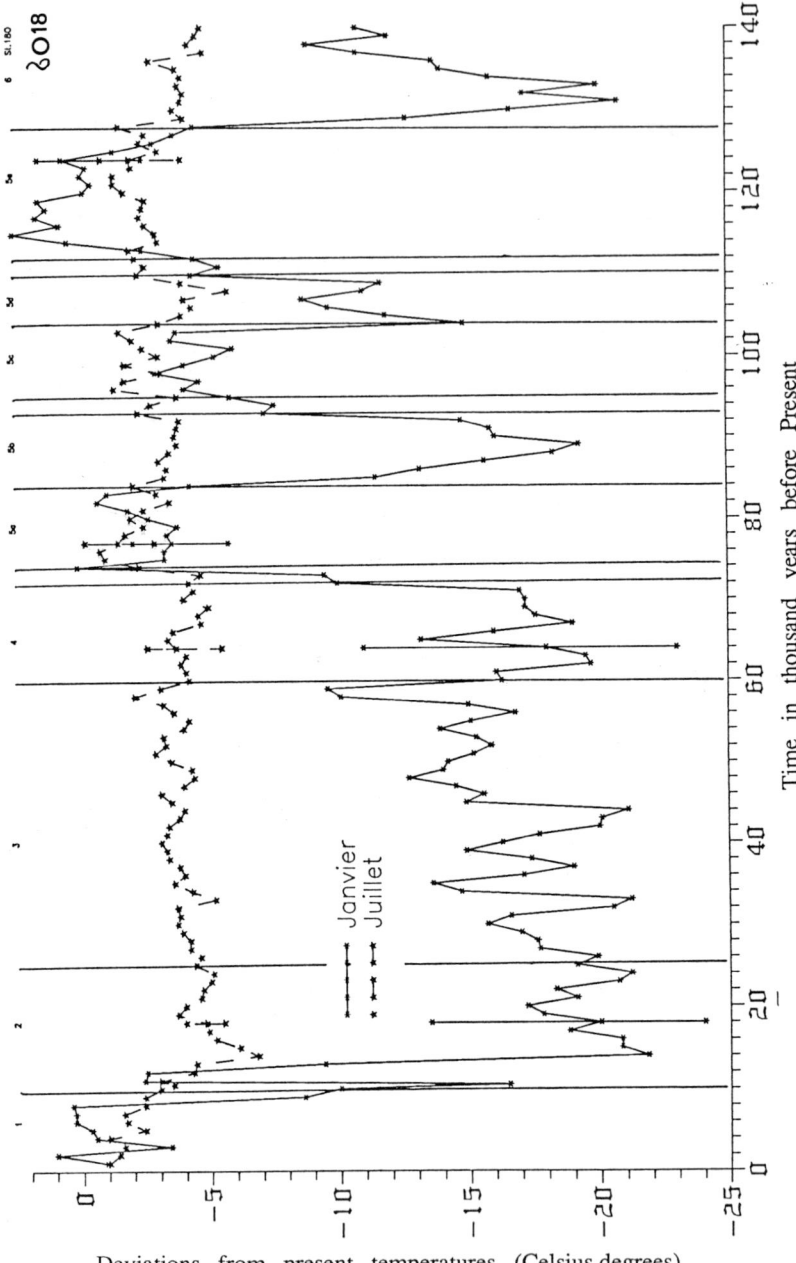

Figure 10 : Reconstruction of the mean temperatures of January and July (mean of Les Echets and La Grande Pile). Only the confidence interval of best characterized episodes is figured.

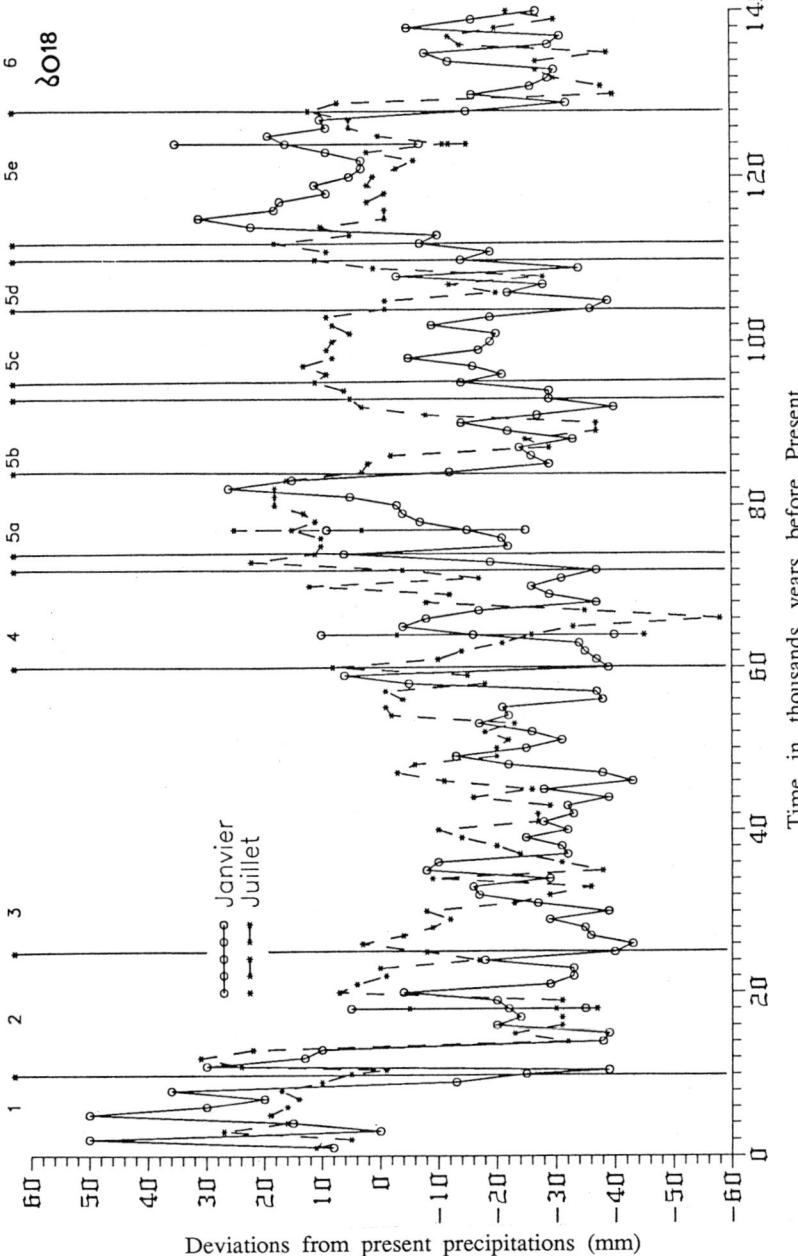

Figure 11 : Reconstruction of the precipitation sum of January and July (mean of Les Echets and La Grande Pile). Only the mean confidence interval of best characterized episodes is figured.

THE GREENHOUSE EFFECT

A. BERGER
Université Catholique de Louvain
Institut d'Astronomie et de Géophysique G. Lemaître
2 Chemin du Cyclotron
B-1348 Louvain-la-Neuve

Summary

The greenhouse effect on the Earth is identified by the difference between the effective radiating temperature of the planet and its surface temperature. The difference between the energy emitted by the surface and that emitted upward to space by the upper atmosphere quantified it; it can therefore be defined as the long wave energy trapped in the atmosphere. Climate forcing and the response of the climate system within which climate feedback mechanisms are contained, will be defined in this review. Quantitative examples will illustrate what could happen if the greenhouse effect is perturbed by human activities, in particular if CO_2 atmospheric concentration would double in the future. Recent measurements by satellites of the greenhouse effect will be given. The net cooling effect of clouds on the Earth and whether or not there will be less cooling by clouds as the planet warms, are discussed following a series of papers recently published by Ramanathan and his collaborators.

1. THE EARTH'S ENERGY BALANCE

Greenhouse gases such as water vapour and CO_2 bring about the greenhouse effect through the property that they absorb strongly in the infrared region of the electromagnetic spectrum. In order to understand clearly this greenhouse effect, we must first look upon the global Earth's energy balance.

Recent satellite measurements show that the so-called solar constant S_0 is about 1365-1372 Wm^{-2} (Willson et al., 1986; Ramanathan et al., 1989a); this solar irradiance is the solar energy per unit area intercepted perpendicularly at the mean Earth-Sun distance per unit time. Changes in this solar constant constitute the primary external forcing of the climate system. The most apparent phenomenon that could cause the solar output to vary at the 10 to 100 years time scale is sunspots. Detailed investigations show that during sunspot maximum, spots may cover 1 to 2 % of the disk and bright areas surrounding the spots known as faculae cover an even larger area than do the spots. From three years of ERBE (Earth Radiation Budget Experiment) solar observations, it was possible to show that solar irradiance varies with sunspot activity, being lowest at the sunspot minimum. Before the sunspot minimum in 1986, it has indeed decreased at a rate of 0.02 %/yr and has increased at the same rate after (Figure 1).

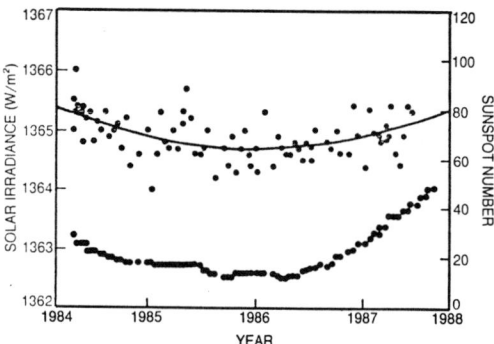

Figure 1. The solar constant from 1984 to 1988 derived from the solar monitor on the Earth Radiation Budget Satellite. The upper curve and dots refer to the solar irradiance and the lower dotted curve to the smoothed sunspot members (from Willson and Hudson, 1988 and Ramanathan et al., 1989a).

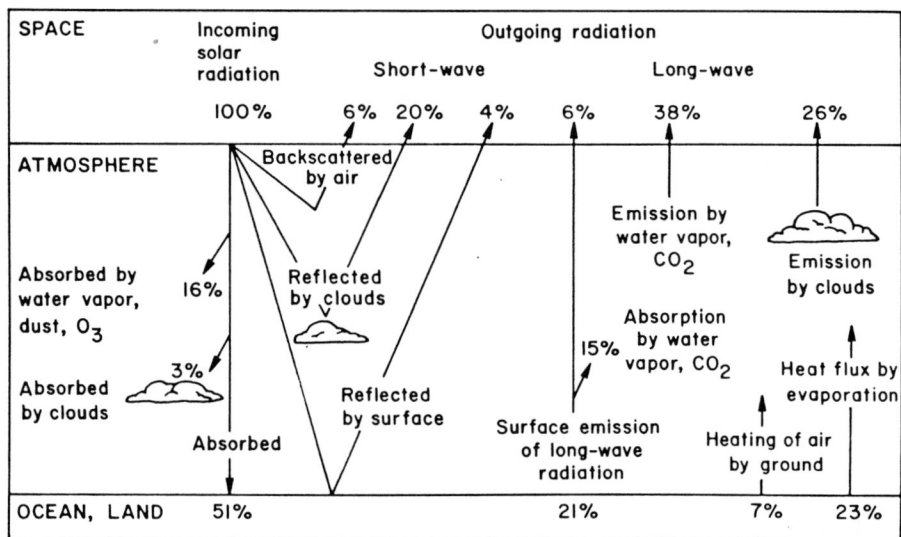

Figure 2. Energy balance of the Earth. 100 % incoming solar radiation represent 342 Wm^{-2}. The net surface emission of long-wave radiation (21 %) is the result of the 6% emission in the atmospheric window and of the balance (15 %) between the surface emission absorbed by the overlying atmosphere (110 %) and the downward long-wave emission by the atmopshere (95 %). All the numbers may differ slightly from one climatology to another due to the uncertainty of the data. The atmospheric window is a term which applies to the region of the absorption spectrum of water vapour existing from about 8.5 to 11 μm. Surface emission in this range of wavelengths is, in contrast to surface emission of other wavelengths, little absorbed by water vapour and, in the absence of cloud, escape to space.

This sunlight, the source of energy for the Earth-atmosphere system, is principally in the visible region, where the Earth's atmosphere is almost transparent. According to the Wien displacement law, the wavelength of maximum emission for a Sun which radiates as a 6,000 K black body in the spectral range from 1.2 to at least 10 μ, is 0.5 μm. Almost 99 % of this sun's radiation is contained in the short wavelengths from 0.15 to 4.0 μm. Of this, 9 % is in the ultraviolet ($\lambda \leq 0.4 \mu m$), 45 % is in the visible ($0.4 \leq \lambda \leq 0.74 \mu m$) and 46 % is in the nea-infrared $\lambda \geq 0.74 \mu m$.

Of the total amount of incoming radiation around 30 % is reflected by clouds (20 %), by particles in the atmosphere (6 %) and by the Earth's surface (4 %), while the other 70 % is absorbed by ozone in the stratosphere (3 %), by water vapour, clouds and aerosols in the troposphere (16 %) and by the Earth's surface (51 %) (Figure 2). These numbers are approximative and may vary slightly from one climatology to another due to the unaccuracy of the observations. Satellite measurements show, for example, that the planetary albedo α (the fraction of the solar irradiance reflected by the planet's surface and atmosphere) is equal to 0.30 $\pm$ 0.03.

To maintain total energy balance, long-wave radiation equivalent to the 70 % incoming solar radiation which was absorbed, needs to be re-emitted to space. This outgoing long-wave radiation is provided by satellite data for the top of the atmosphere (237 Wm^{-2}) along with the estimates of solar radiation (342 Wm^{-2}) and of reflected solar radiation (105 Wm^{-2}). The other quantities of the energy balance are obtained from various published model and empirical estimates. These quantities include the already mentionned atmosphere and surface absorption of solar radiation (68 and 169 Wm^{-2}, respectively), but also the upward long-wave emission by the surface (390 Wm^{-2}, which provides the bulk of the energy emitted to space and among which about 20 Wm^{-2} are directly escaping to space through the so-called atmospheric window), the downward long-wave emission by the atmosphere (327 Wm^{-2}) and the latent and sensible heat fluxes from the surface (90 and 16 Wm^{-2} respectively) (Figure 2). In fact, the atmosphere constantly loses 106 Wm^{-2} of radiating energy because it emits 327 Wm^{-2} to the surface and 217 Wm^{-2} to space and absorbs only 370 Wm^{-2} in the infrared and 68 in the visible. In other words, there is radiative cooling of the atmosphere and a corresponding radiative heating of the Earth's surface, the equilibrium being restored by sensible (16 Wm^{-2}) and latent (90 Wm^{-2}) heat transfer which "convectively" couple the surface to the atmosphere.

2. THE EFFECTIVE RADIATING TEMPERATURE OF THE PLANET EARTH

To illustrate the strong links between the radiation budget, the climate and the circulation of the atmosphere and the oceans, it is useful to examine simple models. The simplest zero-dimensional model of climate considers the long-term average (on a time scale greater than one year) of the global and annual mean temperature. A balance between the absorbed solar energy and the emitted energy governs the temperature T_e defined by :

$$B = \frac{S_0}{4}(1 - \alpha) - \sigma T_e^4 \qquad (1)$$

T_e is the effective radiating temperature of the surface-atmosphere system assumed to emit like a blackbody at equilibrium and σ is the Stefan-Boltzman constant equal to 0.567 10^{-7} Wm^{-2} K^{-4}. If the balance B is zero, T_e is expected to be 255 K and represents the temperature at which satellites see the top of the atmosphere emitting long-wave radiation to space. Although, the atmosphere emits in specific wavelength bands and so, the emission departs significantly from that of a black body, T_e is still a useful parameter provided it is considered as the effective radiating temperature of the Earth.

3. THE GREENHOUSE EFFECT

If the atmosphere did not impede the radiative energy flow, the surface temperature T_s would be the same as T_e. However, T_s is observed to be 288 K presently. The 33 K difference is attributed to the greenhouse effect.

Let us make an illustrative calculation to quantify this greenhouse effect. At a temperature of 288 K, the surface emits E = 390 Wm^{-2}. Satellites tell us that only 237 Wm^{-2} espace to space (which is coherent with S_0 = 1368 Wm^{-2} and α = 0.307 in (1)). If we call it F, the energy trapped in the atmosphere is given by :

$$G \;=\; E \;-\; F \tag{2}$$

(2) defines the greenhouse effect. It is thus the difference between the surface emission and the total energy loss. It is presently 153 Wm^{-2}. Any change in E and/or F, like the only due to the increase in atmospheric concentration of the greenhouse gases for example, leads to a perturbation of the greenhouse effect, ΔG, as it will be explained in section 7.

B can be characterised also by an effective transmissivity factor τ of the atmosphere by relating the absorbed solar radiation to the Earth's surface temperature through :

$$\frac{S_0}{4}\,(1-\alpha) \;-\; \epsilon\,\tau\,\sigma\,T_s^4 \;=\; 0 \tag{3}$$

assuming the Earth's surface emission is close to that of a blackbody (i.e. emissivity, ϵ, is close to 1). Therefore, (3) gives $\epsilon\tau$ = 0.6. (3) shows also that a more efficient trapping (i.e. a lower infrared transmissivity) of the atmosphere will automatically result in a higher surface temperature for a fixed S_0 and α.

In order to illustrate further this greenhouse effect, let us consider Venus. For Venus, S_0 = 2620 Wm^{-2} and α = 0.8. Hence Venus absorbs only 130 Wm^{-2}, only 55 % as much as the Earth despite it receives twice as much solar radiation at the top of its atmosphere as does the Earth. However, Venus has a much hotter surface than the Earth (T_s = 750 K) which emits about 17,900 Wm^{-2}. Thus its atmosphere traps nearly two orders of magnitude greater than the Earth's greenhouse effect. This suggests that there is no conceivable saturation point for the atmospheric greenhouse effect which is limited only by the concentration of gases in the atmosphere. It is also worth to note that because Venus is closer to the Sun, saturation in water vapor was never achieved allowing temperature to run away. On the Earth, the increase of surface temperature halted when the water vapor pressure began to be equal to the saturated vapor pressure and freezing or condensation occured.

4. THE PHYSICAL PROCESSES OF THE GREENHOUSE EFFECT

The physical processes which explain the trapping G defined by (2) are as follows.

1. The atmosphere is quite transparent to solar energy. Various empirical estimates give an atmospheric absorption of solar radiation of 68 Wm^{-2}, the remaining 169 Wm^{-2} being absorbed at the surface.

2. The infrared trapping by the atmoshere is due to greenhouse gases which are primarily water vapor, clouds and CO_2 (Annex 1), with a smaller 5 % contribution from O_3, N_2O and CH_4. Several anthropogenic gases, however, such as the chlorofluorocarbons, like $CFCl_3$ and $CF_2\,Cl_2$, are now beginning to make an appreciable contribution (Hansen et al., 1989).

3. For G to be positive, the temperature must decrease with altitude in the region of the infrared-absorbing gas (Raval and Ramanathan, 1989). Because tropospheric temperatures decrease with increasing altitude, the active atmospheric constituants absorb more upward radiative flux than it emits upwards. The net result of these absorption and emission processes is that part of the infrared radiation emitted by the surface is trapped. If we consider a simple 2-level radiative model such as in Annex 2, we can see that the ground temperature would be 335 K and the calculated radiative temperature profile would be unstable. Thus, a particle disturbed from a location close to the surface would rise and carry energy upwards. The resulting convection currents would mix the atmosphere and alter the temperature profile until the atmosphere is driven toward a neutral thermal lapse rate.

Heating the lower boundary of a fluid while cooling its interior is the classical mechanism for inducing convective instability and turbulence. In the Earth's troposphere, turbulent transfer of heat and condensation of water evaporated from the surface make up for the atmosphere's radiative energy deficit, the combination of these nonradiative processes being basely called convective heat transport. The stratosphere, however, is the region in which the radiative equilibrium lapse rate agrees with the observed lapse rate. Once the lapse rate is prescribed, the surface temperature is therefore the only degree of freedom for the troposphere; it is determined by the net flux of the solar and infrared radiation at the tropopause (the fundamental climate-forcing term).

The radiation fluxes at the upper boundary are influenced strongly by internal parameters such as the distribution of water vapour, clouds and other gases; by the lapse rate; by the surface properties such as ice and snow cover, vegetation types and soil moisture. The dependance of these parameters on the surface temperature T_s gives rise to several feedback loops, of which the interaction between water vapour and T_s is the best understood and that between clouds and T_s the least understood.

It is therefore important to differentiate between what is meant by the greenhouse effect, climate forcing and climate feedback.

5. CLIME FORCING AND CLIMATE FEEDBACKS

The French physicist, Jean-Baptiste Fourier, was probably the first person in 1827 to allude to the greenhouse effect and to suggest that human activities could influence the climate (Ramanathan, 1988, Jones and Henderson-Sellers, 1990).

In the latter half of the nineteenth century, Tyndall (1861) described the greenhouse effect caused by atmospheric water vapour, pointing out that water vapour transmits a large fraction of the incident solar radiation but strongly absorbs infrared radiation emitted by the Earth's surface. Later, Arrhenius (1896) suggested that a climate change may be induced primarily by a change of CO_2 concentration in the atmosphere. One century later, the development of the greenhouse theory took a new dimension. Plass (1956) calculated that the mean global surface temperature would increase by $3.6°C$ if atmospheric carbon dioxide doubled. Möller (1963) provided the first model attempt and suggested that water vapour might also act as an amplifying climate feedback mechanism. In 1967, Manabe and Wetherald provided quantitative results for carbon dioxide induced warming on the basis of a one-dimensional radiative-convective model. Perhaps, one of the most significant advances in the carbon warming was the development by 1975 of a three-dimensional global climatic model reported by Manabe and Wetherald. Since then, the model groups have increased the complexity and hopefully the reliability of their general circulation models (Mitchell, 1989).

In 1989, Raval and Ramanathan cleverly employed satellite measurements to quantify the greenhouse effect and to demonstrate the positive water-vapour feedback.

This provides a convenient means of making the conventional interpretation of climate change as a two stage processes : forcing and response (Cess, 1989).

Let us consider the energy budget,H, of the whole Earth system :

$$H = \frac{S_0}{4} (1 - \alpha) - F \tag{4}$$

For an increase in atmospheric CO_2, the global mean direct radiative forcing of the surface-atmosphere system is the reduction in the top-of-the-atmosphere infrared flux caused solely by the CO_2 increase; it is thus evaluated by holding all the other climate parameters fixed. When the CO_2 increases in the atmosphere, trapping is more efficient and F decreases. Radiative calculations (Dickinson and Cicerone, 1986) has lead to a logrithmic expression of ΔF :

$$\Delta F = - \beta \ln \frac{[CO_2]}{[CO_2]_0} \tag{5}$$

where β is equal to 6.1 ± 0.4 Wm^{-2}. This means for a $2 \times CO_2$: $\Delta F = -4.2 \pm 0.3$ Wm^{-2}.

If we could immobilize the atmosphere and suddenly double its concentration the long wave flux F would decrease by about 4 Wm^{-2} at the tropopause so that the heating increases by 4 Wm^{-2} ($\Delta H = 4$ Wm^{-2}). According to the zero-dimensional climate model, the global situation will restore the radiation energy balance. In other words, the climate system will force H to zero. The planet's surface and troposphere could warm up until it radiates to space the excess 4 Wm^{-2}. The consequent increase in F effected by such as higher temperature balances the decrease in F caused by the increase in CO_2 concentration. The response within which climate feedback mechanisms are contained, is thus the ensuing change in climate that is required to restore the top-of-the-atmosphere radiation balance and corresponds to $\Delta H = -4$ Wm^{-2} (Ramanathan et al., 1989a).

The discussions here ignore the stratosphere (we consider only that part of the atmosphere where the temperature decreases), which in fact will cool because of the increased CO_2 infrared emission.

The infrared emission F is a function of CO_2 and temperature, but also of water vapour because of the evaporation feedback. As the Earth's surface warms, water evaporates more rapidly from the surface. To keep the process near equilibrium, more water must condense, but the net result is increased water vapor in the atmosphere. This vapour will further decrease F and increase H. For these simple models, it is generally assumed that humidity is only a function of temperature. Therefore, (4) becomes :

$$H = \frac{S_0}{4} (1 - \alpha) - F (CO_2, T, e(T)) \tag{6}$$

This energy balance equation changes to become a sum of 5 terms :

$$\Delta H = \frac{1 - \alpha}{4} \Delta S_0 - \frac{S_0}{4} \frac{d\alpha}{dt} \Delta T - \frac{\partial F}{\partial C} \Delta C - \frac{\partial F}{\partial T} \Delta T - \frac{\partial F}{\partial e} \frac{de}{dt} \Delta T \tag{7}$$

The first term expresses the possible change in the solar constant, the second is a function of the albedo-temperature feedback, the third the initial radiative forcing, the fourth one is the direct response and the fifth is the indirect response related to the evaporation-temperature feedback.

One fundamental inference from this model is that climate should be extremely sensitive to small variations in radiative forcing. For example, a 1 % increase in the solar irradiance will increase the absorbed solar radiation by 2.4 Wm^{-2}.

When the system reaches equilibrium, the radiation balance perturbation vanishes ($\Delta H = 0$). Assuming $\Delta S_0 = 0$, the final temperature change is therefore :

$$\Delta T = \frac{- \partial F / \partial C \; \Delta C}{\frac{dF}{dT} + \frac{S_0}{4} \frac{d\alpha}{dt}} \tag{8}$$

where $\frac{dF}{dT} = \frac{\partial F}{\partial T} + \frac{\partial F}{\partial e} \frac{de}{dT}$.

This relation is often rewritten in terms of the forcing, ΔQ, and the feedback parameter, λ, such that :

$$\Delta T = \frac{\Delta Q}{\lambda} \tag{9}$$

The same physical idea has been expressed in different ways, (Manabe, 1983) in particular by using a gain factor (Schlesinger, 1985; Berger and Tricot, 1988).

6. THE FEEDBACK PARAMETER

An upper hypothetical limit of λ (3.7 Wm^{-2} K^{-1}) could be obtained by assuming the climate system is devoid of all feedbacks other than IR radiative damping and that the Earth emits IR radiation as a blackbody with an equilibrium temperature of 255 K :

$$F = \sigma T_e^4 \rightarrow \lambda = \frac{\partial F}{\partial T} = 4 \sigma T_e^3 \tag{10}$$

Current three-dimensional climate models yield global warming by amounts in the range of 1.9 - 5 K (Mitchell et al., 1989; Mitchell, 1989; Schlesinger, 1989). As for a doubling of CO_2, the model estimate of ΔQ is about 4 Wm^{-2}, the theoretical value of λ lies in the range of 0.8 - 2.1 Wm^{-2} K^{-1}.

There have been many empirical estimates of λ from satellite measurements of Earth's radiation budget (Ramanathan, 1987). These lie between 0.7 and 2 Wm^{-2} K^{-1} and are therefore consistent with those estimated from the models. In general, these values derived from observations use the latitudinal and seasonal changes in the observed F, α, and T. This consistency would thus be a satisfactory proof of the models, only if the seasonal climate variations mimics a climate change caused by CO_2. Moreover, the sensitivity of climate to small variations in radiative forcing poses very stringent accuracy requirements on observations (a 1% increase in solar irradiance would lead to a warming of 1.1 to 3°C according to the λ values from GCMs), .

Thus, the measurement of the radiation budget has been a story of increasingly sophisticated instruments and rigorous data processing. After two decades of progress in satellite instrumentation, the Earth Radiation Budget Experiment began in the 1980s. ERBE instruments are carried on three satellites : the Earth Radiation Budget Satellite, NOAA-9 and NOAA-10. A unique feature of ERBE is that it separates the top-of-the-atmopshere fluxes for clear skies from fluxes from cloudy skies, which allows to obtain the greenhouse effect of the atmosphere and that of the clouds separately. Such measurements of F were used recently by Raval and Ramanathan (1989) to quantify the atmospheric greenhouse effect G over the year 1985.

To better estimate the emission E from the surface, they restricted their study to the open oceans. Overall, they expected the error (systematic plus random) in the monthly and regional mean values of G to be 5 to 10 Wm^{-2}.

Globally, for April 1985, the surface emission was 421 Wm^{-2} and the top-of-the-atmosphere emission was 243 Wm^{-2}, resulting in a total trapping of 178 Wm^{-2}. Of this, the atmopsheric greenhouse gases trap 146 Wm^{-2}, whereas clouds trap the remaining 32 Wm^{-2}.

On the other hand, estimates of G over the open oceans for a cloudless atmosphere revealed significant regional variations, generally decreasing from Equator to pole. But the most significant result was that this clear-sky G has a strong positive correlation with sea surface temperature, a correlation that according to Raval and Ramanathan (1989), gives direct evidence of the water vapour feedback.

In order to demonstrate this, they first attempted to eliminate the temperature dependence of G by defining a normalized greenhouse effect $g = \frac{G}{\sigma T^4}$. This definition $g = \frac{G}{E}$ leads to $F = (1 - g)E$.

Their test rested then on three pieces of evidence :

(i) the normalized greenhouse effect increases linearly with temperature :

$$g = g(T_s) = -0.658 + 0.00342 \, T_s \tag{11}$$

which for present day surface temperature gives $g = 0.327$. This feedback is clearly positive as if T_s increases, g and therefore G increases. In fact, E increases more than F with T_s. Moreover, $\frac{dF}{dT_s}$ given by :

$$\frac{dF}{dT_s} = 4 \, \sigma \, T_s^3 \, (1.658 - 0.004275 T_s)$$

is smaller (2.31 Wm^{-2} K^{-1}) than without the water vapour feedback. For this latter case, g being kept constant to its present day value, we have indeed:

$$\frac{dF}{dT_s} = \frac{\partial F}{\partial T_s} = 4 \, \sigma \, T_s^3 \, (1 - g) \simeq 3.65 \; Wm^{-2} K^{-1}$$

(ii) the logarithmic of the measured water vapour concentration varies linearly with surface tempeature, just a dependence that one would calculate from the Clausius-Clapeyron equation governing the saturation vapour pressure.

(iii) g increases logarithmically with the water vapour concentration.

Conclusively, the analysis strongly indicates the presence of the water vapour feedback, which is further supported by the good agreement:

(i) between the sensitivity $\frac{dG}{dT_s}$ of the NCAR Community model and the observed one :

$$\frac{dG}{dT} = 4 \, \sigma \, T_s^3 \, (-0.658 + 0.004275 \, T_s) \sim 3.1 \; Wm^{-2} \; K^{-1}$$

(ii) between the value of the feedback parameter $\frac{dF}{dT_s} = 2.31 \; Wm^{-2} \; K^{-1}$ obtained from ERBE for clear sky conditions and the averages for the 14 GCMs analysed by Cess et al. (1989) : 2.38 ± 0.16 Wm^{-2} K^{-1}, although significant disagreements amongst the models' global mean $\frac{dF}{dT_s}$ occur when clouds are taken into account.

Water vapour feedback is thus clearly detectable from ERBE observations and it can account for most of the observed variation of G with surface temperature. Moreover, the greenhouse perturbation by human activities might be detected by ERBE in the near future (Raval and Ramanathan, 1989). Indeed, the increase in trace gases, if it continues unabated, will globally increase G by 9 to 15.5 Wm^{-2} over the next 50 years (for a direct forcing of 2.5 Wm^{-2} the globe is expected to warm by 2 to 4°C which would lead to a further increase in G of 6.5 to 13 Wm^{-2} due to water vapour feedback).

In summary :

1. $(\frac{\partial F}{\partial T_s})_{g\ fixed} = \frac{dF}{dT_s} = (1 - g)\ 4\ \sigma\ T_s^3 = 3.65\ Wm^{-2}K^{-1}$

   This is clearly the feedback related to the direct response already defined in terms of the temperature at the top of the atmosphere in (10). In such a case,

   $$\Delta T = \frac{-\partial F/\partial C\ \Delta C}{\frac{\partial F}{\partial T}} \qquad (12)$$

   and the rate at which the emission increases with increasing temperature governs the temperature change.

2.

   $$(\frac{dF}{dT_s})_{g\ variable} = 4\ \sigma\ T_s^3\ (1.658\ -\ 0.004275\ T_s) \qquad (13)$$

   This expression which contains both $\frac{\partial F}{\partial T}$ and $\frac{\partial F}{\partial e}\frac{de}{dT}$ as shown in (8), allows to determine the water vapour feedback contribution per se by substracting (12) from (13) :

   $$\frac{\partial F}{\partial e}\frac{de}{dT_s} = 4\ \sigma\ T_s^3\ (0.985\ -\ 0.004275\ T_s) \qquad (14)$$

   which for all $T_s > 230\ K$ is clearly negative, leading to a smaller value of λ and therefore a positive feedback.

3. Finally, the albedo-temperature feedback is also positive:

   $$\frac{S_0}{4}\frac{d\alpha}{dT} = -\ 0.4\ Wm^{-2}K^{-1} \qquad (15)$$

   an increase in T causes indeed a melting of sea ice and snow cover; the decrease in the area of ice and snow lowers the albedo and thereby increases the absorbed sunlight, the whole process amplifying the surface warming.

7. THE RESPONSE TO A 2 X CO_2

A simple calculation can easily show how the climate system is restoring the equilibrium at the top of the atmosphere when it has been forced by doubling the CO_2 concentration in the atmosphere. Because Raval and Ramanathan (1989) refers to clear-sky condition, Cess (1989) has proposed to consider a cloud free Earth with a surface temperature

Table 1: Response of the hypothetical planet's climate system to a 4 Wm^{-2} direct forcing (Cess, 1989)

Process	$\Delta T_s(K)$	$\Delta G(Wm^{-2})$	$\Delta F(Wm^{-2})$
direct forcing (1)	0	4	-4
response with g fixed (2)	1.1	1.9	4
global response (3)	1.7	9.5	0
additional response with g variable (4)	0.6	3.6	0
((4) = (3)-(2))			

of 288 K, where absorption of solar radiation by the surface-atmosphere system is invariant to climate change and no other feedback than water vapour is considered.

Table 1 summarizes the results in 3 steps. The direct infrared forcing simultaneously reduces F and increases G so that $H = 4\ Wm^{-2}$. This imbalance at the top of the atmosphere causes a global warming which is 1.09 K without water vapour feedback. Indeed for g prescribed to its present-day value 0.327 (process 2 in Table 1), we have :

$$\Delta F = 4 = (1-g)\,\Delta E \rightarrow \Delta E = 5.94\ Wm^{-2} \tag{16}$$

which implies, through $\Delta E = 4\,\sigma\,T_s^3\,\Delta\,T_s$, $\Delta\,T_s = 1.09\ K$ and $\Delta\,G = g\,\Delta\,E = 1.9\ Wm^{-2}$.

If the water vapour feedback is considered (process 3) :

$$\Delta F = 4 = (1-g)\,\Delta E - E\,\Delta g \tag{17}$$

which gives   - through $E = \sigma\,T_s^4$ and g given by (11) : $\Delta T_s = 1.73 K$
                - through $G = \sigma\,T_s^4 - F$: $\Delta G = 9.4 - 4 = 5.4\ Wm^{-2}$

This simple illustration demonstrates how the direct greenhouse forcing of 4 Wm^{-2} is first amplified to 5.9 Wm^{-2} through temperature feedback and then to 9.5 Wm^{-2} by water vapour feedback, the further response to variable g being thus 3.6 Wm^{-2}. The combined forcing and feedbacks thus increase the clear sky greenhouse effect from its initial value of 146 Wm^{-2} to nearly 156 Wm^{-2}. Moreover, water vapour feedback amplifies the direct surface warming of 1.1 K by roughly 60 per cent leading to a 1.7 K global warming.

The role of the water vapour feedback has also been described through 3 processes involving ocean-atmosphere thermal interactions as shown in Figure 3 (Ramanathan, 1981 and 1988). For the case of a doubling of CO_2, out of the 4 Wm^{-2} of CO_2 heating, roughly 1 Wm^{-2} is deposited at the surface as increased emission from CO_2 (process 1). The balance, 3 Wm^{-2}, is trapped in the troposphere (process 2). This increase in the infrared energy warms the troposphere and so the surface, including the oceans. More moisture evaporates; the latent heat is released as moisture condenses through precipitation. The warmer troposphere, as a result of the direct greenhouse heating and the release of latent heat holds more water vapour because it is observed that the atmopshere tends to conserve its relative humidity with a change in temperature. The enhanced water vapour in turn traps more infrared radiation and amplifies the greenhouse effect (process 3). In one-dimensional climate models, Ramanathan (1981) has shown that for a global warming of 2.2 K, the 3 processes are responsible for a temperature increase of respectively 0.17, 0.33 and 1.7 K, which shows that the water vapour feedback amplifies the surface warming by a factor of ~ 3.

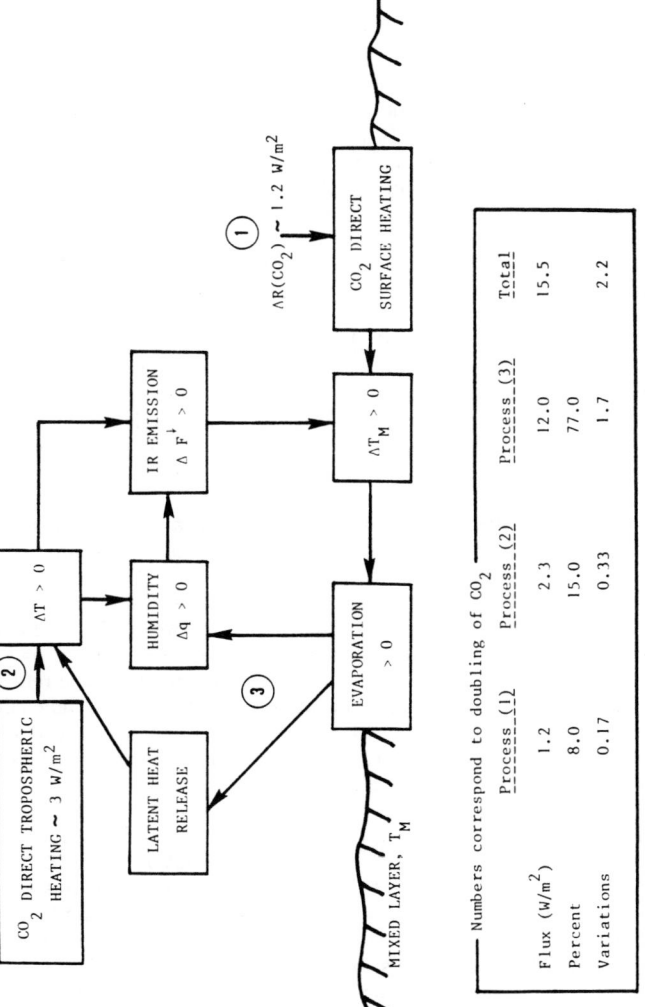

Numbers correspond to doubling of CO_2	Process (1)	Process (2)	Process (3)	Total
Flux (W/m²)	1.2	2.3	12.0	15.5
Percent	8.0	15.0	77.0	
Variations	0.17	0.33	1.7	2.2

Figure 3. Schematic illustration of the ocean-atmosphere feedback processes by which CO_2 increase warms the surface. The table gives the computed contributions by the various processes to the surface radiative heating and to the surface warming. All of the numbers correspond to hemispherically averaged conditions, apply to a doubling of CO_2 and have been obtained with a 1-D radiative-convective model (redrawn from Ramanathan, 1981) (Tricot and Berger, 1986).

The ice-snow albedo feedback is also important, at least locally. For example, sea-ice with overlying snow reflects about 40 to 60 % of the solar radiation, whereas the open ocean at high latitudes reflect only 20 to 30 %. Although the ice albedo feedback amplifies the global warming by only 10 to 20 %, locally near the sea-ice margins and in polar oceans, the warming can be larger than the global warming by factors ranging from 2 to 4.

The cloud feedback, one of the largest sources of uncertainty in the theory of climate change, is investigated in the next section.

8. ALBEDO, CLOUDS AND CLIMATE

Next to water vapour, clouds play an important role in governing the radiation energy balance of the climate system. For example, the reflectivity of clouds increases the albedo of the planet Earth from about 10 % (the clear sky albedo) to the observed value of 30 % (remember than an increase of the planetary albedo of 0.5 % is sufficient to halve the greenhouse effect of CO_2 doubling); this is the short-wave cloud forcing. Clouds are also extremely efficient in trapping the infrared; this will define the long-wave cloud forcing. Moreover, the increased moisture from the warmer oceans should alter cloud distributions and characteristics. The nature of these cloud change and how these changes will affect the net radiative heating is unclear (Ramanathan, 1988).

Hopefully, quantitative estimates of the global distribution of albedo and of clouds-radiative forcing have been obtained from ERBE (Ramanathan et al., 1989b). When atmospheric water vapour condenses to a liquid or a solid, it scatters ultraviolet and visible radiation significantly, compared with cloudless skies. This scattering is both foreward to the surface and backward to space. An individual droplet scatters 85 % of the incident energy in the foreward direction and a cloud of drops can reflect more than 75 %. The resulting enhancement of the planet albedo reduces the solar radiation absorbed by the atmosphere.

The distribution of clear sky albedo for April 1985 reveals that the oceans are the darkest regions of the globe. They have albedoes ranging from 6-10 % in the low latitudes to 15-20 % near the poles (due to low solar elevation). The darkest land surfaces are the tropical rain forest (10-15 %). The brightest parts of the globe are the snow-covered polar areas and the major deserts which reflect between 25 and 40 % of the incident solar radiation. Because clouds reflect more shortwave radiation than the adjacent clear skies, their short-wave forcing, defined as the difference in the reflected solar fluxes between clear and cloudy skies, is negative, leading to a cooling effect. Observations show peaks in the mid- and high-latitudes oceans and over the tropical cirrus systems (less -100 Wm^2). In global average, cloud forcing reduces the absorbed solar radiation by 45 Wm^{-2} (Ramanathan et al., 1989b).

Clouds also significantly enhance the long-wave opacity of the atmosphere, which reduces the radiation emitted to space. This greenhouse effect of clouds (or long-wave forcing) is due to the fact that at their bases they absorb radiation emitted by the warmer surface and at their tops they emit to space at colder temperatures. Deep cirrus clouds, such as the monsoon cloud systems over the Indian Ocean and the jet-stream cirrus clouds at mid-latitudes give a large long-wave forcing (50-100 Wm^{-2}). In average, clouds reduces the long-wave emission to space by 30 Wm^{-2}.

Thus, while the greenhouse effect of clouds warms the planet, their albedo cools it. Unfortunately, the problem is further complicated by the significant dependence of cloud radiation on cloud microphysics (Mitchell et al., 1989; Platt, 1989; Slingo, 1990), which results to the albedo and greenhouse effects being subject to significant variability. The net forcing estimated from ERBE-1985 is varying from -140 to 40 Wm^{-2} roughly (Ramanathan et al., 1989b). In the tropics, the net cloud forcing nearly vanishes, because the two forcing terms cancel each other within the uncertainty of the estimates ($\pm$ 10 Wm^{-2}). Over the midlatitude and polar oceans, persistent stratus and storm-track clouds reduce the radiative heating by as much as 100 Wm^{-2}. As these clouds are very sensitive to surface temperatures and temperature gradients in the troposphere, they can have a significant feedback effect on climate change.

The global averaged long-wave and short-wave forcings for 1985 analysed from ERBE reveal that clouds have a global cooling effect of roughly -15 Wm^{-2}. This should be balanced by a corresponding global mean radiative heating under clear skies to maintain the global energy balance. Without clouds, the planet would be significantly warmer by as much as 10 to 15 K.

As this net cooling is larger than the greenhouse perturbation resulting from a doubling of CO_2 by a factor about 4, small changes (and uncertainties) in the cloud-radiative forcing fields can play a significant role as a climate feedback mechanism (Ingram, 1989).

In the present climate, the balance has tilted toward shortwave forcing - that is net cooling. It is not understood, however, how the balance will shift in an atmosphere where greenhouse gases will be more abundant. Current GCMs suggest, for example, that the balance of cloud forcing is shifting toward less net cooling, i.e. heating, as the planet warms (Ramanathan et al., 1989b; Hansen et al. 1988; Wetherald and Manabe, 1988). For a $\Delta Q = 4 W m^{-2}$ and using the equilibrium CO_2 warming given by GCMs, a λ value of 1.2 $\pm$ 0.3 Wm^{-2} K^{-1} must indeed lead to a positive cloud feedback of (0.7 $\pm$ 0.3), assuming that the reference black-body value for λ is 3.7, the water vapour feedback is 1.4 and the ice albedo feedback is 0.4 :

$$\lambda = 3.7 - 1.4 - 0.4 - (0.7 \pm 0.3) = 1.2 \pm 0.3$$

9. CONCLUSIONS AND THE GEOSPHERE-BIOSPHERE SYSTEM

Although they are very fundamental tool to better understand the greenhouse effect, forcing, response and feedbacks, an important limitation of these radiative-convective models is that they ignore the role of the dynamics, chemistry and biology.

Regions at low latitudes receive more solar energy than do the polar regions. As the long wave emission is much more uniform, there is a net radiative heating at low latitudes accompanied by net cooling at high latitudes. The net radiative heating, H, is balanced by the divergence of heat flow in the atmosphere and oceans and acts, therefore, as the fundamental energy surce for the atmospheric and oceanic circulations. The tropical heating drives the Hadley cell. Strong westerlies generated by the Coriolis force in the upper troposphere on the poleward side of this Hadley cell and mountain barriers give rise to dynamical instabilities that generate eddies in midlatitudes. Oceans transport also heat through wind-driven and thermohaline circulations (Peixoto and Oort, 1984). Including interactions between radiation, convection and planetary-scale dynamics yields to two- and three-dimensional (general circulation) climate models needed to better understand all the physical processes which characterize the climate system.

Moreover, it becomes increasingly apparent that these models should not only account for the interactions among the physics and the dynamics, but also for the interactions of the biosphere with the rest of the system (IGBP, 1988). As a example, Charlson et al. (1987) have suggested the existence of a negative cloud feedback involving marine organisms. The number of condensation nuclei for boundary-layer clouds formed from dimethylsulfide emitted by phytoplanktons would increase in a warmer ocean leading to more cloud drops and to a more negative shortwave cloud forcing which would offset the greenhouse warming due to increase in greenhouse gas concentration. Although, this hypothesis has been criticized (Schwartz, 1989), it deserves great attention (Henderson-Sellers and McGuffie, 1989; Charlson et al., 1989; Gavin et al., 1989; Ghan et al., 1989).

In addition to perturbing Earth's radiation balance, these gases perturb the chemical balance, especially of ozone, hydroxyl radicals, sulfur and nitrogen oxides. Many reactive gases (CO, NO ...) are oxidized by reaction with OH. Several of the reaction products (H_2SO_4, HNO_3, HCl ...) are soluble and are removed by cloud drops leading to acid rain. Others (as chlorofluorocarbons) may play a fundamental role, perturbing the chemistry of the stratosphere leading to the ozone hole. All these chemical reactions have important feedbacks on the climate system.

Unfortunately, no current models accounts for these interactions yet and only attempts to start building advanced climate models are initiated today (CHAMNP, a US Department of Energy initiative, and CSMP at the University Corporation for Atmospheric Research, Boulder, Colorado).

10. REFERENCES

ARRHENIUS S. (1896). On the influence of carbonic acid on the air upon the temperature of the ground. Philos. Mag., 41(5), 237-276.

BERGER A., TRICOT Ch. (1988). Simple climate models. In "Climatic Variations and Impacts, a General Introduction", R. Fantechi (Ed.), Commission of the European Communities, European School of Climatology and Natural Hazards.

CESS R.D. (1989). Gauging water-vapour feedback. Nature, 343, 736-737.

CESS R.D., POTTER G.L., BLANCHET J.P., BOER G.J., GHAN S.J., KIEHL J.T., LE TREUT H., LI Z-X, LIANG Z-Z, MITCHELL J.F.B., MORCRETTE J.J., RANDALL D.A., RICHES M.R., ROECKNER E., SCHLESE U., SLINGO A., TAYLOR K.E., WASHINGTON W.M., WETHERALD R.T., YAGAI I. (1989). Interpretation of cloud-climate feedback as produced by 14 atmospheric general circulation models. Science, 245(4917), 513-516.

CHARLSON R.J., LOVELOCK J.E., ANDREAE M.D., WARREN S.G. (1987). Oceanic phytoplankton, atmospheric sulphur, cloud albedo and climate. Nature, 326 (6114), 655-668.

CHARLSON R.J., LOVELOCK J.E., ANDREAE M.O., and WARREN S.G. (1989). Sulphate aerosols and climate. Nature, 340, 437-438.

DICKINSON R.E. and CICERONE R.J. (1986). Future global warming from atmospheric trace gases. Nature, 319, 109-115.

GAVIN J., KUKLA G., and KARL Th. (1989). Sulphate aerosols and climate. Nature, 340, 438.

GHAN S.J., PENNER J.E., and TAYLOR K.E. (1989). Sulphate aerosols and climate. Nature, 340, 438.

GOODY R.M. (1964). Atmospheric Radiation. I. Theoretical basis. Oxford Monographs on Meteorology, P.A. Sheppard (Ed.), Oxford at the Clarendon Press, 436pp.

HANSEN J., FUNG I., LACIS A., RIND D., LEBEDEFF S., RUEDY R., RUSSELL
G. (1988). Global climate change as forecast by Goddard Institute for Space
Studies three-dimensional model. J. Geophys. Res., 93(D8), 9341-9364.

HANSEN J., LACIS A., PRATHER M. (1989). Greenhouse effect of chlorofluorocarbons and other trace gases. J. Geophys. Res., 94(D13), 16,417-16,421.

HENDERSON-SELLERS A. and McGUFFIE K. (1989). Sulphate aerosols and climate. Nature, 340, 436-437.

IGBP (1988). The International Geosphere-Biosphere Programme : a study of global
change. A plan for Action. International Geosphere-Biosphere Programme, Report n°4, IGBP secretary, Stockholm, Sweden.

INGRAM W.J. (1989). Modelling cloud feedback on climate change. Weather, 44(7),
303-311.

JONES M.D.H., HENDERSON-SELLERS A. (1990). History of the greenhouse effect. Progress in Physical Geography, 14(1), 1-18.

LEE R., BARKSTROM B.R., CESS R.D. (1987). Characteristics of the Earth radiation budget experiment solar monitors. Appl. Opt., 26(15), 3090-3096.

MANABE S. (1983). Carbon dioxide and climatic change. Advances in Geophysics,
25, 39-82.

MANABE S. and WETHERALD R.T. (1975). The effects of doubling the CO_2
concentration on the climate of a General Circulation Model. J. Atmos. Sci.,
32(1), 3-15.

MITCHELL J.F.B. (1989). The "Greenhouse" effect and climate change. Rev. Geophys., 27(1), 115-140.

MITCHELL J.F.B., SENIOR C.A., INGRAM W.J. (1989). CO_2 and climate : a
missing feedback ? Nature, 341, 132-134.

MÖLLER F. (1963). On the influence of changes in CO_2 concentration in air on the
radiation balance of Earth' surface and on climate. J. Geophys. Res., 68(13),
3877-3886.

PEIXOTO J.P., OORT A.H. (1984). Physics of climate. Reviews of Modern Physics,
56(3), 365-429.

PLASS G.N. (1956). Carbon dioxide theory of climatic change. Tellus, 8, 140-154.

PLATT C.M.R. (1989). The role of cloud microphysics in high-cloud feedback effects
on climate change. Nature, 341, 428-429.

RAMANATHAN V. (1981). The role of ocean-atmosphere interactions in the CO_2
climate problem. J. Atmos. Sci., 38(5), 918-930.

RAMANATHAN V. (1987). The role of Earth radiation budget studies in climate
and general circulation research. J. Geophys. Res., 92(D4), 4075-4095.

RAMANATHAN V. (1988). The greenhouse theory of climate change : a test by an
inadvertent global experiment. Science, 240, 293-299.

RAMANATHAN V., BARKSTROM B.R., HARRISON E.F. (1989a). Climate and the Earth's radiation budget. Physics Today, 42(5), 22-33.

RAMANATHAN V., CESS R.D., HARRISON E.F., MINNIS P., BARKSTROM B.R., AHMAD E., HARTMANN D. (1989b). Cloud-radiative forcing and climate : results from the Earth radiation budget experiment. Science, 243, 57-63.

RAVAL A., RAMANATHAN V. (1989). Observational determination of the greenhouse effect. Nature, 343, 758-761.

SCHLESINGER M.E. (1985). Analysis of results from energy balance and radiative-convective models. In "Projecting the Climatic Effects of Increasing Carbon Dioxide", MacCracken M.C., Luther F.M. (Eds), US Department of Energy, DOE/ER-0237, 281-319.

SCHLESINGER M.E. (1989). Model projections of the climatic changes induced by increased atmospheric CO_2. In "Climate and Geo-Sciences", A. Berger, S. Schneider, J.Cl. Duplessy (Eds), 375-416, NATO ASI Series C, vol. 285, Kluwer Academic Publishers, Dordrecht, Holland.

SCHWARTZ S.E. (1989). Sulphate aerosols and climate. Nature, 340, 515-516.

SLINGO A. (1990). Sensitivity of the Earth's radiation budget to changes in low clouds. Nature, 343, 49-51.

TRICOT Ch., BERGER, A. (1986). Modeling the response of temperature to change in trace gases : a review with emphasis on the transient response. Scientific Report 1986/7, Institut d'Astronomie et de Géophysique G. Lemaître, Université Catholique de Louvain, Louvain-la-Neuve.

TYNDALL J. (1861). On the absorption and radiation of heat by gases and vapours, and on the physical connexion of radiation, absorption and conduction. The London, Edinburgh and Dublin Philosophical Magazine and Journal of Science, 4th Series, 22, 169-194, 273-285.

U.K. Department of the Environment (1989). Global climate change. Department of the Environment in association with The Meteorological Office, London.

WHETHERALD R.T., MANABE S. (1988). Cloud feedback processes in a general circulation model. J. Atmos. Sci., 45, 1397-1415.

WILLSON R.C., HUDSON H.S., FROHLICH C. and BRUSA R.W. (1986). Long-term downward trend in total solar irradiance. Science, 234, 1114-1117.

WILLSON R.C. and HUDSON H.S. (1988). Solar luminosity variations in solar cycle 21. Nature, 332 (6167), 810-812.

Annex 1 - The greenhouse gases

Of the radiatively active gases in the current atmosphere, only H_2O, CO_2, CH_4, O_3, N_2O, CFC-11 and CFC-12 are in sufficient concentrations to be important in Earth's overall thermal budget (Table A1). Of these, the strongest absorber by far is water vapour. The vibrational-rotational bands of water vapour block radiation at wavelengths lower than 8 μm and rotational bands block wavelengths larger than 18 μm.

Table A.1. The main properties of the greenhouse gases (Dickinson and Cicerone, 1986; U.K. Department of the Environment, 1989)

Greenhouse gases	pre-industrial concentration (ppmv)	1988 concentration (ppmv)	current rate of change (% per year)	atmospheric life time (years)	relative greenhouse effect per molecule
CO_2	275	350	0.4	60	1
CH_4	0.75	1.7	1	10	30
N_2O	0.285	0.31	0.3	150	160
O_3 (lower atmosphere)	0.02 ?	0.06	1.5	0.2	2,000
CFC-11 ($CFCl_3$)	-	$260\ 10^{-6}$	5	75	21,000
CFC-12 (CF_2Cl_2)	-	$440\ 10^{-6}$	4	110	25,000

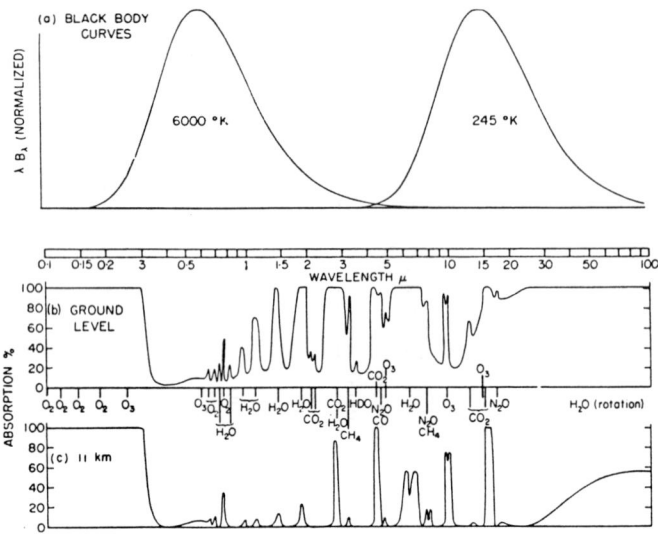

Figure 1: A.1. (a) Blackbody curves for 6000°K and 245°K. (b) Atmospheric gaseous absorption spectrum for a solar beam reaching ground level. (c) The same for a beam reaching the temperate tropopause. The axes are chosen so that area in (a) are proportional to radiant energy. Integrated over the Earth's surface and over all solid angles the solar and terrestrial fluxes are equal; consequently, the two blackbody curves are drawn with equal areas beneath them. An absorption continuum has been drawn beneath bands in (b). This is partly hypothetical because it is difficult to distinguish from the scattering continuum, particularly in the visible and near-red spectrum. Conditions are typical of mid-latitudes and for a solar elevation of 40° or diffuse terrestrial radiation (Goody, 1964).

Water in either vapour or in cloud form absorbs solar radiation and absorbs and emits thermal radiation. Estimates of G over the open oceans for a cloudless (clear sky) and cloudy atmosphere for 1985 (Raval and Ramanathan, 1989) reveal that for a total trapping of 179 Wm^{-2}, 33 Wm^{-2} were trapped by clouds. However, these water vapour states are internal to the climate system, that is, their distributions are controlled by climate processes themselves (the atmospheric hydrological cycle) rather than by sources uncoupled to the climate system (they are not resulting from direct man-made perturbations). Understanding how water vapour concentrations and cloud radiative properties might change in the future thus requires knowledge of how these terms function as feedbacks in the climate system. This is why our primary thrust is concentrated on the role of the other less - abundant trace gases whose concentrations are being directly perturbed by human activities.

Ozone is the only other atmospheric gas which absorbs much solar radiation, mainly in the stratosphere.

CO_2 is but a very weak absorber of solar radiation but it strongly absorbs and emits terrestrial radiation at the wavelength $\sim$ 12 to 18 μm (Figure A1).

The other atmospheric trace gases only affect the atmospheric heat budget through their absorption and emission of the thermal infrared radiation, primarily over the wavelength range 6 to 16 μm.

The remaining spectral region from 8 to 12 μm is known as the window because of the atmosphere relative transparency to radiation over these wavelengths. However, according to the Planck function, the emission of the Earth's surface is maximum in this window and amounts ~ 100 Wm^{-2}. But as this emission varies with temperature approximately as exp (-1500/T) a trace gas at a temperature $\sim 33°C$ less than the surface will only re-emit about half as much energy as it absorbs. Finally, the additional presence of clouds will be responsible for only 20 Wm^{-2} escaping to space from the Earth's surface (Dickinson and Cicerone, 1986). Interestingly enough, CH_4, O_3, N_2O, CFC-11 and CFC-12 all have strong absorption bands in the atmospheric window region. The trapping by O_3 and CFC's whose stronger bands are in the middle of the window region, is affected very little by overlap with water vapour and with carbon dioxide absorption. CH_4 and N_2O whose strongest bands are on the short wavelength edge of the window region, lose about half of their trapping by such overlap.

These trace gases absorb and emit as functions of wavelength in discrete lines with extended wings. The weakest lines absorb radiation significantly only in their line core and this absorption increases essentially linearly with concentration of the absorber gas. Somewhat stronger lines absorb radiation mostly in the wings. Increasing absorber concentration pushes the peak absorption farther into the line wings so that wing absorption increases with the square root of the product of atmospheric pressure and absorber concentration. With even stronger line strengths, the peak absorption is pushed so far into line wings that other lines within the band overlay the absorbing wings and absorption only increases logarithmically with increasing absorber concentration.

The CFC-11, CFC-12 and tropospheric ozone are present in such small concentrations that their absorption of thermal radiation is nearly proportional to their concentration. CH_4 and N_2O, being relatively more abundant, increase their absorption essentially according to the square root of their concentration. CO_2 absorption is proportional to the logarithm of its concentration.

Annex 2 - A simple radiative-convective model

Suppose we divide the Earth's atmosphere into two layers so that each layer just absorbs the infrared radiation incident on it and let each layer have an infrared optical thickness of $\tau = 1$. As the principal absorber in the Earth's atmosphere is water vapour, our 2 layers are centered at heights of 3 and 0.5 km. Both layers radiate as black bodies upwards and downwards and the surface radiates up. Since the planet is emitting at its effective temperature, T_e, $T_1 = T_e$. The energy balance for the 2 layers leads to

$$T_2^4 = 2\,T_e^4 \qquad T_g^4 = 3\,T_e^4$$

From this, one can calculate that the temperature of the top layer, T_1 would be 255 K, of the lower layer, T_2 303 K and of the surface, T_g 335 K. Note that the surface air temperature is lower than the surface temperature and that the radiative temperature profile is unstable.

EUROPEAN SCHOOL OF CLIMATOLOGY AND NATURAL HAZARDS

Course on

"Climate and Global Change"

2. BIOGEOCHEMICAL CYCLES AND THEIR PERTURBATION
   BY HUMAN ACTIVITIES

THE GLOBAL CARBON CYCLE

Ernst Maier-Reimer

Max-Planck-Institut f.Meteorologie Hamburg

Summary

Basic components of the global carbon cycle are described which are treated in existing high resolution global numerical models. With modern supercomputers the models give a fairly realistic reproduction of the observed large scale structure of geochemical tracers in the ocean and thus provide a reliable tool for prognoses.

Introduction

Almost a century ago, S.Arrhenius postulated a dynamical link between the atmospheric CO_2- content and the global climate. In 1938 G.Callendar gave the first warning that human activities could change the CO_2-concentration to an amount which may be sufficient for substantial climatic changes. These warnings were intensified in 1957 by R.Revelle and H.Suess. They pointed out that even after complete equilibration between ocean and atmosphere of the CO_2-concentration, an input of fossil fuel CO_2 will accumulate much less in the ocean than the actual inventories of CO_2. A rise of 1 % in the ocean is in equilibrium with a rise of ten percent in the atmosphere. Consequently, the anthropogenic emissions of CO_2 – seemingly negligibly small compared to the global inventory – are sufficient to raise the atmospheric content substantially. As a consequence of this finding the Scripps institution of Oceanography installed in 1958 the first continuously monitoring station of CO_2 at the Mauna Loa volcano on Hawaii island. Since these measurements are carried out, the atmospheric content of CO_2 has raised from 315 to 350 ppm (parts per million of volume) (Fig.1).

Carbon is the key element of life . The natural food chain is initiated by the ability of plants to transform CO_2 into organic compounds in the presence of light. Carbon exists in three isotopes. 99% occur as the stable ^{12}C, 1 % is in the form of ^{13}C which is stable, too. The radioactive isotope ^{14}C contributes a fraction of only 10^{-12}. It is produced from nitrogen atoms in the upper atmosphere by cosmic rays. The mean life time is 8267 years. Due to the different weight the isotopes behave differently in the incorporation into plant material as well as in the degassing from water. Thus, by careful analyses of isotopic ratios we can get important hints for the emission history and oceanic ventilation rates, which provide crucial tests for constraining and calibrating models.

Fig.2 gives a gross sketch of the natural carbon pools and exchange rates together with the anthropogenic modifications The numbers mean Gigatons and gigatons/year, resp. One gigaton is 10^{12} kg, the weight of roughly one half cube kilometer of pure graphite. The emissions by burning of fossil fuel are known within a very few percent of uncertainty. The estimates of the changes in the terrestrial biosphere, however, are quite uncertain. About two gigatons are emitted from deforestation; the effects of soil erosion, soil formation, and of intensified use of fertilisers is rather poorly known. Another uncertainty lies in the fertilising effect of increased CO_2: carbon is the main contribution to plantal dry mass; plants react on enhanced CO_2 levels with enhanced growth and a better utilisation of moisture. This effect has been clearly found in greenhouses with small plants. It is, however, an open question whether large trees, which represent the main part of terrestrial biomass, react in the same way.

During the past decades the increase of the atmospheric CO_2 content equalled appr. the half of the emissions from fossil fuel burning. The main sink of the other half must have been the ocean. Current estimates of the oceanic circulation exclude the possibility that the ocean has taken up much more than that. Consequently, an upper limit is given for the cumulative effects of all processes of the land biosphere. If we assume that the land biosphere has acted as a significant source of CO_2, we are faced to the "missing sink" problem which has been extensively discussed in the past. Recently, some agreement has been achieved that the net effect of the terrestrial biosphere in the past two decades must have been rather a small sink than a large source.

In the last years the climatologists interest in the carbon cycle has further increased: measurements of air bubbles enclosed in ice clearly revealed that during the last glacial cycle (150,000 years) strong natural variations of atmospheric CO_2 have occured. On a long time scale the variations are closely related to the global ice volume. It is, however, open how this strong correlation reflects any kind of causality. It is unlikely that astronomic variations could influence directly the carbon cycle and thus create climatic variations by the greenhouse effect. On the other hand, the increased solubility in a colder climate of CO_2 in the ocean can account for only a quarter of the observed fluctuations. Several hypotheses have been proposed to explain the connection but they are all ad-hoc assumtions for explaining just one effect. They all imply consequences which are not compatible with other data.

From the same measurements the anthropogenic rise of atmospheric CO_2 has been reconstructed very precisely. In pre-industrial times the concentration was at 280 ±5 ppm. The precise knowledge of the historical increase provides a strong constraint for the net emission from deforestation. The isotopic ratio ^{13}C / ^{12}C in tree rings gives additional information on the emission history. Due to the minor weight, ^{12}C diffuses easier through the membranes of plantal cells than the heavier isotopes. Consequently, plant material has an isotopic ratio 26 permille lower than that of inorganically bounded carbon. In the past 160 years the atmospheric isotopic ratio has decreased by 1.5 permille. By combination of these different data a reliable estimate of the biogenic emission as well as the cumulative uptake of the ocean has been achieved (The emissions from fossil fuel burning are rather precisely known).

Partial pressure and inorganic CO2 chemistry

The ocean contains apr. 37000 Gt of dissolved carbon; the atmosphere contains 750 Gt. From the oceanic carbon only 1 % is gaseous CO_2 which is in direct equilibrium with the atmosphere. 90% are present as bicarbonate (HCO_3^-) and 10 % as carbonate (CO_3^{--}). The partial pressure is related to the concentration of dissolved CO_2

$$[CO_2] = \alpha \ pco_2,$$

where α is the physical solubility. The dissociative equilibrium with the carbonate fractions and the hydrogen ions H^+ is regulated by the law of mass action :

$$[H^+][HCO_3^-] = K_1 [CO_2]$$
$$[H^+][CO_3^{--}] = K_2 [HCO_3^-]$$

(The square brackets denote the concentrations of the ions).
The constants α, K_1, and K_2 vary with temperature. Between 0 and 30^0 C the solubility decreases by more than 50 %, whereas the dissociation constants increase. The overall solubility decreases with increasing temperature but much less than the solubility of the gaseous fraction. At a constant partial pressure the effective solubility decreases by 0.4 % /K.

The differences of the partial pressures of surface waters are compensated by transports in the atmosphere. In upwelling regions, especially around the equator, the water releases CO_2 into the atmosphere; in high latitudes the water is cooled and takes up CO_2. The resulting equator to poles transport in the atmosphere is appr. 2 Gt per year. Fig. 3 shows the global distribution of the partial pressure. It must be emphasized, however, that details of this figure are still in debate. Recent measurements created controversies especially concerning the role of the northern Atlantic.

The equilibrium reaction of the ocean water to a small emission I from fossil fuel burning is easily computed: with the abbreviations A = $[CO_2]$, B = $[HCO_3^-]$, and C = $[CO_3^{--}]$ we get for small changes

$$H\ dB + B\ dH = K_1\ dA$$
$$H\ dC + C\ dH = K_2\ dB$$

the mass balance for the total carbon is

$$dA + dB + dC = I.$$

The fourth equation results from the conservation of electric charges

$$dB + 2\ dC - dH = 0.$$

With realistic values A = 10^{-5}, B = $2\ 10^{-3}$, C=$2\ 10^{-4}$, K_1 = 10^{-6}, K_2 = 10^{-9}, and H = 10^{-8} mol/kg we get approximately dB = 20 dA, dC = -10 dA, and dH = 0.012 dA. The uptake of CO_2 has a completely different relation between the CO_2-fractions than the actual content of CO_2. With increasing CO_2 the dissociation into the ionic forms is reduced, and, consequently, the uptake capacity is reduced, too.

The relation dAdC<0 holds also for reactions from the other end of the dissociation chain; at the formation of calcium carbonate we encounter the -seemingly paradox - fact that taking out carbon ions from the water will increase the partial pressure of CO_2. The ocean contains large quantities of calcium ions. The upper ocean is supersaturated with respect to the formation of calcium carbonate, at the surface by appr. 300 % . It is still an open question why there is no precipitation of calcite crystals. It is suggested that the presence of magnesium or of organic material inhibits the formation, but there is no quantitative formulation of such mechanisms.

In the real ocean these computations turn out to be slightly more complicated due to the presence of other weakly dissociating acids, mainly the borate system which has a concentration of 4 10^{-4} mol/kg. The simple equation for the conservation of electric charges is generalized by the conservation law of "alkalinity" which contains the negative charges of all chemical compounds whose dissociative state may be subject to change in the marine environment (NaCl, for instance, does not contribute to the alkalinity since it will remain completely dissociated). It must be emphasized, however, that this definition is not unique. Many other formulations have been stated to define that charge related property of water which is not changed by physical modifications like heating or compression. Alkalinity and ΣCO_2, the sum of all carbon fractions, are easily to measure; together they determine completely the dissociative state of the water which then can be computed with the temperature dependent coefficients.

For quick estimates it is convenient to consider only the main contributions to ΣCO_2 and Alkalinity. Then:

$$\Sigma CO_2 = HCO_3^- + CO_3^{--}, \text{ and}$$
$$CA = HCO_3^- + 2 CO_3^{--} \quad (\text{"carbonate alkalinity "}).$$

From these reduced definitions it follows immediately that

$$CO_3 = CA - \Sigma CO_2$$

and, applying the law of mass action:

$$PCO_2 = \frac{K_2 (2 \Sigma CO_2 - CA)^2}{\alpha K_1 (CA - \Sigma CO_2)}$$

The adaptation of ocean water to an enhanced CO_2-level in the atmosphere occurs, of course, only at the surface. The effective uptake capacity of the ocean depends crucially on how fast the CO_2-enrichened water is transported into deeper layers and replaced by CO_2-deficient water from below. In a wide range the behaviour of the coupled system with respect to the atmospheric CO_2 - concentration can be described by a linearized impulse-response (Green) function. Fig. 4 shows some empirically determined response functions. The dotted lines stem from an extremely simple ocean model consisting of only three boxes. The solid lines were computed with a high resolution circulation model consisting of 25,000 grid points. In the computations a sudden input of 25, 100, and 300 % of the stationary atmospheric content into the atmosphere has been assumed. Despite the huge differences between the models, the results are rather similar. The curves for 25% and 100% are almost identical which implies that in that range the system indeed behaves linearly. For the 300% input the changes of the system become important; the relative uptake is significantly reduced. Within the thickness of the lines the curves can be described as superpositions of very few exponentials. The solid line for doubling is described by

$$G(\tau) = 0.142 + 0.088 \exp(-\tau/1.7y) + 0.206 \exp(-\tau/19y)$$
$$+ 0.323 \exp(-\tau/80y) + 0.241 \exp(-\tau/319 y).$$

The contributions to this function allow for a firect interpretation: 14 % of the emission remain in the atmosphere, 9% enter quickly the seasonally varying mixed layer of the upper ocean, 20 % enter the warm water lens of the upper kilometer with a time constant of 20 y. The main parts of the ocean takes up CO_2 very slowly with time constants of 80 and 320 y, dependent on how close they are located to the regions of active deep water formation in the northern Atlantic and around Antarctica.

Biotic modifications

In the real ocean, the gradients of CO_2 are much stronger than could be explained by the temperature gradients. In the upper 10 - 100 m where irradiation is sufficient to support photosynthesis ("euphotic" zone) heavy organic material is formed which exports part of the CO_2-content to greater depths. The elementary composition is almost constant in the different species (Redfield ratio):

$$P : N : C : O : Ca = 1 : 17 : 122 : 175 : 33$$

The meaning of oxygen in this expression is the excess of the used CO_2 which is released to the water during production and consumed from the surrounding water during remineralisation. The production is characterized by a very small standing stock compared to the annual gross production which is estimated to be appr. 30 Gt per year. The cumulative production of two years drifts as dead material (Particulate Organic Carbon) and is remineralized by bacteria in regions where oxygen can be consumed. Some 1000 – 3000 Gt of carbon exist in the form of Dissolved Organic Carbon compounds with molecular weights up to a few 100,000. The size of the stock and the role of this DOC in the marine life cycle has just recently become the subject of a very vivid debate. Actually, it may be stated that nobody knows what DOC really means.

The large scale shifts of CO_2 depend mainly on the exports out of the euphotic zone in large aggregates (0.5 μm – 1 mm). These sum up to appr. 5 Gt per year. On a long term, this number is in equilibrium with the large scale upwelling of nutrients, mainly nitrate and phosphate.

Calcareous shells are produced together with soft tissue at a very inhomogeneous relation. When silicate is available, plantal algae seem to prefer to built their skeleta from silicate. Consequently, the rate of calcite production to soft tissue production ranges between 0.01 and 1. In the global average it is appr. 0.25. The effect on the partial pressure of the two forms of production is quite different: in soft tissue production, carbon is extracted from the water as HCO_3^-, and the alkalinity is increased by the replacement of NO_3^- ions by carbon ions. Consequently, the partial pressure is reduced by the production. As mentioned above, the production of calcite shells increases the partial pressure. On the global average the reduction from the formation of soft tissue acts stronger. A sudden switchoff of all biological processes would increase the atmospheric PCO_2 by 200–300 ppm. This consideration often leads to the erroneous statement that the biological pump plays a crucial role in the uptake of fossil fuel CO_2. As long as the oceanic circulation with the upwelling of nutrients and the atmospheric conditions for productivity remain unchanged, the biological pump contributes nothing to the uptake of excess CO_2 by the ocean; it just modifies the basic state which we are now disturbing.

Calcareous shell material in two cristallisation forms of calcium carbonate- calcite and, at minor importance, aragonite – sinks rather quickly to the bottom of the ocean and is able to built up sediment layers. The solubility product increases with the pressure. At a depth of 4 km in the Atlantic and 1.5 km in the Pacific the water becomes undersaturated with respect to calcite; the corresponding solubility horizon for aragonite lies one kilometer closer to the surface. Below these horizons sediment can form only in regions where the rain of carbonates exceeds the dissolution rate. As time goes on, the sediment transforms into solid rock. Only the upper 5 to 10 centimeter ("bioturbated layer") are still in reactive contact with the water. On the time scales of millennia the dissolution will be increased in the acidfied environment and support the net uptake of fossil fuel CO_2 by the ocean. On a very long term, the sediments participate in the game of plate tectonics. The resulting sink of CO_2 is compensated by volcanic emissions and weathering of old carbonate rocks. Crude estimates of this turnover mechanisms are in the order of 0.1 Gt/year. All carbon atoms of the outer globe have undergone several times the transition from the dissolved or gaseous form to the rock state.

The terrestrial biosphere

The turnover rates of the land biosphere are of the same order as of the marine biosphere, but on land, the standing stock is several hundred times higher. The mean life time of plants on land is 10 to 20 years. The production of biomass exhibits even in the global average a pronounced seasonal variation. The peak values are in the northern summer due to the asymmetry of continental distribution with respect to the equator. The Mauna Loa record shows minima in summer, when the binding of CO_2 into leafs has its maximum. In fall and winter the bacterial decomposition of litter dominates the production and CO_2 is released to the atmosphere. The seasonal cycle is strongest at the northernmost station in pt.Barrow, Alaska. Around the equator it exhibits about one half of the amplitude. At the South Pole station, the seasonal cycle is much weaker and reversed in sign.

Deforestation reduces the carbon content of the biosphere by about 2 Gt / y. A small part is used for construction of houses and furniture – there are no data for this type of sink. The major part, however, is lost by burning and enters the atmosphere as a source.

The soil contains 2–4 times more CO_2 than the living plants. Reliable estimates for its turnover are extremely difficult. Rivers transport about one Gt/y of CO_2 from land into the ocean. This amount could be increased in future by increased soil erosion. It is, however, not clear whether this export acts as a source of atmospheric CO_2 or, since this freight is partly oxidized and combined with nutrients, rather as a sink by increasing the biological pump in the ocean.

On the other hand, the land biosphere acts also as a sink for excess CO_2. Measurements on trees gave evidence that during the last 30 years the structures of leaf surfaces have changed. Due to the higher partial pressure of CO_2 plants need a smaller opening of their pores to get the CO_2 needed for growth. Consequently, they have less evaporation and a better usage of water. Experiments in greenhouses with a doubled CO_2 produced plants of almost twice the normal size. The increased agriculture in formerly dry areas, stimulated by increased usage of fertilizers, clearly acts as an additional sink for atmospheric CO_2. The observed increase of the seasonal amplitude in the Mauna Loa record clearly indicates an increase of biological productivity in the northern hemisphere. On the global average these fertilizing effects seem to compensate the losses by deforestation. Without them it would be rather impossible to close the global CO_2 balance.

Carbon Cycle Modelling

Most of the processes described in the previous sections are simulated in the carbon cycle model developed at the MPI Hamburg together with colleagues from Osnabrueck university (land biosphere) and from SIO San Diego. The land biota model is based on FAO data on productivity, vegetation types, climate, and historical land use in 2433 boxes each representing an area of 2.5x2.5 deg. The ocean biota model is based on the current field ofa physical model of the general circulation in a similar horizontal resolution in 15 vertical levels. The tracers of the model are phosphate, alkalinity, oxygen, silicate, ΣCO_2, POC, and calcite. At the bottom the model is closed by interactive layers of organic and calcite sediments. All carbon variables appear in three isotopes. A one layer diffusive atmospheric model is in contact with the upper layer of the ocean and with the land biosphere.

The functional formulation of the biological processes is not unique. All data from measurements exhibit enormous scatter. A validation is possible primarily by comparison of the resulting tracer fields with the observed distributions. This, of course, will not give a validation for the formulation of an isolated process but only for the system as a whole. It turns out that all tracer fields in the ocean depend critically on the structure of the velocity field. They provide much stronger validation criteria for the physical model than the genuine physical tracers like temperature and salinity. Some of the tracers, especially phosphate and radiocarbon, are almost independent of the details of the biological model. The most critical distributions for the biological part are the surface PCO_2 and the solubility horizons of calcite.

Fig. 5 shows the models distribution of phosphate as meridional sections of Atlantic and Pacific which can be compared with the profiles measured during the GEOSECS experiment. The characteristic differences between the circulation fields of the major oceans, which play a key role in the climate system, come out very clearly.

Fig. 6 shows the reconstruction of the historical increase of the atmospheric CO_2 together with the data from Mauna Loa and from ice cores. In this model run, the terrestrial biosphere acted as a source until 1960. In the time until 1984 it removed up to 0.5 Gt/y from the atmosphere due to the different fertilizing effects.

Suggested reading:
-Bolin, B., E.T.Degens, S.Kempe, and P.Ketner: The global carbon cycle, 491 pp. SCOPE 13, Wiley & sons, 1979.
-Bolin,B., B.R.Döös, and J.Jäger (eds.): The Greenhouse effect, climatic change, and ecosystems, 541 pp. SCOPE 29, Wiley&sons, 1986.
-Broecker, W.S. and T.H.Peng. Tracers in the Sea. Eldigio Press, Lamont - Doherty Geological Observatory , Palisades, N.Y., 690pp., 1982.
-Degens, E.T.: Perspectives on Biogeochemistry, 423 pp. Springer, 1989.
-Sundquist,E.T. and W.S.Broecker(eds.): The Carbon Cycle and AtmosphericCO2: Natural Variations Archean to Present. AGU Geophysical Monograph 32, 627 pp. Washington D.C. 1985.
-Tellus 39 B. Proceedinges of the 2 nd international conference on atmospheric carbon dioxide, its sources, sinks and transport. 1987
-Trabalka,J.R. and D.E.Reichle (eds.). The changing carbon cycle - a global analysis. 592 pp., Springer 1986.

Figure captions

Fig. 1 Increase of CO_2 since 1958 measured at Mauna Loa

Fig.2 Major carbon fluxes and pools of the global system (Degens)
      Units are Gt and Gt/y

Fig.3 Air – Sea difference of PCO_2 (Takahashi). Units are μbar.

Fig.4 empirical impulse-response (Green) functions for a sudden release of CO_2 into the atmosphere from the full circulationmodel (solid lines) and from Siegenthalers outcrop – diffusion model (dotted lines). The curves were computed from a sudden increase on the atmosphere by 25 % (a), 100% (b), and 300% (c),resp.

Fig. 5a Phosphate in the model Atlantic   (μ mol/kg)

Fig. 5b Phosphate in the model Pacific   (μ mol/kg)

Fig. 5c Phosphate in the real Atlantic (GEOSECS, 1973) (μ mol/kg)

Fig. 5d Phosphate in the real Pacific (GEOSECS, 1973) (μ mol/kg)

Fig. 6 Reconstruction of the increase of atmospheric PCO_2 since 1750

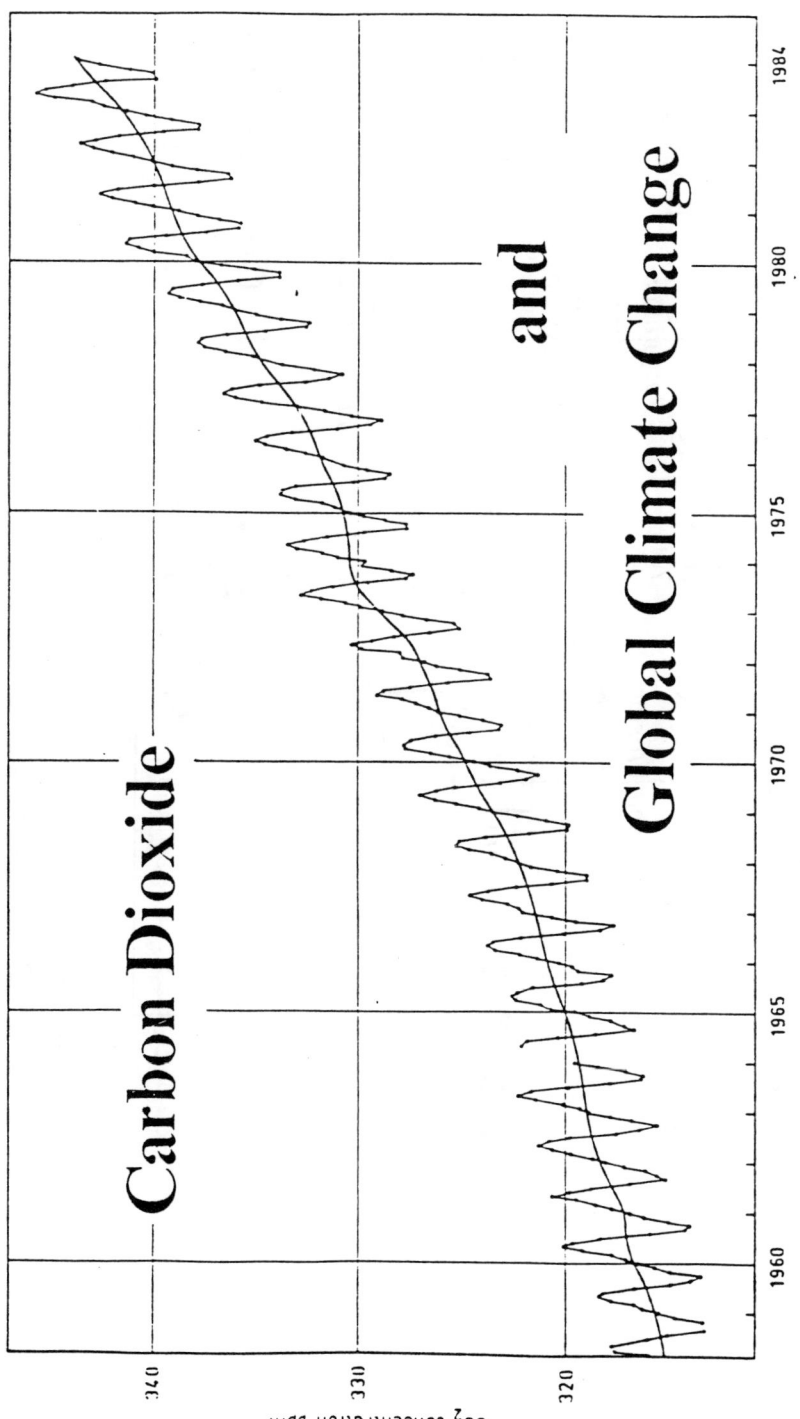

Carbon Dioxide and Global Climate Change

Figure 1

Concentration of atmospheric CO_2 at Mauna Loa Observatory, Hawaii. Dots indicate monthly averages determined from continuous measurements. Based on data reported by Bacastow and Keeling (1981), supplemented by data from recent years supplied by personal communication

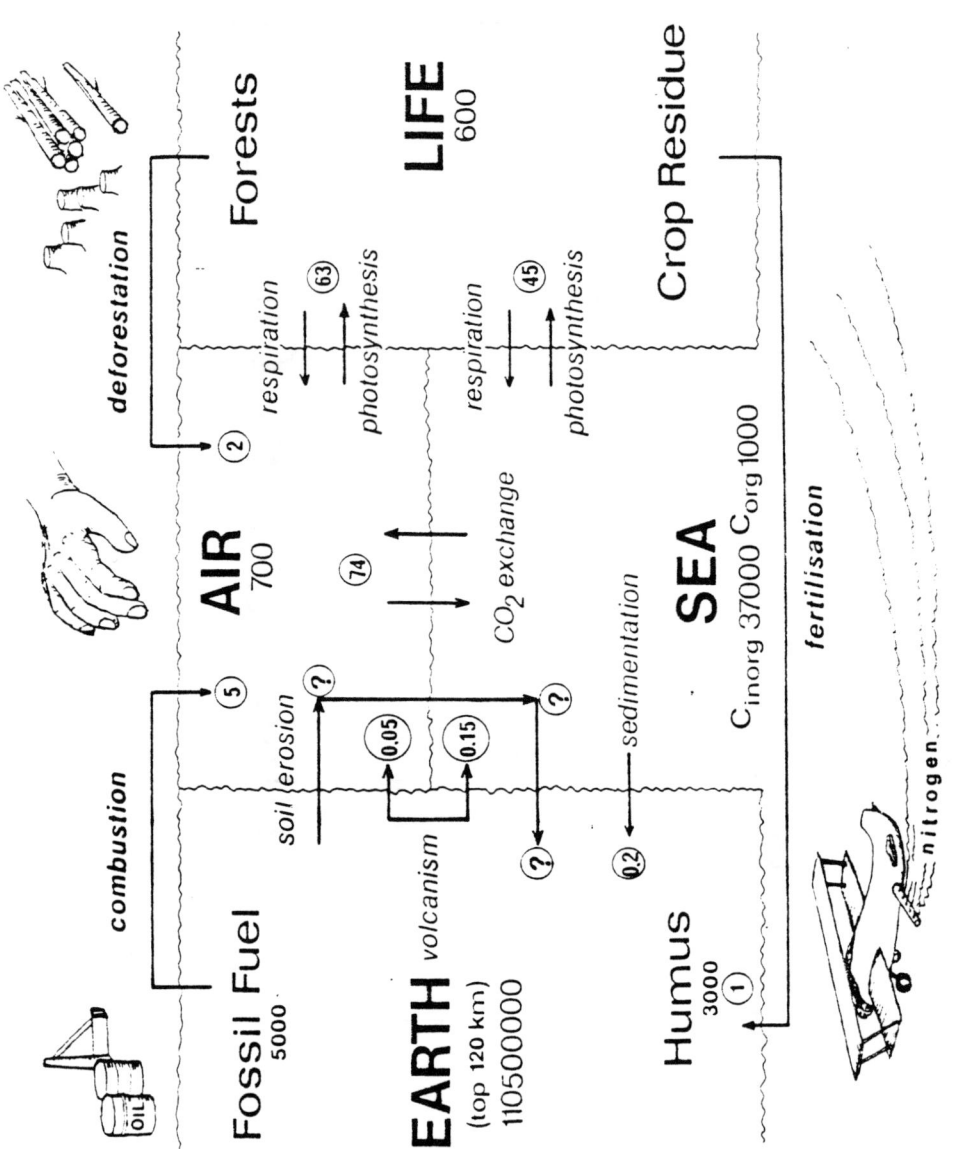

Figure 2

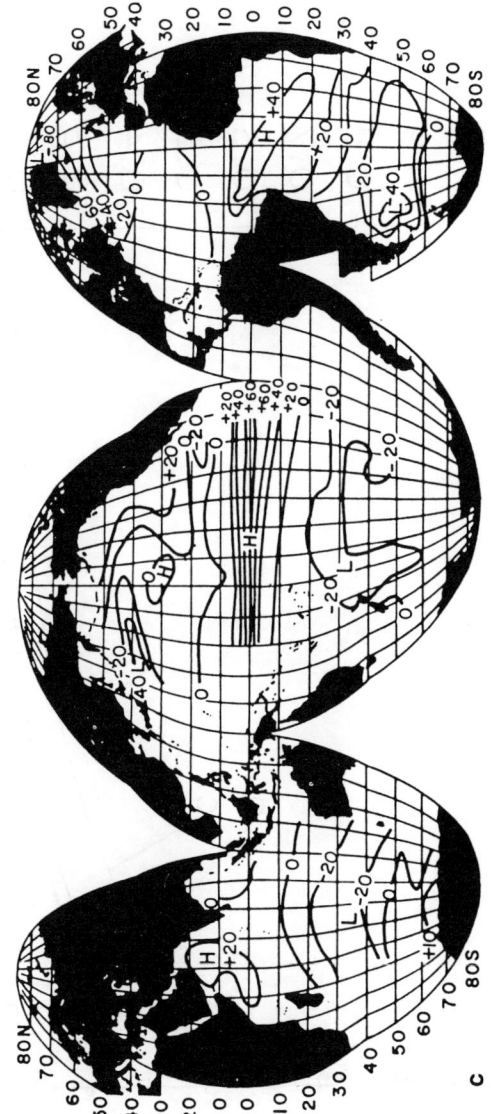

Figure 3

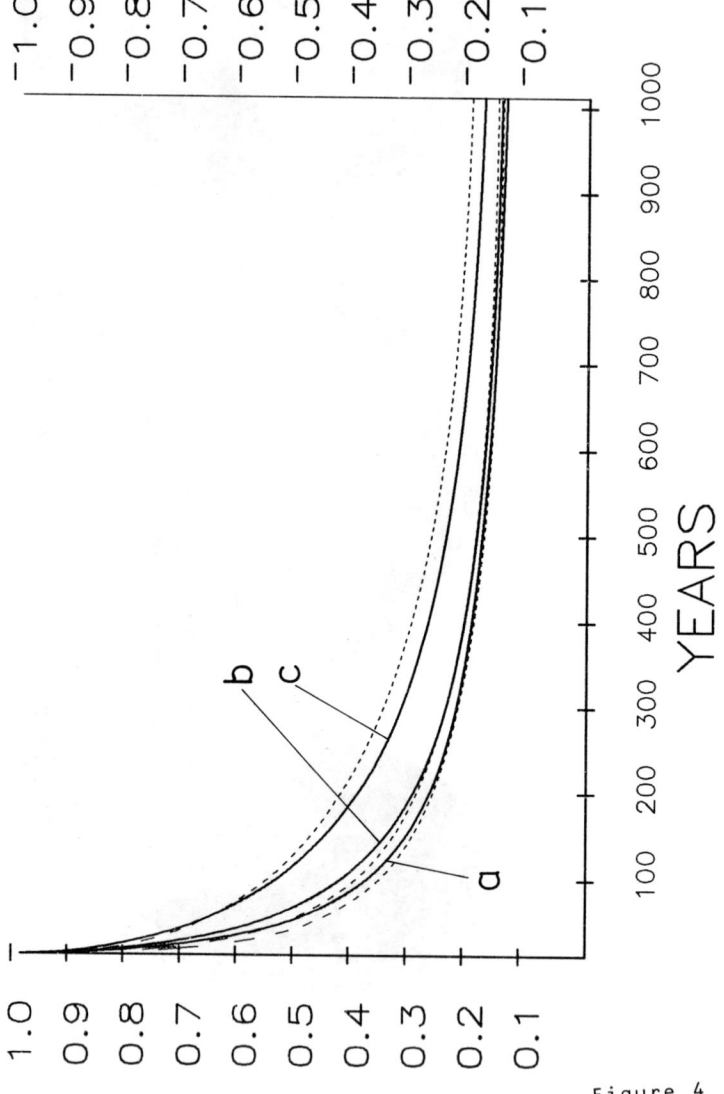

Figure 4

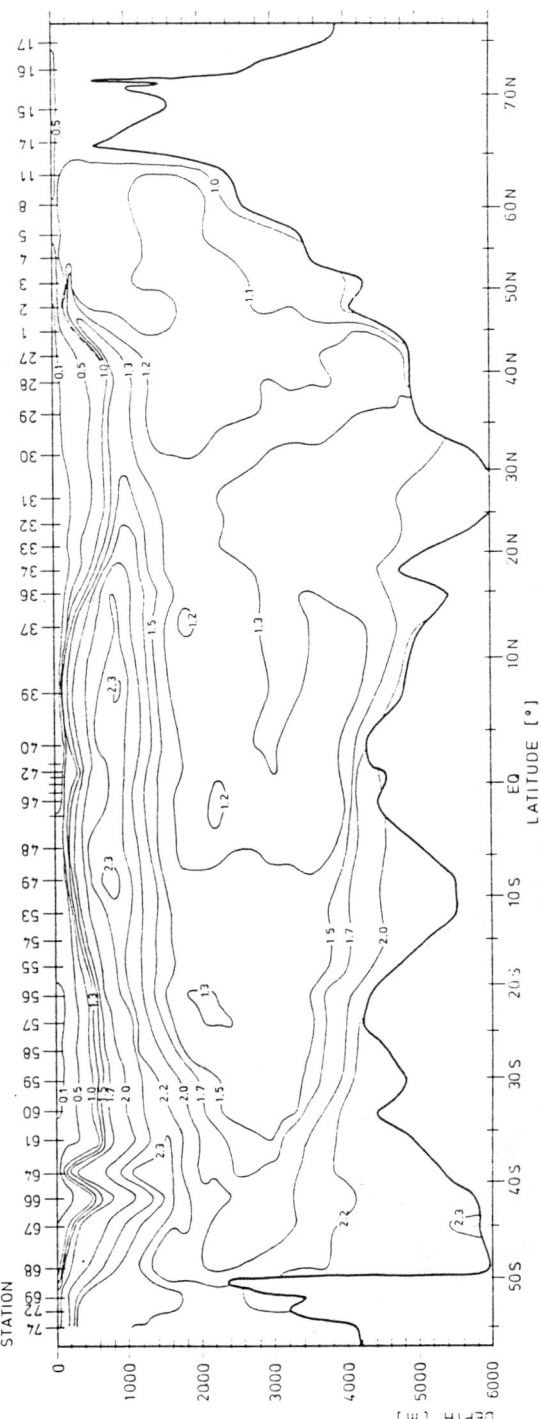

WESTERN ATLANTIC

PHOSPHATE [µM/kg]
JUL.-DEC. 1972

Figure 5a

Figure 5b

WESTERN PACIFIC

PHOSPHATE [μM/kg]
SEP. 1973 - MAR. 1974

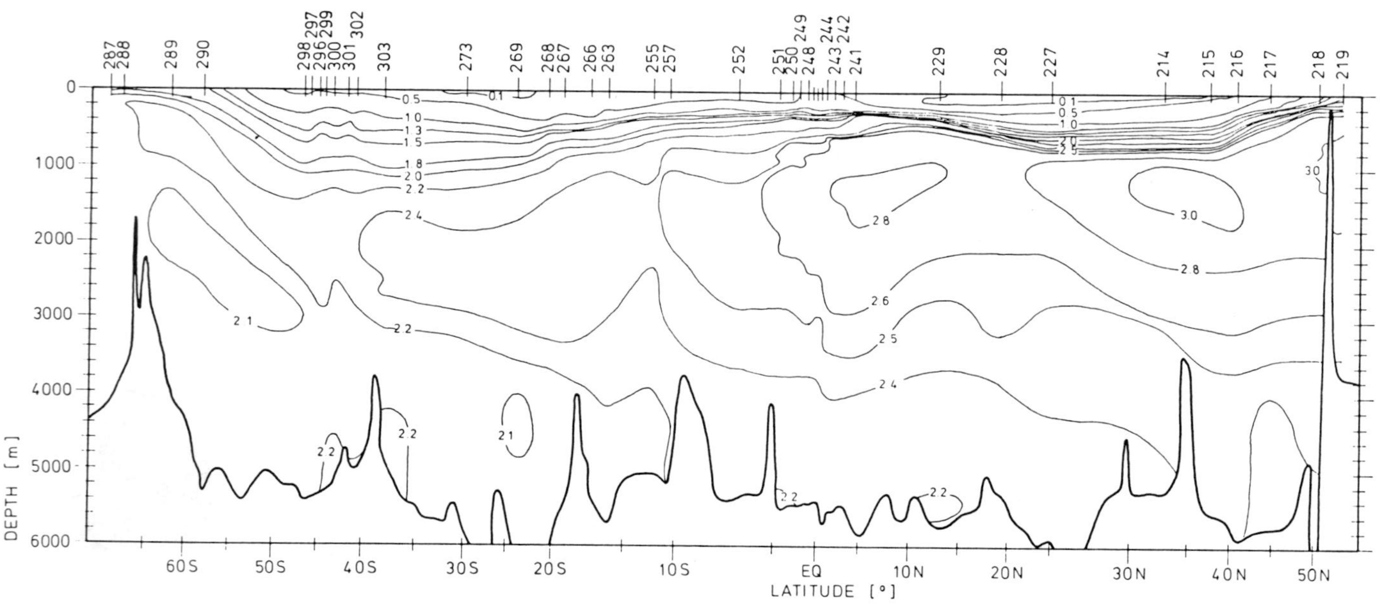

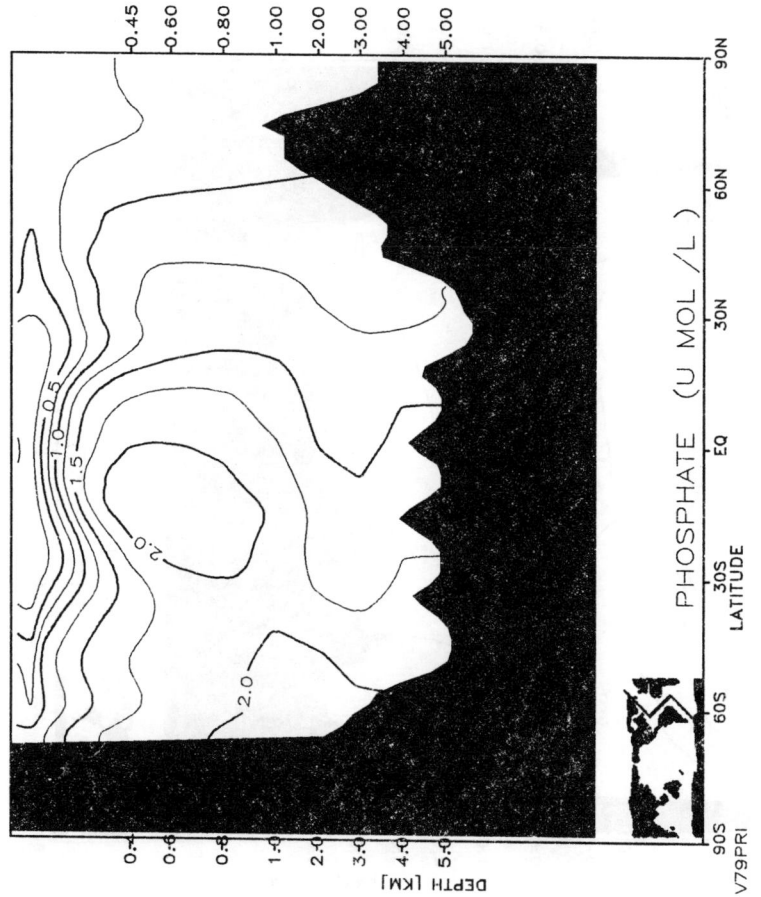

Figure 5c

71

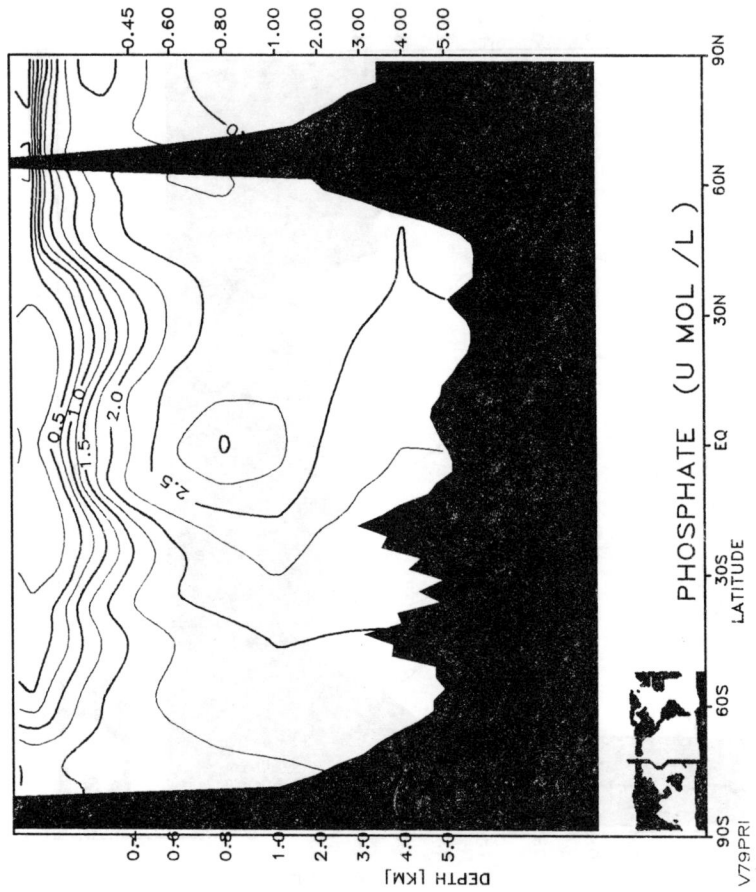

PHOSPHATE (U MOL / L)

Figure 5d

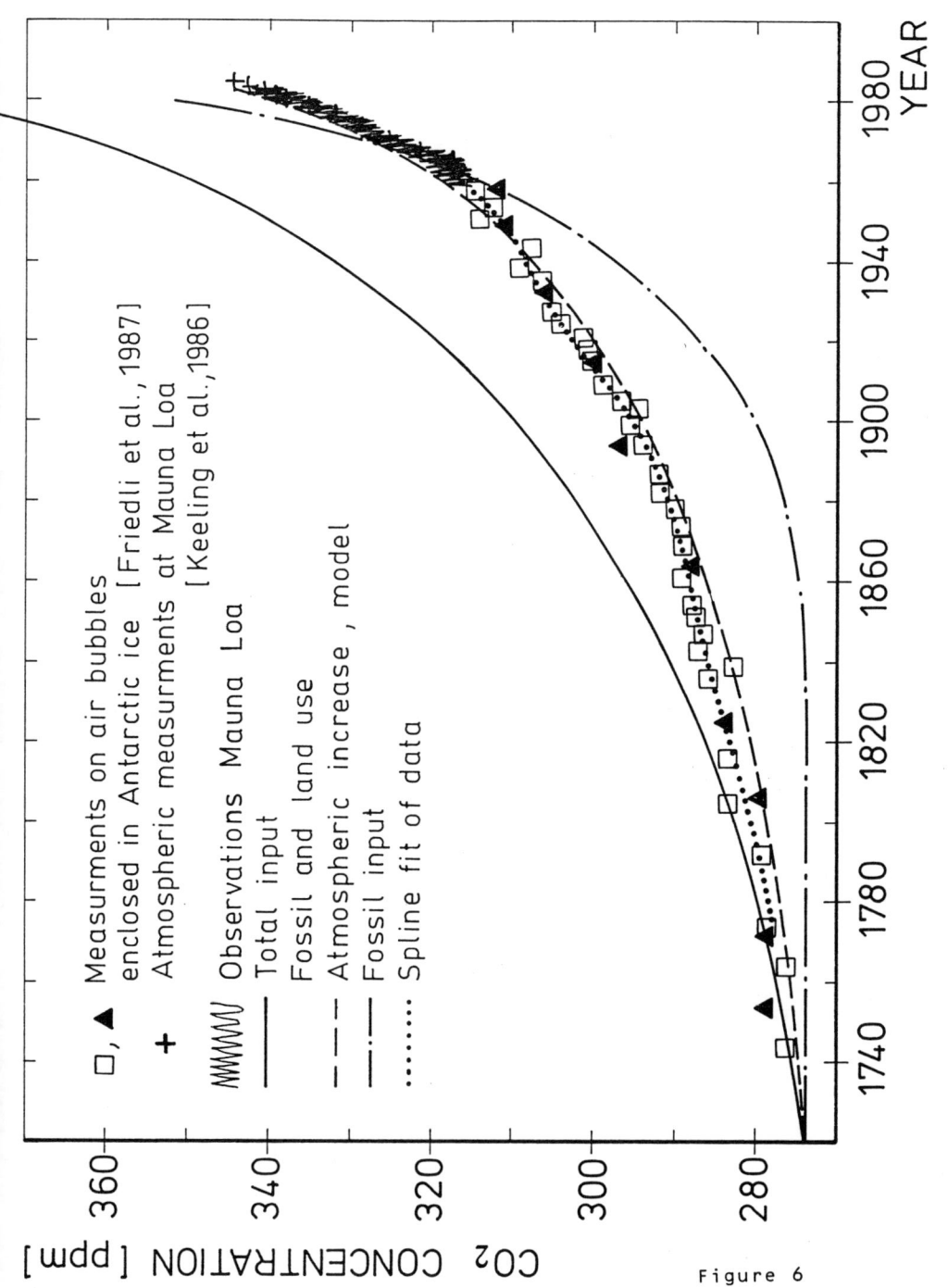

Figure 6

THE SULPHUR CYCLE

P.S. LISS
School of Environmental Sciences
University of East Anglia
Norwich NR4 7TJ, United Kingdom

Summary

The impact of sulphur from fossil fuel combustion on the environment is well recognised. This is particularly so in Europe, North America and other regions of high fossil fuel consumption. On the other hand the role of sulphur gases, especially dimethyl sulphide (DMS), produced by marine phytoplankton on environmental processes is less well established. Although it is recognised that the atmospheric oxidation of marine derived DMS is important for the acidity of rain and aerosols in remote areas, the significance of this route is less well established for industrialised regions. We are largely ignorant of the role that increased anthropogenic nutrient inputs into coastal marine areas may have had on DMS production. Similarly, we are only just beginning to study the possible atmospheric interactions of other marine biogenic trace gases, such as methyl iodide and bromoform, on the oxidation of DMS and other facets of atmospheric chemistry. The suggestion that sulphate aerosol particles derived from marine DMS may play a role in regulating climate has added a new and provocative dimension to these studies.

1. INTRODUCTION

The cycling of the element sulphur in the form of its various compounds plays many important environmental roles, particularly with respect to the atmosphere. For example, sulphur dioxide (SO_2) from both natural and fossil fuel sources is important in determining the acidity of rain and atmospheric aerosols. Carbonyl sulphide (COS) and carbon disulphide (CS_2) following oxidation in the atmosphere are important sources of stratospheric particles, and sulphate aerosols, formed by oxidation of marine-derived dimethyl sulphide (DMS), are the main constituents of cloud condensation nuclei (CCN) in the troposphere. This list is far from complete; for more examples and a comprehensive treatment of many aspects of the topic the interested reader is referred to the recently published book on the Global Biogeochemical Sulphur Cycle (Brimblecombe and Lein, 1989).

In this chapter the global cycle of sulphur is first reviewed both as it is now and how it is thought to have been prior to major anthropogenic perturbation. Then the role of both natural and fossil fuel derived SO_2 on rain and aerosol acidity in remote and industrialised regions is discussed. Finally the proposal that marine algae may play a part in climate regulation is examined.

2. THE GLOBAL SULPHUR CYCLE

Following the pioneering attempt by Eriksson (1960), several authors have presented global budgets of sulphur, including Junge (1963), Robinson and Robbins (1968), Kellogg et al. (1972), Friend (1973), Granat et al. (1976), Zehnder and Zinder (1980) and Ivanov and Freney (1983). The most recent attempt is that of Brimblecombe et al. (1989). This last group's results are shown in Figures 1 and 2, which respectively refer to the current (mid-1980s) budget as influenced by man's activities, and a prediction of how it was prior to major industrialisation. All fluxes given on the figures and elsewhere in this chapter are in units of Tg S a^{-1} i.e. 10^{12} g S per year. A comparison of the Figures 1 and 2 reveals some interesting apparent changes in the sizes of the inter-reservoir fluxes and also some for which there is little or no evidence for change.

There is no reason to believe that volcanic emissions of sulphur have changed significantly over this short time period for either marine or land volcanoes. Similarly there is no evidence for significant alterations in the sea-to-air fluxes of either volatile sulphur (mainly DMS, with smaller amounts of COS, CS_2 and possibly H_2S) and sea-salt sulphate or terrestrial emissions of sulphur gases (a similar suite of compounds to those for marine emissions but with H_2S playing a major, possibly the dominant role). It is important to note that these gaseous fluxes are important components in the geochemical cycle for sulphur, so that the budgets cannot be balanced without them, and that the total of marine plus terrestrial emissions to the atmosphere is 70% of the size of the amount of sulphur put into the atmosphere by fossil fuel burning (Figure 1).

Important parts of the cycles which have changed as a result of human activities include:

i) Aeolian emissions of sulphur-containing soil dust particles are thought to have increased by a factor of about two, from 10 to 20 Tg S a^{-1}. This is largely as a result of man-induced changes in farming and agricultural practice, particularly pasturing, ploughing and irrigation.

ii) By far the most significant impact on the system has been the input of sulphur (largely as SO_2) direct to the atmosphere from the burning of fossil fuels, metal smelting and other industrial activities. Such emissions have increased approximately twenty-fold over the last 120 years. It is not sure that this upward trend will continue since there are now important moves in the most advanced industrial nations to restrict emission, by, for example, burning sulphur-poor fuels and removal of SO_2 from power station stack gases. Because of its large size in relation to the natural cycle of sulphur, this input has substantial impacts on other parts of the cycle, some of which are discussed below.

iii) The deposition flux of sulphur from the atmosphere onto the oceans and land surfaces has increased by approximately 25% and 163% respectively. Although this input has essentially no effect on the chemistry of seawater, due to its buffer capacity and the large amount of sulphate it contains, its impact on poorly buffered soils and freshwaters can be profound. In areas with thin soil cover and soft water rivers and lakes significant changes in water chemistry and biology are well documented.

iv) Comparison of Figures 1 and 2 indicates that the amount of sulphur entering the oceans in river runoff has probably more than doubled due to man's activities. This has come about in part from sulphur-rich waste waters and agricultural fertilizers entering river courses, although another major factor is material deposited from the atmosphere finding its way into rivers etc. Lest the question be asked as to whether the enhanced inputs given in iii) and iv) may be having a measurable effect on the

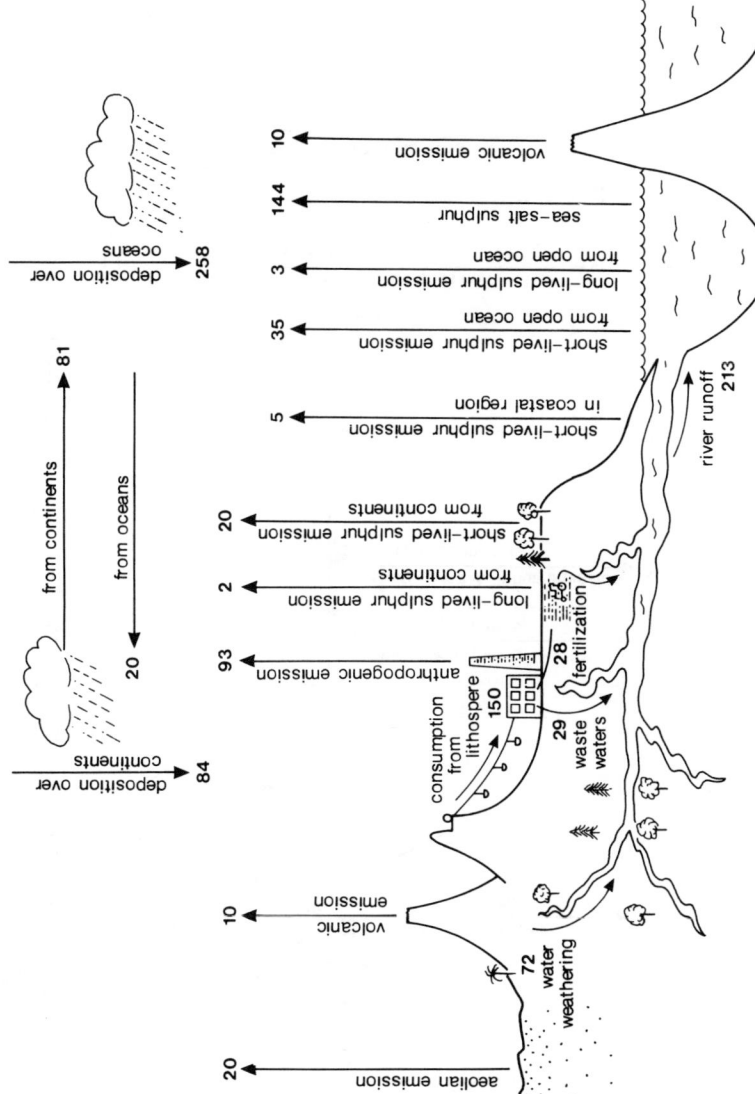

FIGURE 1. Global sulphur cycle in the mid-1980s. Units are Tg S a⁻¹. From Brimblecombe et al. (1989).

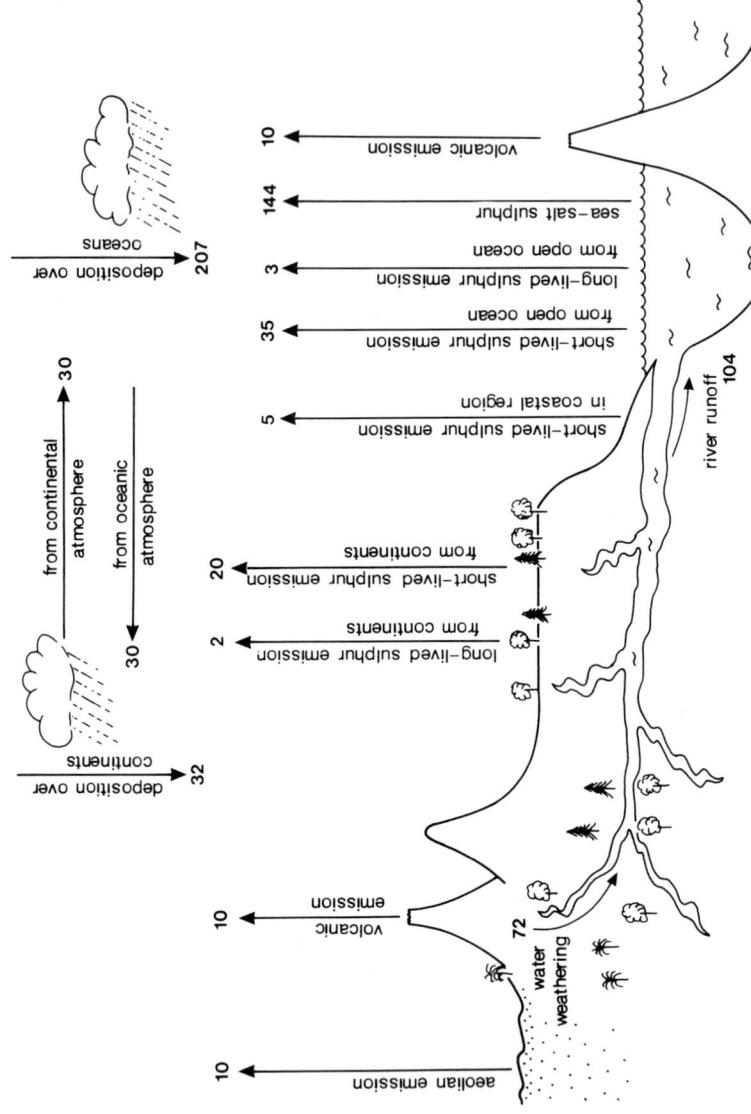

FIGURE 2. Global sulphur cycle prior to major anthropogenic influence. Units are Tg S a^{-1}. From Brimblecombe et al. (1989).

amount of sulphur (as sulphate) in the oceans, a simple calculation indicates a potential increase of about 10^{-5}% per annum. This estimate is probably an upper limit since it assumes that removal processes taking sulphur into ocean sediments remain the same as previously.

v) A final difference highlighted by the two figures is in the balance of sulphur flows between the continental and marine atmospheres. In the unperturbed cycle sulphur in air coming from the land just about balanced that in air masses moving in the opposite direction. Now the situation is out of balance, with about four times as much sulphur being exported from the continental to the marine atmosphere than flows in the opposite direction.

Although some interesting differences are apparent from the above comparison of pre- and post industrialisation budgets, a note of caution should be sounded concerning the absolute accuracy of the fluxes. Since no confidence limits or ranges are given the uncertainties in the flux estimates are not immediately apparent. However, inspection of other attempts at global sulphur budgets (listed earlier) reveals considerable differences between the various estimates, with factors of two divergences not being uncommon. Uncertainties in the individual fluxes are discussed in detail in Brimblecombe et al. (1989) - the source of Figures 1 and 2.

3. SULPHUR AND THE ACIDITY OF THE ATMOSPHERE

If carbon dioxide was the only atmospheric gas which controlled the acidity of rain then the pH of rainwater would be close to 5.6 (Raiswell et al., 1980). However, the vast majority of measurements of rain pH are below this value, indicating other sources of acidity. Obvious sources of this extra acid are SO_2 and methane sulphonic acid. Sulphur dioxide, whether from marine or anthropogenic sources, can be further oxidised to sulphuric acid, largely within cloud water drops. Backing-off some of these sources of acid is ammonia, for which there are strong sources on land, especially where intensive agriculture is being practised. Even from this oversimplified description, it is clear that the pH of a rain sample is dependent on the wide variety of sources of acid and alkali and the large number processes which affect these substances once they are in the atmosphere. For a more detailed discussion of the factors controlling the acidity of natural (largely unpolluted) rain the interested reader should see Charlson and Rodhe (1982). In broad terms rain unaffected by manmade emissions has a pH in the region of 5 (see for example Galloway and Gaudry's (1984) data from Amsterdam Island in the middle of the Indian Ocean), whereas rain in Europe and other highly industrialised regions regularly has pH values in the region of 4.5 or lower. Since pH is a logarithmic scale, this pH difference corresponds to at least a three-fold change in acidity between clean and polluted areas. From the above discussion, it should be clear that sulphur compounds play a vital role in determining the pH of rain and atmospheric aerosols.

In remote regions of the globe, particularly over and close to the oceans, it is generally considered that atmospheric oxidation of marine DMS is the dominant source of acidity for rain and particulate samples. This is in part because of their proximity to marine DMS sources but also because of rapidity with which land-derived acids and ammonia are removed from the atmosphere. This removal is often in rain but can also be by deposition of 'dry' particles. Thus, natural and man-made emissions from terrestrial sources are removed from the atmosphere largely on a local or regional scale.

As an example of a situation with respect to sulphur acidity in a remote location, Figure 3 shows plots of measurements made on atmospheric aerosols at a coastal site in Antarctica during the period 1983-1985 by

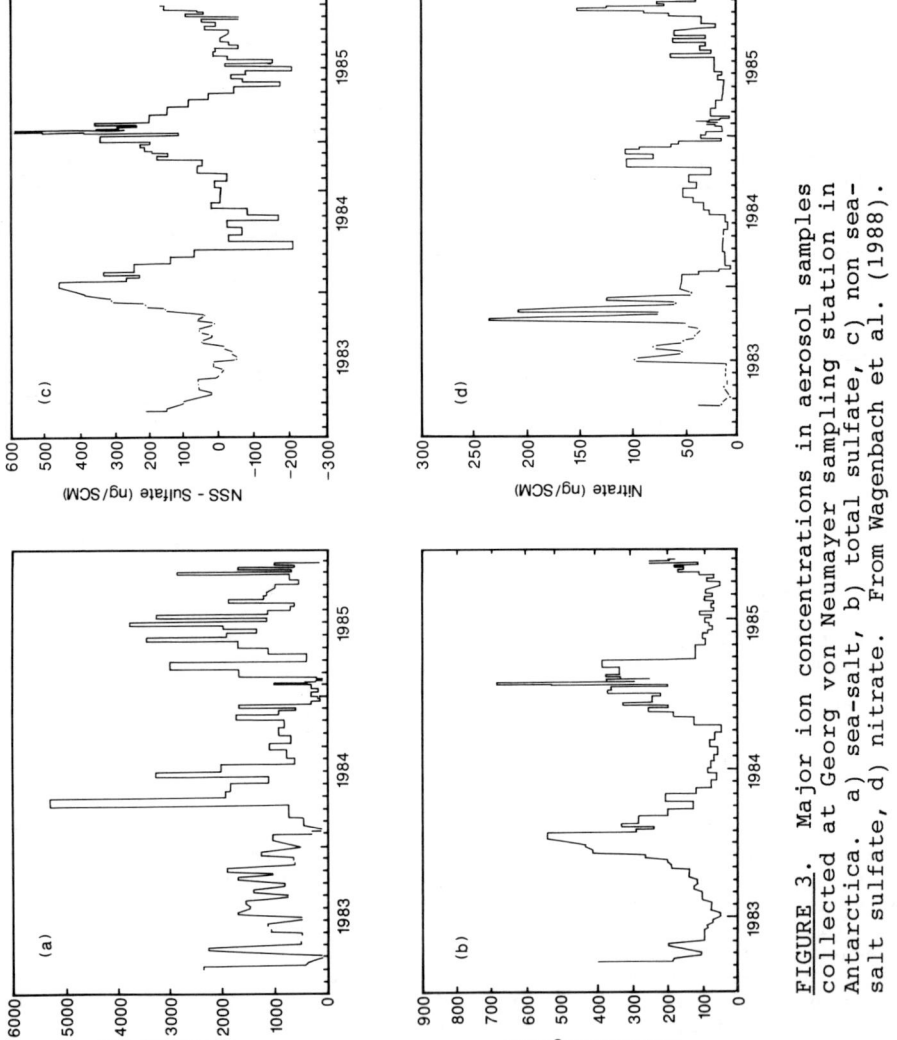

FIGURE 3. Major ion concentrations in aerosol samples collected at Georg von Neumayer sampling station in Antarctica. a) sea-salt, b) total sulfate, c) non sea-salt sulfate, d) nitrate. From Wagenbach et al. (1988).

Wagenbach et al. (1988). As well as total sulphate and non sea-salt sulphate (NSS sulphate, which is defined as total SO_4^{2-} minus SO_4^{2-} directly from sea water), nitrate and sea salt in the aerosol are also plotted. For both total and non sea-salt sulphate there is a clear seasonal pattern of concentration, with highest values in local summer and very low (sometimes even negative for NSS sulphate) values in winter. Sea salt concentrations show no similar seasonal distribution. (Although aerosol nitrate appears to show a seasonal pattern, with high values in local spring, the authors cannot find a satisfactory explanation for it). For the total and NSS sulphate a ready explanation is from the proximity of the site to the Antarctic ocean - an obvious source of the precursor DMS. A similar conclusion was reached by Savoie and Prospero (1989) in a study of aerosol composition over the North Pacific. They concluded that 80% of the NSS sulphate in samples from that region was of marine biogenic origin via the DMS oxidation route, the remaining 20% being from land-based sources.

In the marine atmosphere closer to industrialised land areas the importance of DMS derived sulphur becomes of less consequence. The extent of this decrease in importance is related not only to the proximity of the location considered to industrialised land areas but also to the degree of biological activity in the water with respect to DMS production.

As an example of a situation potentially intermediate between the remote Antarctic and North Pacific ones discussed above and a heavily polluted industrial environment, we can examine the budget for atmospheric sulphur in the North West Atlantic, specifically in the area between the eastern seaboard of North America and the island of Bermuda. Galloway and Whelpdale (1987) present such a budget, which is shown in Figure 4. These authors estimate that anthropogenic emissions to the atmosphere from eastern North America are 11.7 Tg S a^{-1}. Of this 3-4 Tg S a^{-1} is transported into the North Atlantic marine atmosphere, leaving 8-9 Tg S a^{-1} to be deposited on land. Once over the sea about 2 Tg S a^{-1} is deposited onto the area to the west of Bermuda; roughly equal amounts of this deposition being by wet (rain) and dry processes. A further 1-2 Tg S a^{-1} is transported out of the study area to the east of Bermuda. Natural marine input of DMS derived sulphur is estimated at 0.2 Tg S $^{-1}$, i.e. about 10% of the man-made S being deposited into the sea. The conclusion is thus that for this region the atmospheric budget is largely controlled by anthropogenic emissions from North America. Although the Galloway and Whelpdale study was done at a time when there were very few direct estimates of sea-to-air DMS fluxes for this particular area, and indeed more recent work has shown somewhat higher fluxes particularly at times of high biological activity in the seawater, the general conclusion stated above would appear to be robust.

Turning now to a case where it might be thought that man-made sulphur sources would be totally dominant, we examine the situation of atmospheric acidity in northwestern Europe. Figure 5 shows the distribution of rainwater pH for the area and gives some idea of its spatial distribution. Values are generally lower (more acidic) in the industrial heartlands of Germany, the Low Countries and eastern United Kingdom and higher (less acidic) in the less populated/industrialised areas to the north and south.

Several mathematical models have been constructed to simulate the deposition (both wet and dry) of sulphur over Europe from industrial/urban sources. The output from such models shows maximum deposition occurring in the broad industrial band stretching from England across Germany and into Eastern Europe. the general agreement between the model predicted distribution of deposition and the observed pattern of rainfall pH shown in Figure 5 is indicative of man-made sources of sulphur being dominant in the

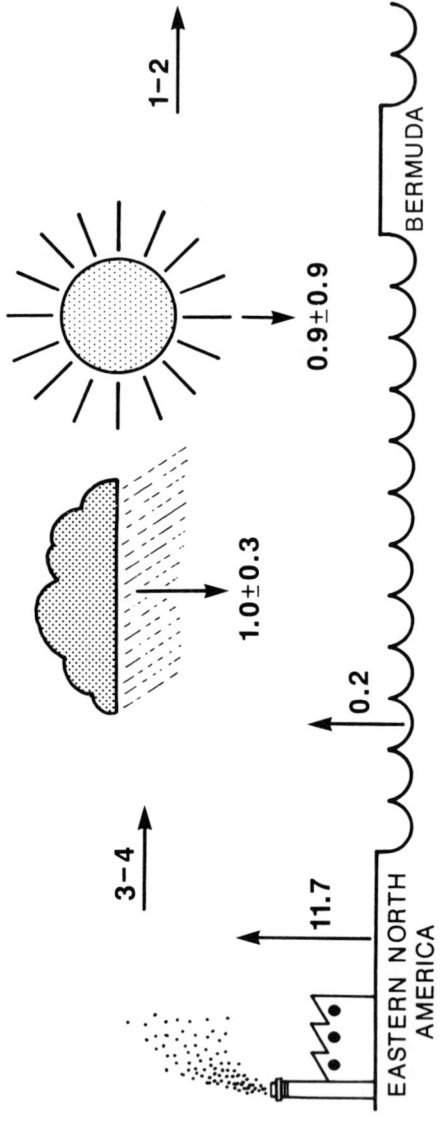

FIGURE 4. A sulphur budget for the Western North Atlantic Ocean. Units are Tg S a^{-1}. From Galloway and Whelpdale (1987).

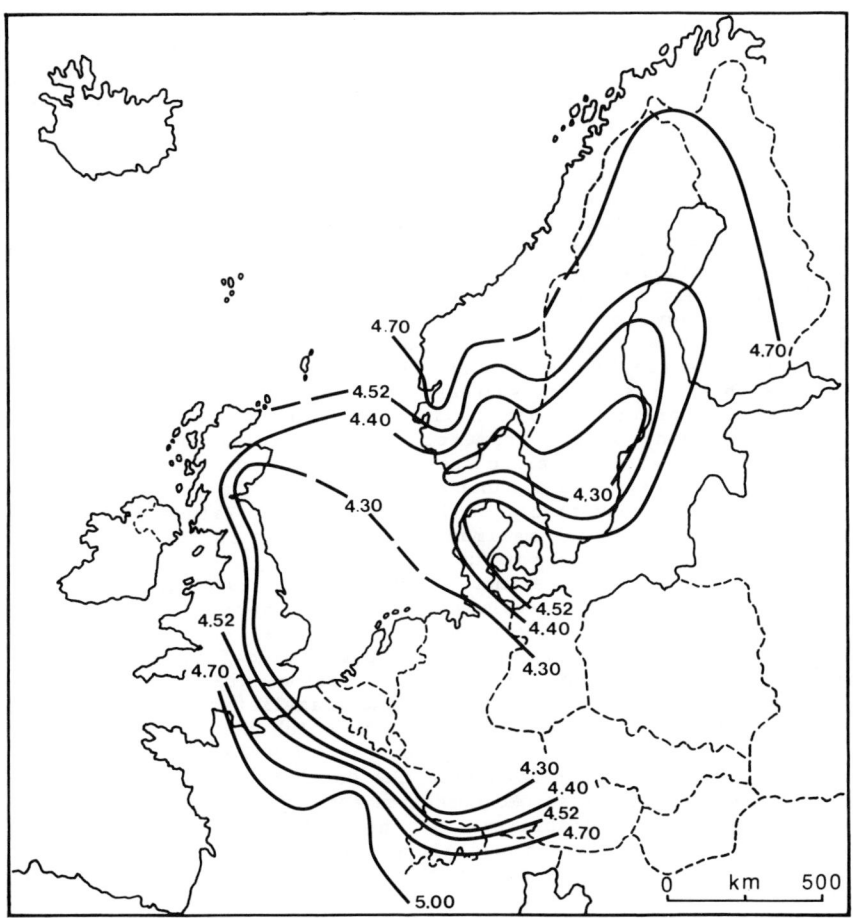

<u>FIGURE 5</u>. Variations in pH of rainfall over Europe,
1974. From Park (1987), based on OECD data.

region, as might be expected. However, this general conclusion must be treated with some caution. For example, it is known that the models tend to underestimate the amount of sulphur deposited in rain in the wetter regions of Europe. Also, model predictions do not quantitatively balance with measured rates of deposition. This mis-match is usually attributed to 'background' sources of sulphur not included in the models, which invariably deal only with industrial/urban emissions. The origin of this background sulphur is unclear but it could be due inter alia to very long-range transport (e.g. pollutant sulphur from North America) or biogenic sulphur emissions from seawater. Some support for this latter suggestion comes from the fact that the greatest discrepancy between the observed depositions in Europe and the model predictions occurs in the spring when marine phytoplankton activity is generally at its maximum.

Further information on the processes controlling DMS production and its importance for atmospheric sulphur chemistry in Europe can be obtained from recent work we have been carrying out as part of the U.K. Natural Environment Research Council's North Sea Programme. We have participated in 9 research cruises in the southern North Sea in 1989 and measured DMS, as well as its precursor dimethylsulphoniopropionate (DMSP), plant nutrients and many other parameters, on a regular survey track with approximately 100 water sampling stations on each cruise. The preliminary results show the expected pronounced seasonal cycle for DMS (and DMSP) concentrations in these highly productive waters. Values are low during autumn and winter and show maximum concentrations in May and June, with some evidence for a secondary maximum in August.

Previous results (for example, Turner et al., 1988) indicate that any relationship between DMS levels and general measures of phytoplankton activity such as chlorophyll concentration are likely to be weak. This is because different organisms appear to have very different abilities to produce DMSP and hence DMS. For example, diatoms are thought to be poor DMS producers whereas coccolithophores and flagellates seem to make large amounts. Thus, chlorophyll determinations, which measure carbon fixed by the plankton, are likely to be poor predictors of DMS production.

It is clear from our data from the North Sea that, as well as the large seasonal variations in DMS concentrations already discussed, there is also a large spatial variability. In the spring and early summer months highest concentrations are found along the Dutch/German/Danish coasts, probably associated with blooms of the organism Phaeocyctis pouchetti, which is known to be common in this region at that time. Although broad features such as this can be identified, close inspection of the data shows a high degree of variability from point to point. This is not unexpected since 'patchiness' in plankton distributions is generally the rule, in this case enhanced by the varying abilities of different classes of plankton to synthesise DMSP and ultimately produce DMS. A consequence of this patchiness, coupled with large seasonal cycle, is that in order to properly quantify the flux of marine biogenic sulphur to the atmosphere a large data base of DMS concentration measurements must be available.

Although DMS levels are clearly affected by the particular organisms present in the water, the question of the role of plant nutrients in DMS production is also pertinent. For, if anthropogenic activities are increasing nutrient levels in coastal waters, does this lead to enhanced production of DMS in these areas? There is some field evidence from the North and Baltic Seas (Turner et al., 1988 and Leck et al., 1990), together with some preliminary results from studies in culture (Turner et al., 1988) that at least for nitrate there may be an inverse relationship. For an explanation of this possibly unexpected result it is necessary to look at the function of DMSP in plankton cells, which is generally assumed to be as an osmolyte. However, DMSP is not the only osmolyte that plant cells synthesise. Another is glycine betaine, which is chemically the analogue of DMSP but with nitrogen as the central atom rather than sulphur. The hypothesis is that under conditions where there is abundant nitrate or other forms of nitrogen in the water the organisms will use glycine betaine as the osmolyte and will only switch over to forming DMSP when nitrogen is in short supply. It should be stressed that this is only a concept at the present and more work needs to be done before its validity can be assessed. However, if it is correct then the conclusion has to be that although increased nutrient inputs to marine waters will in general lead to increased productivity, this will not necessarily produce a proportional increase in the formation of DMS.

Turner et al. (1988) have made an attempt to calculate the seasonal sea-to-air flux of sulphur, as DMS, from British coastal and shelf waters in order to assess this natural source of atmospheric sulphur relative to the input of man-made SO_2. The calculation method they adopted was that outlined by Liss (1983) and Liss and Merlivat (1986). Basically, it is necessary to know the sea-air DMS concentration difference, which is the driving force for the net interfacial flux, and the rate at which the gas exchange occurs (expressed as a transfer velocity, k). The net flux is then obtained from the product of these two terms.

For the DMS concentration difference term Turner et al. (1988) use the concentration data they had for the study area at that time. The transfer velocity was obtained from the relationships given by Liss and Merlivat (1986) for the variation of k with wind speed. The calculated winter and summer fluxes of DMS out of the water were close to 16 and 935 ug S m^{-2} d^{-1}. The equivalent emissions to the atmosphere of man-made sulphur as SO_2 are 11.4 and 5.7 x 10^3 ug S m^{-2} d^{-1} for winter and summer, respectively (using OECD, 1977 emission figures). These authors therefore conclude that emissions per unit area from the North Sea are trivial compared to land-based sources. However, in summer the areal emission rate is about 16% of equivalent anthropogenic emissions. Thus, although in summer fossil fuel burning is still the dominant source of sulphur to the atmosphere, natural emissions from these coastal and shelf waters are not insignificant. Further, the calculation probably underestimates the importance of natural emissions in some areas of low industrialisation since it assumes that anthropogenic emissions are evenly distributed over Europe. In addition, since the 1977 OECD report significant reductions in SO_2 emissions have taken place.

The likely significance of DMS as a source of sulphur to remote areas of Europe which are downwind of marine waters is confirmed by a computer modelling exercise conducted by Fletcher (1989). He predicts that the DMS oxidation product methanesulphonic acid (MSA) could account for 30-50% of the total sulphur acidity in air over Scandinavia when plankton blooms occur in the North Sea between the beginning of May and early August. Such effects may be more widespread than just in the North Sea. For example, recent measurements of MSA in atmospheric aerosols at Mace Head on the west coast of Eire exhibit a strong seasonal signal and in the early summer show very high values relative to measurements made elsewhere (J. Prospero, University of Miami, personal communication). This finding implies that the North East Atlantic is probably also a strong and extensive source for atmospheric sulphur during times of high plankton activity.

DMS is only one of a suite of biologically-produced trace gases in seawater, which subsequently degas to the atmosphere and may affect its chemistry. Other examples include methyl iodide (CH_3I) and bromoform ($CHBr_3$). The latter has been invoked, via its photochemical transformation, in the breakdown of tropospheric ozone in the Arctic atmosphere in spring (Barrie et al., 1988). Methyl iodide is of interest since it has been invoked as a catalyst in the atmospheric oxidation of dimethyl sulphide. The reaction scheme involves photolysis of methyl iodide in the atmosphere to form I atoms, which then react with ozone molecules to form the IO radical. In the final step of the chain, IO reacts with a DMS molecule to form the compound dimethyl sulphoxide (DMSO), with the release of an I atom (Barnes et al., 1987). In an analogous way to the role that DMS plays in the global sulphur cycle, methyl iodide fulfils a vital function in balancing the environmental cycle of iodine (Liss, 1986).

4. THE SULPHUR CYCLE AND CLIMATE REGULATION

As well as its importance for rain and aerosol acidity, the sulphur cycle also plays an important role in controlling the radiation balance of the atmosphere. Shaw (1987) pointed out the efficiency of sulphate aerosols in interacting with solar radiation and hinted at a climate regulating mechanism acting through changes in the optical depth of the atmosphere arising from variations in aerosol numbers. Such variations would be driven by changes in DMS production by plankton in the surface of the oceans; marine DMS, after oxidation in the atmosphere, being the dominant source of sulphate aerosols in remote areas. The idea was developed by Charlson et al. (1987) and their hypothesised mechanism for such climate regulation is shown in Figure 6.

Perhaps the easiest way to understand how the proposed mechanism works is to take the following example. Suppose the temperature of the atmosphere increases due to some perturbation from outside the scheme shown in the figure, for example a change in solar constant or an increase in atmospheric CO_2 whether natural or man-made. An increase in air temperature will lead to a concomitant increase in ocean surface water temperatures. A key links in the chain is that this will produce enhanced production of DMS by marine plankton. If it does then, all other things being equal, this will give rise to a greater flux of DMS to the atmosphere and so to enhanced sulphate aerosol numbers. This implies greater numbers of cloud condensation nuclei, more scattering of solar radiation by cloud droplets and so an increase in cloud albedo. All these changes result in a greater amount of incident solar radiation being reflected back to space, which will tend to cool the atmosphere, thus counteracting the initial perturbation.

If correct, the Charlson et al. hypothesis gives ocean plankton a key role in climate regulation. As might be expected, an idea as dramatic as this has received considerable amounts of critical comment, both positive and negative (see, for example, Schwartz, 1988; Slingo, 1988; Wigley, 1989; and letters to the Editor of Nature published on the 10 and 17 August, 1989). The matter seems far from resolved for the moment. One item worthy of comment here, in view of earlier discussion, is what factors control DMS production by plankton. Charlson et al. (1987) assume that in warmer seawater more DMS will be produced. This is based on the somewhat larger concentrations of DMS currently found in tropical and equatorial areas, compared with cooler waters further north and south. An alternative explanation is that lack of nitrogen in the warmer waters leads to the enhanced DMS levels, as discussed earlier in the context of DMS formation in the North and Baltic Seas. To resolve this matter, it is clearly necessary to carry out some careful research to try to unravel the individual roles of water temperature and nutrient status, together with other factors such as light availability, in the production of dimethyl sulphide.

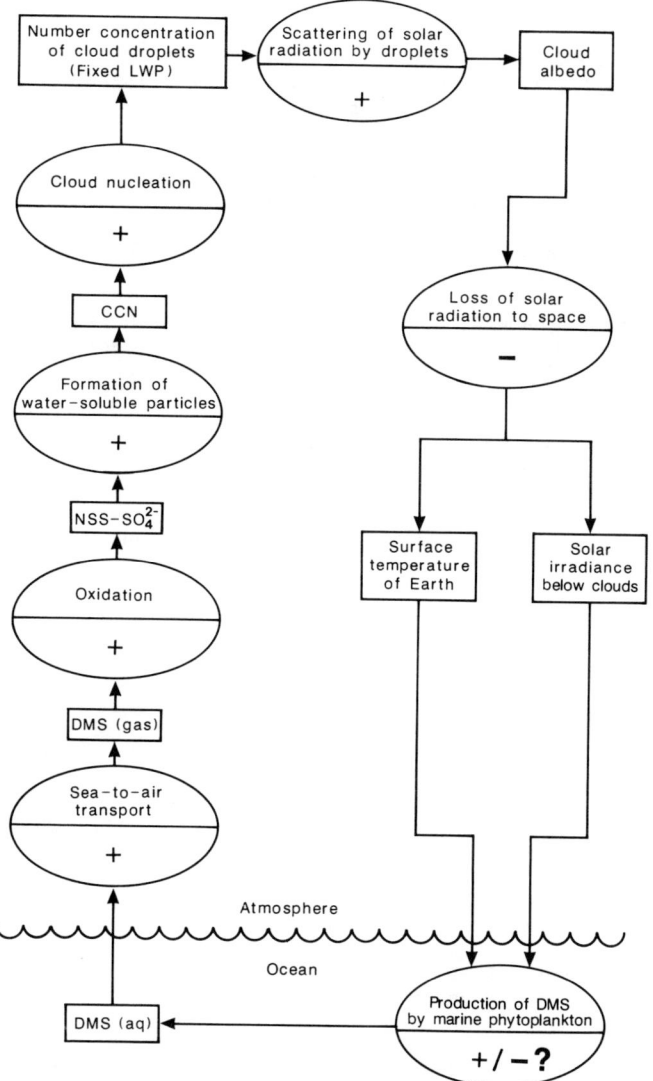

FIGURE 6. Conceptual diagram of a possible climatic feedback loop. The rectangles are measurable quantities, and the ovals are processes linking the rectangles. The sign (+ or -) in the oval indicates the effect of a positive change of the quantity in the preceding rectangle on that in the succeeding rectangle. From Charlson et al. (1987).

REFERENCES

BARNES, I., BECKER, K.H., CARLIER, P. and MOUVIER, G. (1987). FTIR study of the DMS/NO_2/I_2/N_2 photolysis system: the reaction of IO radicals with DMS. International Journal of Chemical Kinetics 19: 489-501.

BARRIE, L.A., BOTTENHEIM, J.W., SCHNELL, R.C., CRUTZEN, P.J. and RASMUSSEN, R.A. (1988). Ozone destruction and photochemical reactions at polar sunrise in the lower Arctic atmosphere. Nature 334: 138-141.

BRIMBLECOMBE, P. and LEIN, A. YU. (Eds) (1989). Evolution of the Global Biogeochemical Sulphur Cycle. John Wiley, 241pp.

BRIMBLECOMBE, P., HAMMER, C., RODHE, H., RYABOSHAPKO, A. and BOUTRON, C.F. (1989). Human influence on the sulphur cycle. In: Evolution of the Global Biogeochemical Sulphur Cycle (ed. P. Brimblecome and A. Yu. Lein). John Wiley, p.77-121.

CHARLSON, R.J. and RODHE, H. (1982). Factors controlling the acidity of natural rainwater. Nature 295: 683-685.

CHARLSON, R.J., LOVELOCK, J.E., ANDREAE, M.O. and WARREN, S.G. (1987). Oceanic phytoplankton, atmospheric sulphur, cloud albedo and climate. Nature 326: 655-661.

ERIKSSON, E. (1960). The yearly circulation of chloride and sulphur in nature: meteorological, geochemical and pedological implications. Part 2. Tellus 12: 63-109.

FLETCHER, I. (1989). North Sea dimethyl sulfide emissions as a source of background sulfate over Scandinavia: A model. In: Biogenic Sulfur in the Environment (ed. E.S. Saltzman and W.J. Cooper). American Chemical Society, p.489-501.

FRIEND, J.P. (1973). The global sulphur cycle. In: Chemistry of the Lower Atmosphere (ed. S.I. Rasool). Plenum Press, p.177-201.

GALLOWAY, J.N. and GAUDRY, A. (1984). The composition of precipitation on Amsterdam Island, Indian Ocean. Atmospheric Environment 18: 2649-2656.

GALLOWAY, J.N. and WHELPDALE, D.M. (1987), WATOX-86 overview and western north Atlantic Ocean S and N atmospheric budgets. Global Biogeochemical Cycles 1: 261-281.

GRANAT, L. RODHE, H. and HALLBERG, R.O. (1976). The global sulfur cycle. In: Nitrogen, Phosphorus and Sulfur - Global Cycles (ed. B.H. Svensson and R. Soderlund). Ecological Bulletins 22: 89-134.

IVANOV M.V. and FRENEY, J.R. (Eds) (1983). The Global Biogeochemistry Sulfur Cycle. John Wiley, 470pp.

JUNGE, C.E. (1963). Air Chemistry and Radioactivity. Academic Press, 381pp.

KELLOGG, W.W., CADLE, R.D., ALLEN, E.R. LAZRUS, A.L. and MARTELL, E.A. (1972). The sulfur cycle. Science 175: 587-596.

LECK, C., LARSSON, U., BAGANDER, L.E., JOHANSSON, S. and HAJDU, S. (1990). Dimethyl sulfide in the Baltic Sea: Annual variability in relation to biological activity. Journal of Geophysical Research 95: 3353-3363.

LISS, P.S. (1983). Gas transfer: Experiments and geochemical implications. In: Air-Sea Exchange of Gases and Particles (ed. P.S. Liss and W.G.N. Slinn). D. Reidel, p.241-298.

LISS, P.S. (1986). The air-sea exchange of low molecular weight halocarbon gases. In: The Role of Air-Sea Exchange in Geochemical Cycling (ed. P. Buat-Menard). D. Reidel, p.283-294.

LISS, P.S. and MERLIVAT, L. (1986). Air-sea gas exchange rates: Introduction and synthesis. In: The Role of Air-Sea Exchange in Geochemical Cycling (ed. P. Buat-Menard). D. Reidel, p.113-127.

OECD (1977). The OECD programme on long range transport of air pollutants.

PARK, C.C. (1987). Acid Rain: Rhetoric and Reality. Methuen, 272pp.

RAISWELL, R.W., BRIMBLECOMBE, P., DENT, D.L. and LISS, P.S. (1980). Environmental Chemistry. Edward Arnold, 184pp.

ROBINSON, E. and ROBBINS, R.C. (1968). Sources, Abundance and Fate of Gaseous Atmospheric Pollutants. Stanford Research Institute, 110pp.

SAVOIE, D.L. and PROSPERO, J.M. (1989). Comparison of oceanic and continental sources of non-sea-salt sulphate over the Pacific Ocean. Nature 339: 685-687.

SCHWARTZ, S.E. (1988). Are global cloud albedo and climate controlled by marine phytoplankton? Nature 336: 441-445.

SHAW, G.E. (1987). Aerosols as climate regulators: A climate - biosphere linkage? Atmospheric Environment 21: 985-986.

SLINGO, T. (1988). Can plankton control climate? Nature 336: 421.

TURNER, S.M., MALIN, G., LISS, P.S., HARBOUR, D.S. and HOLLIGAN, P.M. (1988). The seasonal variation of dimethyl sulfide and dimethylsulfoniopropionate concentrations in nearshore waters. Limnology and Oceanography 33: 364-375.

WAGENBACH, D., GORLACH, U., MOSER, K. and MUNNICH, K.O. (1988). Coastal Antarctic aerosol: The seasonal pattern of its chemical composition and radionuclide content. Tellus 40B: 426-436.

WIGLEY, T.M.L. (1989). Possible climate change due to SO_2-derived cloud condensation nuclei. Nature 339: 365-367.

ZEHNDER, A.J.B. and ZINDER, S.H. (1980). The sulfur cycle. In: The Handbook of Environmental Chemistry 1A (ed. O. Hutzinger). Springer-Verlag, p.105-145.

THE TROPOSPHERIC CYCLES OF METHANE AND OTHER HYDROCARBONS

J. RUDOLPH

Institut für atmosphärische Chemie, Forschungszentrum Jülich GmbH, Postfach
1913, D-5170 Jülich, F. R. G.

Summary

Methane and the numerous nonmethane hydrocarbons (NMHC) are
important participants in the photochemical reaction cycles in the
troposphere. Methane is with a global average mixing ratio of 1.7
ppm the by far most abundant and also most important hydrocarbon
compound in the troposphere. As a result of the long atmospheric
residence time of methane, its distribution is relatively uniform
and the gradient between the northern and southern hemisphere is
only about 0.1 ppm. During the past years the tropospheric methane
concentration has been increasing at a relative rate of nearly
1 %/yr. The other hydrocarbons have much shorter atmospheric
lifetimes, between a few month and less than one day. Their mixing
ratios are typically in the range of several ppt to a few ppb. As
consequence of the short lifetimes, the NMHC exhibit a substantial
variability which can only partly be ascribed to systematic
vertical, latitudinal or seasonal trends. For the less reactive
NMHC (e.g. ethane, propane, acetylene) it has been possible to
establish the existence of systematic seasonal and latitudinal
trends. However, the presently available data do not allow any
identification of possible long term trends.

1. INTRODUCTION

     Methane and the other hydrocarbons belong to the key compounds in the
photochemical reaction cycles of the troposphere. Hydrocarbon oxidation
leads to the formation of several other important atmospheric trace gases
such as ozone, carbon monoxide, and hydrogen. The formation of tropospheric
ozone from hydrocarbon (RH) oxidation proceeds via the following, extremely
simplified, reaction sequence.

$$RH + OH \longrightarrow R^{\cdot} + H_2O$$
$$R^{\cdot} + O_2 \longrightarrow ROO^{\cdot}$$
$$ROO^{\cdot} + NO \longrightarrow RO^{\cdot} + NO_2$$
$$NO_2 + h\nu \longrightarrow NO + O$$
$$O_2 + O \longrightarrow O_3$$

There are competing reactions such as $ROO^{\cdot} + O_3 \longrightarrow RO + 2O_2$, reactions of
peroxyradicals with each other etc. which do not result in ozone formation.
Consequently the NO concentration plays a key role in determining wether
the oxidation of NMHC results in ozone production or not. Very roughly,
above 10 ppt the reaction sequence above starts to produce ozone. Thus the
production of ozone from hydrocarbon oxidation can be enhanced not only by
increasing hydrocarbon (and CO) levels but also by increasing NO_x

concentrations. In the latter case, the effect of any NO_x change will strongly depend on the NMHC levels.

The reactions of hydrocarbons with OH-radicals also have a substantial impact on the concentration of OH- and HO_2-radicals in the troposphere. Furthermore methane acts as a greenhouse gas and changes in atmospheric methane concentrations contribute to climatic changes. During the photooxidation of hydrocarbons a variety of more or less complex compounds is formed, e.g. aldehydes, ketones, carboxylic acids, peroxides and peroxyacetyl nitrate. Especially the NMHCs can act as precursors of a variety of compounds which have no or only marginal direct atmospheric sources.

For essentially all hydrocarbons the dominant removal mechanism is the reaction with OH-radicals. However, the reaction rate constants differ by several orders of magnitude and consequently the various hydrocarbons can have very different atmospheric residence times. Not only the reactivity of the hydrocarbons but also their atmospheric concentrations and the variability in their distributions can be very different. This is illustrated in Table 1. The Table lists the average mixing ratios and the relative standard deviation of the most important hydrocarbons measured at a remote continental site in Germany (Deuselbach) during three campaigns of 2-5 days in 1983. Also included are the rate constants for the reaction with OH-radicals and the resulting removal rates due to this reaction.

Table 1. AVERAGE HYDROCARBON MIXING RATIOS AT A REMOTE CONTINENTAL
         SITE IN GERMANY (DEUSELBACH)

Compound	Mixing Ratio ppb	Relative Standard Deviation	OH-Rate Constant $cm^3 molecule^{-1}s^{-1}$	Removal Rate* ppb/d
Methane	1740	2%	$8.0*10^{-15}$	0.79
Ethane	2.05	18%	$2.9*10^{-13}$	0.034
Propane	0.60	39%	$1.5*10^{-12}$	0.050
n-Butane	0.50	42%	$2.5*10^{-12}$	0.070
i-Butane	0.22	45%	$2.4*10^{-12}$	0.030
n-Pentane	0.17	53%	$3.7*10^{-12}$	0.035
i-Pentane	0.32	47%	$4.6*10^{-12}$	0.096
n-Hexane	0.08	59%	$4.8*10^{-12}$	0.022
Ethene	0.54	54%	$8.1*10^{-12}$	0.25
Propene	0.16	63%	$1.5*10^{-11}$	0.13
Acetylene	0.57	30%	$1.0*10^{-12}$	0.003
Benzene	0.14	40%	$1.2*10^{-12}$	0.009
Toluene	0.17	54%	$6.4*10^{-12}$	0.060
Xylenes	0.06	51%	$1.4-2.4*10^{-11}$	0.070
Ethylb.	0.06	63%	$7.5*10^{-12}$	0.025

* Based on an average OH-radical concentration of $6.5*10^5$ molecules/cm^3
+ Rate constants from ref. (1)

The data in Table 1 are based on only about 25 measurements made under relatively clean air conditions at a single station. Therefore they should not be considered to be representative for the continental background troposphere. Nevertheless, these data demonstrate several points which can be observed for most hydrocarbon measurements in the continental background. Methane is the by far most abundant hydrocarbon and its turnover exceeds that of every other individual hydrocarbon, although its rate constant for the reaction with OH-radicals is relatively slow. The other hydrocarbon mixing ratios are typically 3-4 orders of magnitude less abundant. This is partly compensated by their higher reactivity and therefore the NMHC can contribute significantly to the overall hydrocarbon turnover. In the example in Table 1 the sum of all NMHC even exceeds methane in its molecular turnover. However, none of the individual NMHC contributes more than 15% to the total of all hydrocarbon reactions (methane plus NMHC).

There is another important point which can be derived from Table 1. The relative variability of the hydrocarbon mixing ratios generally increases with increasing reactivity (or decreasing atmospheric lifetime). Indeed, the distribution of methane shows only relatively small systematic and unsystematic variations, whereas it is impossible to give a representative average value for the large scale concentrations of the alkenes or the xylenes. For this reason and since methane is the most important individual hydrocarbon, we will discuss the budgets and distributions of methane and the NMHC separately.

2. METHANE SOURCES AND SINKS

There is a considerable number of emission inventories for methane published by different authors. Table 2 summarizes some of the more recent estimates of the global methane source strength. In general the total global emission rates range around $500*10^{12}$ g/yr, but the contributions of the various sources sometimes differ significantly. Most of the authors give a range of uncertainty of about a factor of two for their estimates. It should be mentioned, that the C-14 content of the atmosphere and the various emissions as well as their C-13/C-12 isotopic composition allow to put some constraints on the methane budget. However, a discussion of the isotopic composition of atmospheric methane would be beyond the scope of this paper.

Table 2. SUMMARY OF GLOBAL METHANE EMISSION RATES (10^{12} G/YR)

Type of source	Ref(2) (1978)	Ref(3) (1983)	Ref(4) (1984)	Ref(5) (1987)	Ref(6) (1988)
Ruminants	160	120	85	75	80
Rice paddy fields	280	95	53	136*	110
Swamps & marshes	245	150	38		115
Oceans & lakes	22	23	4		15
Landfills				50	40
Biomass burning		25	75	65	55
Coal mining	18		30	35	35
Tundra	2	12			
Natural gas losses		40	25	18	45
Others	15	88	10	15	40
Total	742	553	320	394	535

* including wetlands

93

Since the atmospheric methane concentration is changing only at a rather slow rate (see below) these emissions have to be balanced by removal mechanisms. There exist two sinks for tropospheric methane, reaction with OH-radicals, including transport into the stratosphere and subsequent photooxidation in the stratosphere, and uptake by soils. Based on a global OH-radical distribution scaled to the global budget of CH_3CCl_3 a photochemical removal rate of $425\pm125 * 10^{12}$ g/yr can be calculated for methane (7). The uptake by soils is less important, only $10\pm5 * 10^{12}$ g/yr (7). Finally we have to consider that the increase of the tropospheric methane mixing ratio (see below) results in an accumulation in the atmosphere of about $60\pm15 * 10^{12}$ g/yr (7). Within the uncertainty of the different budgets, the removal balances the global emissions. Therefore it seems plausible to accept that the total methane emissions estimates are reasonably correct. The largest uncertainties lay probably in the apportionment of the total flux to the different types of sources.

3. METHANE DISTRIBUTIONS AND TRENDS

In the past few years there has been a large number of investigations of the global distribution, the seasonal cycle, and long term trend of the tropospheric methane mixing ratios (3,8-14) and the distribution, trends, and seasonal cycles of tropospheric methane are now reasonably well established. The global average mixing ratio of tropospheric methane in 1984 was 1.625 ppm (8). The largest systematic gradient within the troposphere is found between the northern and the southern hemisphere with a north/south ratio of 1.07 corresponding to 0.11 ppm (3,8). There exists a small, but significant north-south gradient within the northern hemisphere, but only a marginal latitudinal dependence can be found for the southern hemisphere. Within the troposphere no significant vertical gradient is observed. A clear seasonal cycle with a summer/winter difference of 0.035 ppm is reported for the northern hemisphere (8). This is mainly due to the seasonal change in the tropospheric removal rate due to the seasonal cycle of the OH-radical concentration, but also transport effects probably influence the shape of the seasonal variation of the methane concentration. The amplitude of the southern hemispheric seasonal cycle is smaller and shifted in phase by 6 month.

There is a significant and well established secular trend in the tropospheric methane concentration. The relative rate of increase between the mid seventies and the mid eighties is about 1 %/yr (8,9,11,14). The present day (1989-1990) global average of methane mixing ratio is about 1.7 ppm. The analyses of air bubbles trapped in old ice from glaciers show that the methane level prior to 1700 was only about 0.7 ppm (10,15). There are indications, that the relative north/south gradient was possibly slightly larger than today (15). Possible reasons for the change of the methane concentration are the increase in the number of cattle and the area of rice fields, possibly to some extent also increasing losses of natural gas.

4. NONMETHANE HYDROCARBON SOURCES AND SINKS

The emission rates of the main sources for atmospheric NMHC on a global scale have been evaluated by Ehhalt and Rudolph (16), Table 3. There are a few other estimates of the global emission rates for several different types for NMHC sources (17,18) but on a global scale the different NMHC sources and their emission rates are still poorly known. From these estimates we can see that the global source strength of NMHCs (not including isoprene and terpenes, see below) is about 25% of the methane source strength. This comparison underestimates the importance of

Table 3. ESTIMATES OF GLOBAL HYDROCARBON EMISSIONS INTO THE ATMOSPHERE (10^{12} G/Y)*

Source	C_2H_6	C_3H_8	C_4H_{10}	C_5H_{12}	C_2H_4	C_3H_6	C_2H_2	C_6H_6	C_7H_8	C_8H_{10}	$\Sigma NMHC^+$
Engine exhaust	0.2	0.02	0.4	1.3	1.9	0.8	1.4	0.8	1.7	1.5	19
Evaporation losses	--	--	1.0	2.6	--	--	--	0.14	0.7	0.5	10
Natural gas losses	1	1	1	--	--	--	--	--	--	--	5
Oil and coal burning	--	--	--	--	--	--	--	--	--	--	4
Chemical industry	0.1	--	--	--	0.3	--	--	0.1	0.1	--	3
Solvent use	--	--	--	--	--	--	--	--	1	0.5	15
Biomass burning	4.1	1.7	--	--	4	3.8	0.6	--	--	--	14
Emissions from foliage	0.5	--	--	--	1.0	--	--	--	--	--	1.5
Microbial production	0.7	--	--	--	4.4	0.6	--	--	--	--	5.7
Ocean emissions++	1 (3)	1.3 (3)	-- (3)	-- (3.4)	7 (15)	11.6 (11)	--	--	--	--	21 (52)
Total	7.6 (9.6)	4 (5.7)	2.4 (5.4)	3.9 (7.3)	15 (23)	16.8 (16.2)	2	1	3.5	2.5	98 (129)

* Data from ref. (16)
+ not including isoprene and terpenes
++ Data in parenthesis are from ref. (17)

Since the list of inidividual NMHC is not complete, the Σ NMHC sometimes exceeds the sum from the listed compounds.

these NMHCs for two reasons. The intermediate products of the photooxidation of NMHC are considerably more complex than the methane oxidation products. This allows other, more complex pathways and reactions in the photochemical reaction chains, e.g. the formation of peroxyacetyl nitrate. Furthermore, due to the, compared to methane relatively short atmospheric residence time of most NMHC, they often dominate in regions with strong NMHC emitters, e.g. Europe or North America.

The main removal mechanism for NMHC from the atmosphere is the reaction with OH-radicals. For alkenes also the reaction with ozone has to be taken into account. At mid northern latitudes the alkene-ozone reaction amounts on the average to roughly 30% of the total atmospheric removal. The average atmospheric lifetimes of NMHC range from a few month for ethane to several hours for the reactive alkenes (cf. 19,20).

5. THE TROPOSPHERIC DISTRIBUTION OF NONMETHANE HYDROCARBONS

The current picture of the distribution of nonmethane hydrocarbons (NMHC) in the troposphere is still far from being complete. The two main reasons are the still rather limited number of investigations and the large, to a considerable part seemingly unsystematic variability of the background NMHC concentrations. An example of the variability of the NMHC concentrations in rural areas of continental Europe is given in Figure 1 (taken from ref. 21). The measurements were made under varying meteorological conditions, ranging from stagnant conditions with high pollution levels to near background conditions. For comparison the emission pattern (scaled to 10 ppb of ethane) of continental NMHC sources is included. Evidently, with decreasing pollution levels not only the absolute mixing ratios show a general decrease, but the relative pattern also changes systematically. With decreasing hydrocarbon levels, the more reactive NMHC decrease faster than the less reactive NMHC. This depletion of reactive NMHC is the result of the faster photochemical removal.

Large scale distributions of NMHC in the remote troposphere have been published by several investigators (16,19-33). An overview over various measurement series of light NMHC in the remote marine boundary layer is shown in Figure 2. The distribution of the longer-lived NMHC (τ > 1 week) show some systematic features which can be ascribed to some extend to representative latitudinal profiles. On the average the highest mixing ratios are observed at mid to high northern latitudes, roughly 1.5 – 2.5 ppb of ethane, 0.2 – 0.6 ppb of propane and n-butane, 0.1 – 0.5 ppb of acetylene, 0.05 – 0.25 ppb of benzene and 0.02 – 0.1 ppb of i-butane. As an example a series of measurements of light n-alkanes over the remote Atlantic is shown in Figure 3.

In general all NMHC show a considerable decrease towards lower latitudes and the slope is more pronounced for the shorter-lived NMHC. This reflects both the source distribution with high emissions in the industrialized zone of the northern hemisphere and the faster removal at low latitudes due to the higher OH radical concentrations in tropical latitudes. Average south hemispheric mixing ratios are roughly 0.15 – 0.4 ppb ethane, 0.03 – 0.1 ppb propane, 0.01 – 0.05 ppb n-butane, 0.01 – 0.05 ppb acetylene, 0.01 – 0.03 ppb benzene and 0.005 – 0.03 ppb i-butane. There seems to be a slight, but systematic decrease from low southern latitudes towards mid and high southern latitudes.

For ethane and probably acetylene there seems to be a significant systematic seasonal cycle. In both hemispheres the minima are in the respective late summer, maxima in late winter (24,33). Outside of areas with strong ground level sources, no systematic vertical gradient is observed for these compounds.

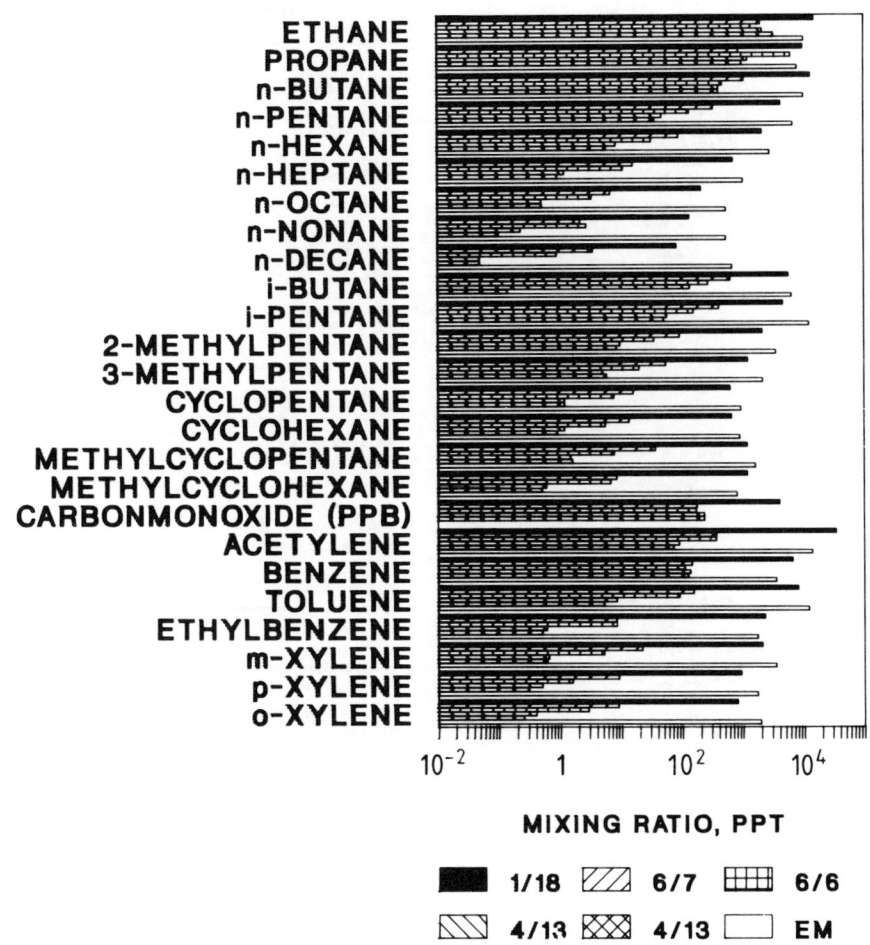

Fig. 1: Hydrocarbon mixing ratios measured in a rural area in West Germany in 1985. 1/18: Stagnant weather conditions, 6/6 and 6/7: Moderate wind from southwest, 4/13: Strong winds from southwest, E. M. (open bars): Emission pattern scaled to an ethane mixing ratio of 10 ppb.

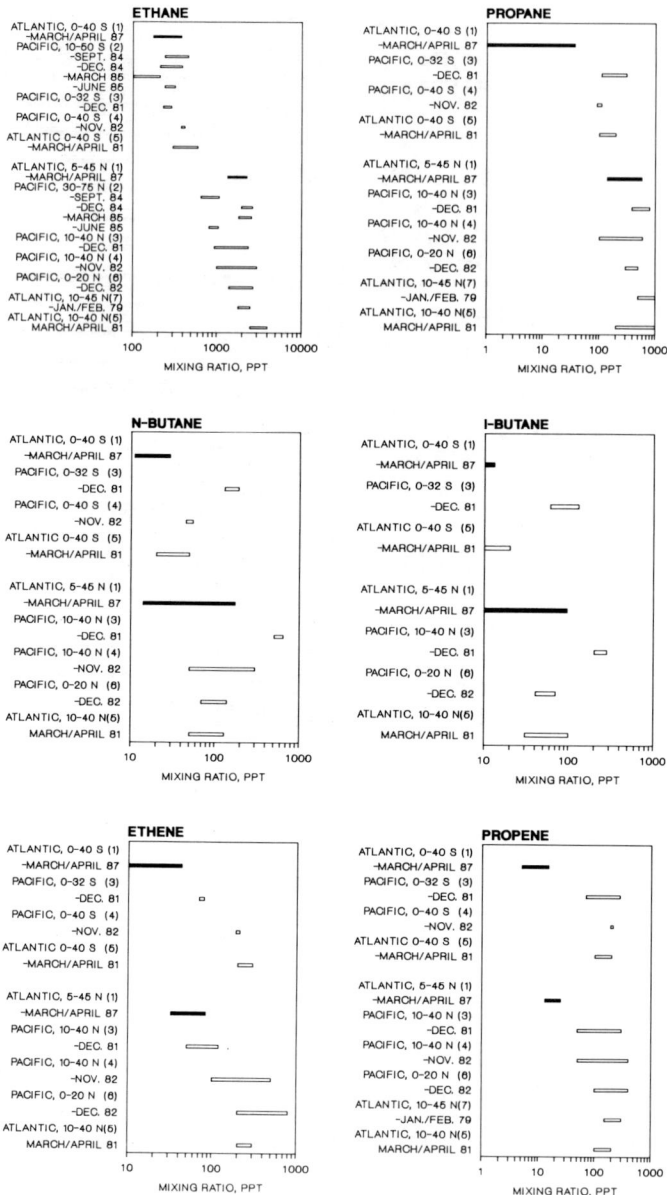

Fig. 2: Overview over measurements of light NMHC in the remote marine atmosphere, from ref. (31)

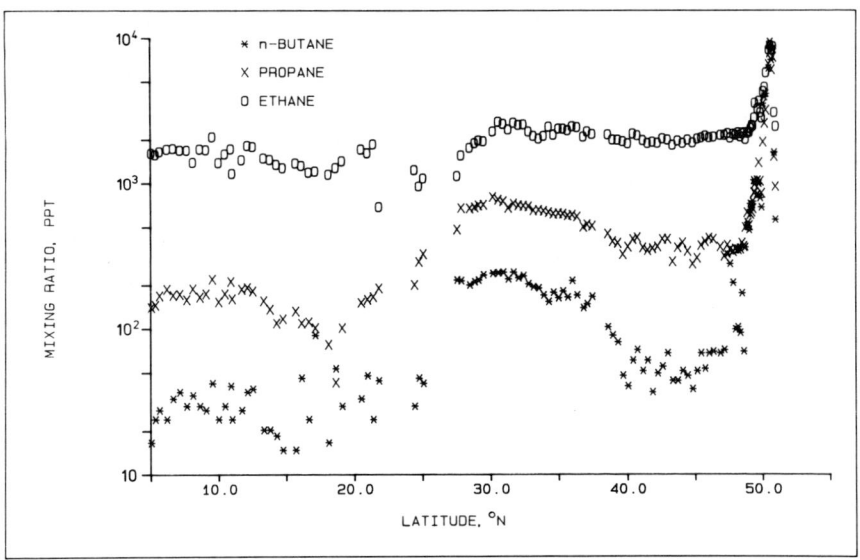

Fig. 3: Latitudinal distribution of some light alkanes in the northern hemisphere over the remote Atlantic. Data from ref. (31)

For more reactive NMHC with average atmospheric residence times of less than a few days, it is no longer justified to consider their distributions as systematic with latitude. The atmospheric concentrations of these species are more or less determined by emissions, removal reactions, and transport on a local or regional scale. Still, for NMHC with average atmospheric lifetimes of more than half a day transport can be of considerable importance and the impact of large sources often is visible in distances of several hundred kilometres. The most important NMHC in this category (1 week > τ > 0.5 days) are light alkenes and, somewhat less important, the C_5 and heavier alkanes and alkylbenzenes. Since the sources of light alkenes include biomass burning, emissions from vegetation and from oceans, substantial concentrations in the range from 0.1 ppb to a few ppb of ethene and propene have been observed over non industrialized continental regions and the oceans. Simplified model calculations showed that light alkenes can potentially contribute to the production of tropospheric ozone in the order of some ten ppb/day on a regional scale (34). However in regions without significant alkene sources only a marginal impact of alkenes on the photochemistry of the troposphere is found (34). In contrast to model predictions (35,36) measurements of the vertical profiles of reactive alkanes and alkenes show no or little systematic decrease in their mixing ratios with increasing altitude below the tropopause (37-41). This is most probably the result of fast convective vertical exchange processes which are not considered in the 1-D eddy diffusional models.

The existing measurements of NMHC in the troposphere do not allow to recognize any systematic long-term trend. Moreover, the observed variability of the NMHC most probably will prevent the recognition of any global trend of atmospheric NMHC in the near future. On local or regional scales changes of emission rates (e.g. increasing oil and natural gas production and consumption or control actions to reduce emissions from automobiles) can have clearly visible effects. On a global scale a sufficiently dense monitoring network for NMHC would require very large efforts. More promising seems to estimate future (and past) changes of atmospheric NMHC concentrations from possible changes in the magnitude of their main sources. However, our current knowledge about the emission rates of NMHC from the various types of sources is far from being satisfactory and not yet sufficient to allow predictions of future NMHC emission rates.

6. ISOPRENE AND TERPENES IN THE ATMOSPHERE

In contrast to the light alkenes and reactive alkanes, the extremely reactive terpenes and isoprene allways show a very strong decrease with increasing altitudes and can generally be observed in significant concentrations only within the atmospheric boundary layer (42-45). These compounds can only be found near to their sources (exclusively emissions from vegetation) and substantial concentrations of isoprene and terpenes can often be observed over the continents. Over areas with dense vegetation, isoprene mixing ratios around 0.5 - 2 ppb and monoterpene mixing ratios approaching 1 ppb are frequently found. At those levels the biogenic NMHC often have a large impact on the photochemical reaction chains in the atmosphere. However it should be noted that the emissions of many biogenic NMHC strongly depend on temperature, relative humidity and light-intensity (cf. 46,47). Also the types of emitted compounds depend on the type of vegetation (cf. 48) e.g. isoprene is primarily emitted from deciduous plants whereas coniferous trees mainly act as terpene sources. The total annual global emission rates are estimated to be $450 * 10^{12}$ g/yr for isoprene (cf. 18) and roughly the same amount of terpenes (cf. 16). However, these estimates are highly uncertain.

REFERENCES

(1)   Hampson, R. F. and Garvin, D. (1977). Reactionrate and photochemical data for atmospheric chemistry. National Bureau of Standards Special Publication 513. U.S. Government Printing Office, Washington, D.C.

(2)   Ehhalt, D. and Schmidt, U. (1978). The atmospheric cycle of methane. PAGEOPH 116, 452-464

(3)   Khalil, M.A.K. and Rasmussen, R.A. (1983). Sources, sinks and seasonal cycles of atmospheric methane. J. Geophys. Res. 88, 5131-5144.

(4)   Seiler, W., Conrad, R., and Scharffe, D. (1984). Field studies of methane emissions from termite nests into the atmosphere and measurements of methane uptake from soils, J. Atm. Chem. 1, 171-186.

(5)   Bingemer, H.G. and Crutzen, P.J. (1987). The production of methane from solid wastes. J. Geophys. Res. 92, 2181-2187.

(6)   Cicerone, R.J. and Oremland, R.S. (1988). Biogeochemical aspects of atmospheric methane. Global Biochemical Cycles 2, 299-327.

(7)   McElroy, M.B. and Wofsy, S.C. (1985). Tropical forests: Interactions with the atmosphere. In Symposium volume on "Tropical forests and World Atmospheres" edited by Prance, G.T.

(8)   Steele, L.P., Fraser, P.J., Rasmussen, R.A., Khalil, M.A.K., Conway, T.J., Crawford, A.J., Gammon, R.H., Masarie, K.A., and Thoning, K.W. (1987). The global distribution of methane in the troposphere. J. Atmos. Chem. 5, 125-171.

(9)   Rasmussen, R.A. and Khalil, M.A.K. (1986). Atmospheric trace gases: Trends and distributions over the last decade. Science, 232, 1623-1624.

(10) Khalil, M.A.K. and Rasmussen, R.A. (1987). Atmospheric methane: Trends over the last 10,000 years. Atmos. Environ.21, 2445-2452.

(11) Blake, D.R. and Rowland, F.S. (1986). World-wide increase in tropospheric methane, J. Atmos. Chem. 4, 43-62.

(12) Blake, D.R,Mayer, E.W., Tyler, S.C., Makide, Y., Montague, D.C., and Rowland, F.S. (1982). Global increase in atmospheric methane concentrations between 1978 and 1980. Geophys. Res.Lett. 9, 447-480.

(13) Fraser, P.J., Khalil, M.A.K., Rasmussen, R.A., and Crawford, A.J. (1981). Trends of atmospheric methane in the southern hemisphere. Geophys. Res. Lett. 8, 1063-1066.

(14) Zander, R., Demoulin, Ph., Ehhalt, D.H., and Schmidt, U. (1989). Secular increase of the vertical column abundance of methane derived from IR solar spectra recorded at the Jungfraujoch station. J. Geophys. Res. 94, 11029-11039.

(15) Rasmussen, R.A. and Khalil, M.A.K. (1984). Atmospheric methane in the recent and ancient atmospheres: Concentrations, trends and interhemispheric gradient. J. Geophys. Res. 89, 11599-11605.

(16) Ehhalt, D.H. and Rudolph, J. (1984). On the importance of light hydrocarbons in multiphase atmospheric systems, Ber. Kernforschungsanlage Jülich, JÜL-1942, pp. 1-43.

(17) Bonsang, B., Kanakidou, M., Lambert, G., and Monfray, P. (1988). The marine source of C_2-C_6 aliphatic hydrocarbons. J. Atmos. Chem. 6, 3-20.

(18) Rasmussen, R.A. and Khalil, M.A.K. (1988). Isoprene over the Amazon basin. J. Geophys. Res. 93, 1417-1421.

(19) Rudolph, J. and Ehhalt, D.H. (1981). Measurements of C_2-C_5 hydrocarbons over the North Atlantic. J. Geophys. Res. 86, 11.959-11.964.

(20) Greenberg, J.P. and Zimmerman, P.R. (1984). Nonmethane hydrocarbons in remote tropical, continental and marine atmospheres. J. Geophys. Res. 89, 4767-4778.

(21) Rudolph, J. and Koppmann, R. (1990). Sources and atmospheric distribution of light hydrocarbons. Proceedings of the 28th Liège International Astrophysical Colloquium "Our Changing Atmosphere", June 26-30, 1989, Liège, Belgium, 435-447.

(22) Singh, H.B., Viezee, W., and Salas, L.J. (1988). Measurements of selected C_2-C_5 hydrocarbons in the troposphere: Latitudinal, vertical and temporal variations. J. Geophys. Res. 93, 15861-15878.

(23) Bonsang, B. and Lambert, G. (1985). Nonmethane hydrocarbons in an oceanic atmosphere. J. Atmos. Chem. 2, 257-271.

(24) Blake, D.R., and Rowland, F.S. (1986). Global atmospheric concentrations and source strength of ethane. Nature 321, 231-233.

(25) Nutmagul, W. and Cronn, D.R. (1985). Determination of selected atmospheric aromatic hydrocarbons at remote continental and oceanic locations using photoionization/flame-ionization detection. J. Atmos. Chem. 2, 415-433.

(26) Ehhalt, D.H., Rudolph, J., Meixner, F.X., and Schmidt, U. (1985). Measurements of selected C_2-C_5 hydrocarbons in the background troposphere: Vertical and latitudinal variations. J. Atmos. Chem. 3, 29-52.

(27) Singh, H.B., and Salas, L.J. (1982). Measurements of selected light hydrocarbons over the Pacific Ocean: Latitudinal and seasonal variations. Geophys. Res. Lett. 9, 842-845.

(28) Rudolph, J. (1988). Two-dimensional distribution of light hydrocarbons: Results from the STRATOZ III experiment. J. Geophys. Res. 93, 8367-8377.

(29) Rudolph, J., Jebsen, C., Khedim, A., and Johnen, F.J. (1984). Measurements of the latitudinal distribution of light hydrocarbons and halocarbons over the Atlantic. In Proceedings of the 3rd European Symposium on Physico-Chemical Behaviour of Atmospheric Pollutants, edited by B. Versino and G. Angeletti, pp. 492-501, D. Reidel, Hingham, Mass..

(30) Rudolph, J., Ehhalt, D.H., Khedim, A., and Jebsen, C. (1982). Latitudinal profiles of some C_2-C_5 hydrocarbons in the clean troposphere over the Atlantic. Paper presented at 2nd Symposium on Composition of the Nonurban Troposphere, Am. Meteorol. Soc., Boston, Mass..

(31) Rudolph, J. and Johnen, F.J. (1990). Measurements of light atmospheric hydrocarbons over the Atlantic in regions of low biological activity. J. Geophys. Res., in press.

(32) Hov, O., Penkett, S.A., Isakson, I.S.A., and Semb, A. (1984). Organic gases in the Norwegian Arctic. Geophys. Res. Lett. 11, 425-428.

(33) Rudolph, J., Khedim, A., and Wagenbach, D. (1989). The seasonal variation of light nonmethane hydrocarbons in the Antarctic Troposphere. J. Geophys. Res. 94, 13039-13044.

(34) Rudolph, J., Koppmann, R., Johnen, F.J., and Khedim, A. (1990). The distribution of light nonmethane hydrocarbons in the troposphere and there potential impact on photochemical ozone formation. Proceedings of the Quadrennial Ozone Symposium, 8-14 August 1988, Göttingen, Germany, in press.

(35) Chameides, W.L. and Cicerone, R.J. (1978). Effects of nonmethane hydrocarbons in the atmosphere. J. Geophys. Res. 83, 947-952.

(36) Fishman, J. and Carney, T.A. (1984). A one-dimensional photochemical model of the troposphere with planetary boundary layer parameterization. J. Atmos. Chem. 1, 351-376.

(37) Tille, K.J.W., Savelsberg, M., and Bächmann, K. (1985). Airborne measurements of nonmethane hydrocarbons over western Europe: Vertical distribution, seasonal cycles of mixing ratios and source strength. Atmos. Environ. 11, 1751-1760.

(38) Rasmussen, R.A., Khalil, M.A.K., and Fox, R.J. (1984). Altitudinal and temporal variation of hydrocarbons and other tracers of arctic trace. Geophys. Res. Lett. 10, 144-147.

(39) Rudolph, J., Ehhalt, D.H., and Tönnissen, A. (1981). Vertical profiles of ethane and propane in the stratosphere. J. Geophys. Res. 86, 7267-7272.

(40) Rudolph, J., Ehhalt, D.H., Schmidt, U., and Khedim, A. (1982). Vertical distributions of some C_2-C_5 hydrocarbons in the nonurban troposphere. Paper presented at 2nd Symposium on Composition of the Nonurban Troposphere, Am. Meteorol. Soc., Boston, Mass..

(41) Rudolph, J., Ehhalt, D.H., and Khedim, A. (1984). Vertical profiles of acetylene in the troposphere and stratosphere. J. Atmos. Chem. 2, 117-124.

(42) Zimmerman, P.R., Greenberg,J.P., and Westberg, C.E., (1988). Measurements of atmospheric hydrocarbons and biogenic emission fluxes in the Amazon boundary layer. J. Geophys. Res. 93, 1407-1416.

(43) Gregory, G.L., et al. (1986). Air chemistry over the tropical forest of Guyana. J. Geophys. Res. 91, 8603-9612.

(44) Hov, O., Schjoldager, J., and Wathne, B.M. (1983). Measurement and modeling of the concentrations of terpenes in coniferous forest air. J. Geophys. Res. 88, 10.697-10.688.

(45) Ayers, G.P., and Gillett, R.W. (1988). Isoprene emissions from vegetation and hydrocarbon emissions from bushfires in tropical Australia. J. Atmos. Chem. 7, 177-190.

(46) Lamb, B., Westberg, H., and Allwine, G.(1986). Isoprene emission fluxes determined by an atmospheric tracer technique. Atmos. Environ. 20, 1-8.

(47) Lamb, B., Westberg, H., and Allwine, G. (1985). Biogenic hydrocarbon emissions from deciduous and coniferous trees in the United States. J. Geophys. Res. 90, 2380-2390.

(48) Graedel, T. (1979). Terpenoids in the atmosphere. Rev. Geophys. and Space Phys. 17, 937-947.

EUROPEAN SCHOOL OF CLIMATOLOGY AND NATURAL HAZARDS

Course on

"Climate and Global Change"

3. MONITORING AND FORECASTING GLOBAL CHANGES

CLIMATIC EVOLUTION DURING THE LAST CENTURY

C.J.E. Schuurmans
Royal Netherlands Meteorological Institute
De Bilt, The Netherlands
and
Institute of Meteorology and Oceanography
Utrecht University, Utrecht, The Netherlands

Summary

In most areas of the world temperature and precipitation are percei-
ved as the key climatic elements. However, these elements are not
necessarily the best indicators of climatic evolution. In search for
such indicators, we have to broaden our concept of climate change, to
not only include temperature and precipitation changes (locally,
regionally and globally), but also the associated changes of the
atmospheric circulation. Related changes in e.g. sea level and
glacier extend are also of interest. An attempt is made to list the
most important changes of the coupled climate system over the last
100 years. The most obvious of these is a small global, though non-
uniform, warming. Some of the changes clearly result from a specific
mode of behaviour of the system, like the Southern Oscillation cycle.
Others might be the result of external forcing, which is of solar,
volcanic or antropogenic origin. Except for some of the latter, e.g.
the increasing greenhouse gas concentration of the atmosphere, the
quantitative time history of these forcings is unknown. This hampers
the determination of the fraction of the observed warming that may be
attributed to the enhanced greenhouse radiation. Another outstanding
problem is how to account for some of the apparently jumplike changes
of climate which occurred in the present century. Is the climate
system with its feedbacks able to produce these, all by itself?

0. INTRODUCTION

Any attempt to describe the evolution of climate over the last
century inevitably will be incomplete, due to the vastness of the subject.
See e.g. Kukla and Robinson, (13). Climate of course can be studied from a
practical point of view - its impact on living conditions - or from a
theoretical viewpoint. It has statistical and physical aspects and as far
as its description is concerned, we have to do with a variety of climatic
elements (temperature, precipitation, etc.), each of them having its
characteristic variations on specific time and space scales (local,
regional or global).

In this lecture I therefore limit the scope of the subject to only
10 aspects:
1. the impact of climate change
2. climate change under constant external forcing
3. climate change due to external forcing
4. short term climate changes (less than 10 years)
5. decadal variations
6. changes of the atmospheric circulation
7. changes in variability
8. recent climate in longer perspective
9. climatic jumps as a mode of variation
10. detection of the greenhouse warming.

Before embarking on these points I introduce in figure 1 the variation of hemispheric and global mean temperature for the period 1861-1988, just as a point of reference.

The curves make clear that even the averages over large areas vary from year to year by an appreciable amount. Furthermore the curves show some coherence in time over periods of several years and finally in both hemispheres an upward trend is present in the early part of this century (up to about 1940), which after some interuption (till 1970) continues until the present day.

The reliability of the curves shown in Fig. 1 has sometimes been doubted because of the non-uniformity of the observations in space and time. Also the so-called city-effect on temperatures has been mentioned as a possible source of unreliability. By the latter we mean that temperature observations in cities will be affected by the large heat capacity of the surrounding buildings and probably also by the heat production, which is mostly concentrated in cities. Both the non-uniformity and city-effect on hemispheric and global mean temperature curves have been studied by Hansen and Lebedeff (8). They come to the conclusion that the gross features of the temperature curves remain the same when the selection and number of stations is varied, or when cities are excluded in the derivation. They however admit that the city effect may be responsible for some 0.2 °C of the general trend shown by curves like in Fig. 1.

1. IMPORTANCE OF CLIMATE CHANGE

In most areas of the world temperature and precipitation are percei-ved as the key elements of climate. Changes in these elements may have a strong impact on the environment and living conditions of man. Just a few examples from the last 100 years may illustrate the point (Wallén, (27)).
1. In Finland, due to climate warming in 1920-1940 summer wheat could be cultivated north of the polar circle (66 °N), at the end of the 1930's. Due to cooling off in the 1950's and 1960's however, many settlers had to leave their farms.
2. In Canada in 1920-1940 crop limits also moved northward due to higher temperatures, but southward in the Canadian Prairies and the Great Plains of the USA, at the same time repeated crop failures occurred due to the heat and wide-spread drought.

Though on average less clear, strong local impacts also may result from year-to-year changes in temperature and amount of precipitation. Convincing examples of this e.g. may be found in Lamb (15).

This means that changes of temperature and precipitation may be important on all timescales. Let us call this the first important conclu-sion: impact of climate change occurs on all timescales. So, from the point of view of climate change impact all possible changes may be worth looking at. Clearly the examples given in 1. and 2. above have to do with the decadal time scale of climate evolution. For the different time scales of variation the areas involved are not necessarily the same, however. This means that it is virtually impossible to construct a meaningful time series of temperature and/or precipitation without a specific impact in mind.

2. UNDERSTANDING CHANGES BY INTERNAL FEEDBACKS

We now turn to the subject of understanding of climate change, leaving from now on the impact issue. Climate being the time average of a number of individual weather situations will necessarily be variable like weather itself. For instance, it is extremely unlikely that two different years produce 365 equal weather situations, at a single location. So individual years will be different from each other, just by a different set of weather patterns, occurring under one and the same seasonably variable solar forcing. These weather patterns are part of the atmospheric circulation, which apart from solar irradiation is influenced by mechani-cal and thermal forcing at the earth's surface. In this way a weather pattern may just be seen as the atmospheric part of the state of the coupled earth-atmosphere-ocean system at a certain time.

Physical models of this system, even with a limited number of variables, do show that the time variation of such a system is not limited to just interannual differences. Without any change in the external solar irradiation the system may produce long-period variability of considerable amplitude (Lorenz, (16); James and James, (11)). Of course, if we have interannual variability we will also have variability on longer time scales, simply by the statistical sampling process. E.g. if the annual values of a certain parameter are normally distributed with standard deviation σ, the distribution of the N-year averages of this parameter will have $\sigma_N = \sigma\ N^{-\frac{1}{2}}$. The long-term variations referred to above, however are much larger than can be explained in this way, at least for some periods. The real cause of these (quasi)-periodic changes therefore must be sought in certain feedback processes between temperature and other quantities in the (model)system.

The point is that if these feedbacks occur in a model of the climate system, they almost certainly also occur in the real system. This would mean that basically all or perhaps most of the climate changes we observe can be due to such internal feedbacks between parts of the system. At least in principle. The second important conclusion therefore is that the benign climate of the 1930's in high Northern latitudes just could be the result of specific mode of behavior of the climate system under constant external forcing. Also on a time scale of months to seasons internal feed backs may contribute significantly to climate variability (Madden and Shea, (17)).

3. CHANGES IN EXTERNAL FORCING

But we know that external forcing may vary and over the last century certainly has varied. What climate changes have been caused by these variations? Let us first look at the time histories of these forcings. There are three of them: solar radiation, volcanic dust and the concentration of greenhouse gases. The first two are in fact unknown for most of the century and we have to rely on proxy data. Only the last one is relatively well-known, if we accept the indirectly estimated concentrations of e.g. CO_2, before regular measurements began, in 1958.

These forcing functions can be fed into a time dependent climate model and the response can be studied. It is found that certain features of temperature evolution in this century can be simulated succesfully. See Gilliland and Schneider (7) and Oerlemans (19). The models used for this purpose are quite simple and can only simulate the crudest aspects of the temperature evolution, usually only the global mean temperature, over the last 100 years. Therefore we have no guarantee that a close correspondance between the simulated and observed temperature curve means that indeed the variations of the external forcing functions are the real cause of the long-term temperature changes. In fact, and this is the third main conclusion, we have a number of possible explanations of the observed variation of global mean temperature over the last century, but we have as yet no possibility to determine the real cause(s).

This situation more or less invites to a further investigation of the pattern of interrelated changes of the global system, for each of the timescales of evolution, regardless the cause of the associated changes.

4. SHORT-TERM VARIATIONS

On the time scale of less than 10 years we have the well-known El Niño-Southern Oscillation (ENSO) cycle. This is an irregular oscillation of duration 3-7 years, involving the (tropical) oceans and global atmosphere. The average evolution of the phenomenon is described by e.g. Horel and Wallace, (10). Of major concern in our study of climate over the past century is the continuity of the ENSO-phenomenon. Based on an analysis of sea surface temperatures in the tropical Pacific and of variations of sea level atmospheric pressure in certain key areas the following list of warm and cold event years has been produced.

Table I List of warm and cold event years (according to Bradley et al., (3).

Warm events:
1884, 1888, 1891, 1896, 1899, 1902, 1904, 1911, 1913, 1918, 1923, 1925, 1930, 1932, 1939, 1951, 1953, 1957, 1963, 1965, 1969, 1972, 1976, 1982 1986.

Cold events:
1886, 1889, 1892, 1898, 1903, 1906, 1908, 1916, 1920, 1924, 1928, 1931, 1938, 1942, 1949, 1954, 1964, 1970, 1973, 1975, 1988.

We may conclude that ENSO has been with us all over the past 100 years and most probably for many centuries earlier. This does not mean that the atmospheric response pattern at the extratropical latitudes always has been the same. This response depends on the structure of the tropical wind field and this may change on time scales of decades or centuries, parallel to changes of global climate. However possible differences of the extratropical response pattern to ENSO's between early and recent decades of this century have not been fully documented. Nevertheless, conclusion number four reads that over the past 100 years our climate system maintained an oscillatory fluctuation of 3-7 years duration which caused appreciable temperature and precipitation variability at this time scale, all over the globe.

5. CHANGES ON DECADAL TIMESCALE

Our main interest is in the changes of the climate system on the 10-100 years time scale. We already referred to the warm period in the 1930's, the subsequent cooling in the years 1950-1970 and the temperature rise afterwards. This is the well-known behavior of global mean temperature and as far as the cooling in 1950-1970 is concerned, more specifically of the Northern Hemisphere mean temperature.
A more detailed description of the decadal changes over the past century has to include the following:
1. differences of the global mean behavior of temperature for different regions of the globe
2. related changes in precipitation in different parts of the world
3. associated changes of the atmospheric circulation
4. changes of the slowly varying components of the climate system like changes in sea level and changes in the extent of sea ice and mountain glaciers.
This list could be at infinity extended by inclusion of seasonal differences in long term behavior, distinction between day and nighttime temperatures, changes in yearly amplitude of temperature etc.
For reasons of availability of data and the required conciseness of the description I had to restrict myself to the following remarks. The temperature increase in the last century is indeed a worldwide phenomenon. In the Northern Hemisphere extratropics the cooling in the 1950-1970 period is more marked than elsewhere. See Fig. 2. From a regional study for Europe we may conclude that this cooling also here was mainly a mid- to highlatitude phenomenon, which was much less the case for the warming early in the century (Schuurmans, (21)).
For northern latitudes the changes in winter are much larger than in any of the other seasons. This means that for those areas the use of annual averages can be misleading.
On the other hand we have time shifts in the start of a certain change, between different seasons. E.g. in Europe the warming in winter started already early in the century (around 1910), while summer warming only started about 1930. So when studying changes of seasonal mean temperatures for fixed intervals of time it may happen that the changes for winter are smaller than for the other seasons. See Fig. 3. From the same figure we get the impression that the geographical pattern of change in different seasons might be the same.
Precipitation over the Northern Hemisphere land areas north of the

subtropics is positively correlated with temperature. Compare Fig. 4 with Fig. 2. In the low latitude zone precipitation decreases during the recent 40 years, but this decrease seems to be strongly dominated by the well-known Sahelian drought, which possibly could have been strengthened by human factors.

In terms of the mass balance of glaciers the increase of precipitation over the past century had less effect than the increased melting due to higher temperatures. According to Oerlemans (19), glaciers in Iceland, Norway and in the Alps, all have been retreating during the last 100-150 years.

Glacier retreat during this period could even be a world-wide phenomenon, which together with thermal expansion of the ocean due to higher temperatures, explains part of the observed sea level rise in the last 100 years. The global average of this rise is 10-15 cm per century.

When studying the climatic changes on a decadal and longer time scale it is not difficult to find some consistency and coherence of changes at different places and between different quantities. However, <u>without considering the important role of the atmospheric circulation</u>, not only at the surface but also in the free atmosphere, <u>we will never make real progress in our understanding of such changes</u>. This of course does not mean that all climatic changes on these time scales are necessarily caused by changes of the atmospheric circulation.

6. ATMOSPHERIC CIRCULATION CHANGES

The changes in the atmospheric circulation of the Northern Hemisphere over the last century are difficult to quantify. In most cases the so-called zonal index of the circulation, being a measure of the intensity of the westerlies, has been used. Others have defined an index of meridionality, indicating a tendency for meridional or blocking circulations. It is not certain if the latter physically makes sense. The zonal index certainly does, and also the meridional heat transport by transient and stationary eddies (van Loon, (25)).

In the latter paper and also in van Loon and Williams (26) the relationship between Northern Hemisphere temperature in winter and the heat transports by transient and stationary waves is studied. They come to the rather surprising conclusion that mid-latitudes are colder when the poleward heat fluxes by stationary waves are stronger. The explanation is that reduced heat transports by the planetary waves favor increased temperatures by reducing the effectiveness of the Arctic cold sink in winter. This mechanism can only work in winter since in other seasons meridional heat transport is always dominated by the transient waves. See Oort (20). Consequently, changes in the planetary waves affect Northern Hemisphere temperature in winter locally and by affecting the horizontal temperature gradients. These in turn affect the transient waves, giving rise to further temperature and precipitation changes. This is just one example of how a change in one component of the atmospheric circulation - the planetary waves - determines the winter climate of the whole hemisphere.

The zonal index, or at least its substitute the Azores-Iceland pressure difference, has recently been used to study its influence on long-term temperature trends in north-western Europe (Moses et al., (18)).

It is clear that in winter temperatures in north-western Europe are low when the normal north-south pressure gradient is reversed. Oppositely, in the more normal situation that we have an Iceland low and an Azores high - the high index situation - winter weather is relatively mild. Leaving out the extreme months however, did not lead to a full reduction of the observed trends in winter temperature. In particular the early 20th century upward trend is not at all eliminated by eliminating high index years. This led the authors to conclude that some true net change of winter temperature occurred, not associated with changes in circulation regime.

Whether or not this is the case has to be found out by further study. Anyhow, changes in atmospheric circulation do have an effect, which sometimes can be quantified. Unfortunately, upper air data do not go further back than about 1950. This causes a problem even if in a study is only made use of circulation types. For Europe e.g., we have the well-known German system of Grosswetterlagen, introduced by Hess and Brezowsky (9).

On the basis of sea level pressure distributions daily circulation types have been defined for all days from January, 1, 1881 onwards. As soon as upper air data became available (around World War II) the circulation at 500 mb is also taken into account in the classification. One of the striking changes of the Grosswetterlagen over the last century is a strong increase in the frequency of occurrence of the south-westerly circulation type, mainly at the expense of the north-westerly type.

The change however is most pronounced between the period 1881-1945 and the period thereafter. This at least suggests that the incorporation of upper air data in the (subjective) classification could have played a role.

Bearing such difficulties in mind, much more can be learned about the role of the atmospheric circulation in shaping climate and climate change, than hitherto has been tried. It might be possible however that definitive answers concerning this role can only be obtained from a more complete reconstruction of the behaviour of the atmospheric system over e.g. the last 10 years. (the period at least that sufficient data is available).

7. CHANGES IN VARIABILITY

Changes of climatic averages, whether or not directly related to changes of the atmospheric circulation are only one aspect of climate change. There is also the issue of changes in variability and in the occurrence of extremes. Only a few authors have studied this subject in any depth (van Loon and Williams, (24), Schuurmans and Coops, (22)).

The point is that in popular belief changes of climate go along with a more erratic behavior of climate. This is certainly not true. The true situation is much more complex. Just as an example let us look at seasonal temperatures at De Bilt, The Netherlands over the period 1875-1975. We define interannual variability just as the difference in absolute sense between two consecutive winter (or spring, summer, autumn) temperatures. This series is correlated with the series of seasonal mean temperatures itself. We find the following correlation coefficients:

winter	- 0.27
spring	- 0.06
summer	- 0.12
autumn	- 0.17

The 99% level of significance is at 0.25 which means that only in winter we probably have a negative correlation between interannual variability and the average temperature level. Interpretation of this correlation is that when temperatures on average are lower, they will be more variable. For north-western Europe this is exactly what we expect. Periods with low winter temperatures must have winds from the north or from the east. By advection temperature may reach very low values, depending on the source of cold air and the strength of the advection. Intermittent mild winters in such period will cause a very large variability. Periods with higher winter temperatures, however, have mainly winds from the (south)-west. Temperatures are limited then by the sea surface temperature, because the main source area is over the sea. For the whole of north-western Europe for winter temperatures the same argument applies.

Although this is only one example we tend to conclude that the relation between climate averages and variability, no matter how the latter is expressed, is a complicated one. In many cases there might be no relation and when there is a relation it may vary from place to place.

Apart from its relation to average climate the question may be asked if recently or over the last century "climate variability" has increased. It is difficult to answer this question in general. From my own studies I may conclude however that at a selection of 13 European stations studied, interannual variability of seasonal mean temperatures in the period 1875-1975 did not show any significant trend. An update and extension of this study to other areas might however be useful.

8. LONG-TERM TREND

For most areas of the globe climatological observations data go back less than 150 years. This means that for those areas the evolution of climate over the past century is the only detailed description of climate we have. Whether or not for these places the last century was representative for some longer period, cannot with certainty be determined. Fortunately, for Europe we have some longer series of climate observations, especially for temperature. Among the longest are those of Central England (since 1659) and De Bilt (The Netherlands) (since 1706, for winter temperatures even going back to 1645).

All of these long series show what has been called by Lamb (14) "a one-way story, a trend towards greater warmth that was interrupted only by various shorter term variations".

The curve of non-overlapping 10-year averages of winter temperatures in Holland, illustrates the point (Fig. 5). The reconstruction and quality of the series has been discussed by Van den Dool, Schuurmans and Krijnen (23). Updated and including the very mild winters of 1988/89 and 1989/90 the linear trend is + 0.0025 C per year. Interuptions as Lamb calls them are found in the periods of winter cooling which run from about 1740-1780, 1860-1890 and from 1920-1960. Note that the 10-year average around 1740 was about equally high as the maximum in the present century. Deviations from the trendline, i.c. the detrended time series shows that the last two very mild winters where about equally mild as some winters in the 18th century (1737 and 1739). Not only in view of the detection of the greenhouse warming (see paragraph 10), but also from a more general point of view, the cause of this long upward trend is something of utmost importance in relation to our subject. However, we do not know this cause. It could be the slow recovery from the Little Ice Age steered by internal feedbacks as discussed in paragraph 2, but long term variations in external forcing (increase of solar activity or decrease in volcanic activity) cannot be ruled out either.

9. JUMPLIKE CHANGES

Up till now in this lecture we have considered climate variations of three types: irregular interannual changes, coherent changes over a number of years (say 3-7) and long-term trends.

However, some evidence exists that the latter is not as regular and smooth as the term suggests. In Fig. 5 we saw already these so-called interruptions of the long-term warming trend. In closer view these interruptions sometimes have a jumplike character. They start rather abruptly and may finish abruptly when the general upward trend is resumed. Such a time series therefore may also looked upon as a sequence of alternating subperiods of low and high average values. The change points between the subperiods simply can be determined from the cumulative deviations of the long term average of the series. In Fig. 6 this procedure has been applied to data for summer climate at De Bilt, The Netherlands. The series as a whole do not show a trend, but most of the differences between succesive subperiods have a very low probability of occurrence, which probably means that the interruptions are not related to the trend and that their respective causes are independent of each other.

Abruptness of climate change on time scales of 10-100 years is something which has not received much attention, neither by statisticians, nor by theoreticians. Further examples may be found in Flohn (6) and Berger and Labeyrie (2).

10. GREENHOUSE WARMING DETECTED?

In 1990 a lecture on climate evolution over the past century has to say something explicit about the greenhouse warming signal. As already stated above, enhanced greenhouse warming is only one of the possible explanations of the global temperature increase of the past century.

A quantative estimate of its possible contribution can be obtained as follows. The radiative forcing ΔQ of increasing CO_2-concentration can be expressed as

$$\Delta Q = 6.3 \ln C/C_o$$

In this formula, derived by Augustsson and Ramanathan (1), C is the actual concentration of CO_2 and C_o some reference concentration, usually taken to be the pre-industrial level.

Since not only CO_2 but also greenhouse gases like CH_4, N_2O, O_3 and CFC's are changing in atmospheric concentration, the total radiative forcing ΔQ must be considered. This can be expressed by the same formula, when we define C as an equivalent CO_2-concentration, to be denoted by C..

Based on the observed increases of the atmospheric concentration of greenhouse gases since pre-industrial times and their theoretically computed contributions to the increased downward longwave radiation, Wigley (29) derived for 1985 ΔQ = 2.2 W/m^2.

According to the above formula this is equivalent with an equivalent CO_2-concentration C. = 400 ppm which means a 45% increase of the pre-industrial value C_o = 275 ppm.

Climate sensitivity to enhanced greenhouse radiation is usually expressed in terms of the expected increase of global mean temperature at the doubling of the CO_2-concentration. Symbolically: T_{2x}. This so-called equilibrium warming is poorly known, its range of uncertainty being

$$1.5 < T_{2x} < 4.5 \text{ K}$$

This means that the equilibrium warming at present is less than half of the above values, which nevertheless would mean an increase of at least 0.7 K since pre-industrial times.

However, equilibrium responses are very much delayed by the thermal inertia of the oceans. Taking this in account, Wigley and Raper (28) predict that the greenhouse warming over 1880-1985 lies in the range 0.4 - 1.1 K.

When compared with the observed warming over this period of 0.5 K, we tend to conclude that the climate sensitivity lies at the lower end of the indicated range. However, we cannot rule out other possibilities. For instance, other climate forcings could have partially offset the greenhouse warming. Therefore we have to conclude that from global mean temperatures alone we cannot say that we have detected the greenhouse warming, nor can we say that the high T_{2x}-values are impossible.

To overcome the problem of attribution, detection methods must make use of more variables than global mean temperature only. However, also multivariate "fingerprint"-methods up till now did not proof, that the 20th century warming, nor the more recent warming, is likely caused by the enhanced greenhouse radiation. On the other hand, multivariate methods, by which several aspects of the enhanced greenhouse signal are compared with observed climate changes, ultimately must show the real quantitative impact. Evidently, when in the forthcoming decade or so global mean temperature will increase by such an amount that no known factor reasonably may cause such an increase, climatologists confidently will conclude that the greenhouse warming has been detected.

11. REFERENCES

(1) AUGUSTSSON, T. and RAMANATHAN, V. (1977). A radiative-convective model study of the CO_2 climate problem. J. Atm. Sci., 34, 448-451.
(2) BERGER, W.H. and LABEYRIE, L.D. (eds.) (1987). Abrupt climate change. Reidel Publ. Comp., 425 pp.
(3) BRADLEY, R.S. et al. (1987). ENSO signal in continental temperature and precipitation records. Nature, 327, 497-501.
(4) BRADLEY, R.S. et al. (1987). Precipitation fluctuations over Northern Hemisphere land areas since the mid-19th century, Science, 237, 171-175.
(5) COOPS, A.J. and SCHUURMANS, C.J.E. (1986). Detection of the CO_2-effect upon the climate of western Europe. Theor. Appl. Climatol., 37, 111-125.
(6) FLOHN, H. (1986). Singular events and catastrophes now and in climatic history. Naturwissenschaften, 73, 136-149.

(7) GILLILAND, R.L. and SCHNEIDER, S.H. (1984). Volcanic, CO_2 and solar forcing of Northern and Southern Hemisphere surface air temperatures. Nature, 310, 38-41.

(8) HANSEN, J. and LEBEDEFF, S. (1987). Global trends of measured surface air temperature. J. Geo. Res., 92, 13345-13372.

(9) HESS, P. and BREZOWSKY, H. (1977). Katalog der Grosswetterlagen. Ber. Dtsch. Wetterdienst, Offenbach, 113, Bd. 15, 39 pp.

(10) HOREL, J.D. and WALLACE, J.M. (1981). Planetary-scale atmospheric phenomena associated with the Southern Oscillation. Mon. Wea. Rev., 109, 813-829.

(11) JAMES, I.N. and JAMES, P.M. (1989). Ultra low frequency variability in a simple atmospheric circulation model. Nature, 342.

(12) JONES, P.D. (1988). The influence of ENSO on global temperatures. Climate Monitor, 17, 80-89.

(13) KUKLA, G.J. and ROBINSON, D.A. (1981). Temperature changes in the last 100 years, Climatic Variations and Variability: Facts and Theories. A. Berger (ed.), Reidel Publ. Comp., p. 287-301.

(14) LAMB, H.H. (1977). Climate: Present, Past and Future. Vol. 2, Methuen, 835 pp.

(15) LAMB, H.H. (1982). Climate, history and the modern world. Methuen, 387 pp.

(16) LORENZ, E.N. (1986). The index cycle is alive and well. Namias Symp., Scripps Inst. of Ocean. Ref. Series 96-17, 188-196.

(17) MADDEN, R.A. and SHEA, D.J. (1978). Estimates of the natural variability of time-averaged temperatures over the United States. Mon. Wea. Rev., 106, 1695-1703.

(18) MOSES, T. et al. (1987). Characteristics and frequency of reversals in mean sea level pressure in the North Atlantic sector and their relationship to long-term temperature trends. J. of Climatol., 7, 13-30.

(19) OERLEMANS, J. (1988). Simulation of historic glacier variations with a simple climate-glacier model. J. of Glaciol., 34, 333-341.

(20) OORT, A.H. and RASMUSSON, E.M. (1971). Atmospheric Circulation Statistics. NOAA Professional paper 5, 323 pp.

(21) SCHUURMANS, C.J.E. (1984). Climate variability and its time changes in European countries, based on instrumental observations. The climate of Europe: Past, Present and Future. H. Flohn and R. Fantechi (eds.), Reidel Publ. Comp. p. 65-101.

(22) SCHUURMANS, C.J.E. and COOPS, A.J. (1984). Seasonal mean temperatures in Europe and their interannual variability. Mon. Wea. Rev., 112, 1218-1225.

(23) VAN DEN DOOL, H.M., SCHUURMANS, C.J.E. and KRIJNEN, H.J. (1978). Average winter temperatures at De Bilt (The Netherlands): 1634-1977. Clim. Change, 1, 319-331.

(24) VAN LOON, H. and WILLIAMS, J. (1978). The association between mean temperature and interannual variability. Mon. Wea. Rev., 106, 1012-1017.

(25) VAN LOON, H. (1979). The association between latitudinal temperatures gradient and eddy transport. Pt.I: Transport of sensible heat in winter. Mon. Wea. Rev., 107, 525-534.

(26) VAN LOON, H. and WILLIAMS, J. (1980). The association between latitudinal temperature gradient and eddy transport. Pt.II: Relationships between sensible heat transport by stationary waves and wind, pressure and temperature in winter. Mon. Wea. Rev., 108, 604-614.

(27) WALLEN, C.C. (1984). Present century climate fluctuations in the Northern Hemisphere and examples of their impact. WMO/TD-No. 9, 85 pp.

(28) WIGLEY, T.M.L. and RAPER, S.C.B. (1987). Thermal expansion of sea water associated with global warming. Nature, 330, 127-131.

(29) WIGLEY, T.M.L. (1988). Expected greenhouse-gas-induced climatic change: transient response and regional effects. Carbon dioxide and other greenhouse gases: climatic and associated impacts (R. Fantechi, ed.). Reidel Publ. Comp., p.

FIGURE CAPTIONS

Fig. 1 Hemispheric and global surface air temperatures for the 1861-1988 period. The annual values are plotted as anomalies from the 1950-79 period, the filtered curve is a 13-term Gaussian filter designed to suppress variations on time scales less than 10 years (according to Jones (11)).

Fig. 2 Surface air temperature change for different latitude zones (adapted from Hanssen and Lebedeff (8)).

Fig. 3 Geographical distribution of average temperature difference between 1925-1975 and 1875-1925 for winter (upper left), spring (upper right), summer (lower left) and autumn (lower right). Units: degree centigrade.

Fig. 4 Precipitation indices for different latitude zones (according to Bradley et al. (4)). Dashed lines indicate starting year of 50% coverage of observations.

Fig. 5 Non-overlapping 10-year averages of winter temperature at De Bilt, The Netherlands, since 1634 (update of Van den Dool et al. (23)).

Fig. 6 Cumulative deviations from the long term average temperature (1) in °C, precipitation amount (2) in dm and air pressure (3) in hPa at De Bilt, The Netherlands. Transition points of the cumulative series are indicated by dots, while averages of the climatic elements (temperature, precipitation and air pressure) for sub periods are given by the block diagram in tenths of the scales indicated. The length of the series is, respectively, for temperature (1709-1984), precipitation (1896-1984) and air pressure (1902-1984). (according to Coops and Schuurmans (2)).

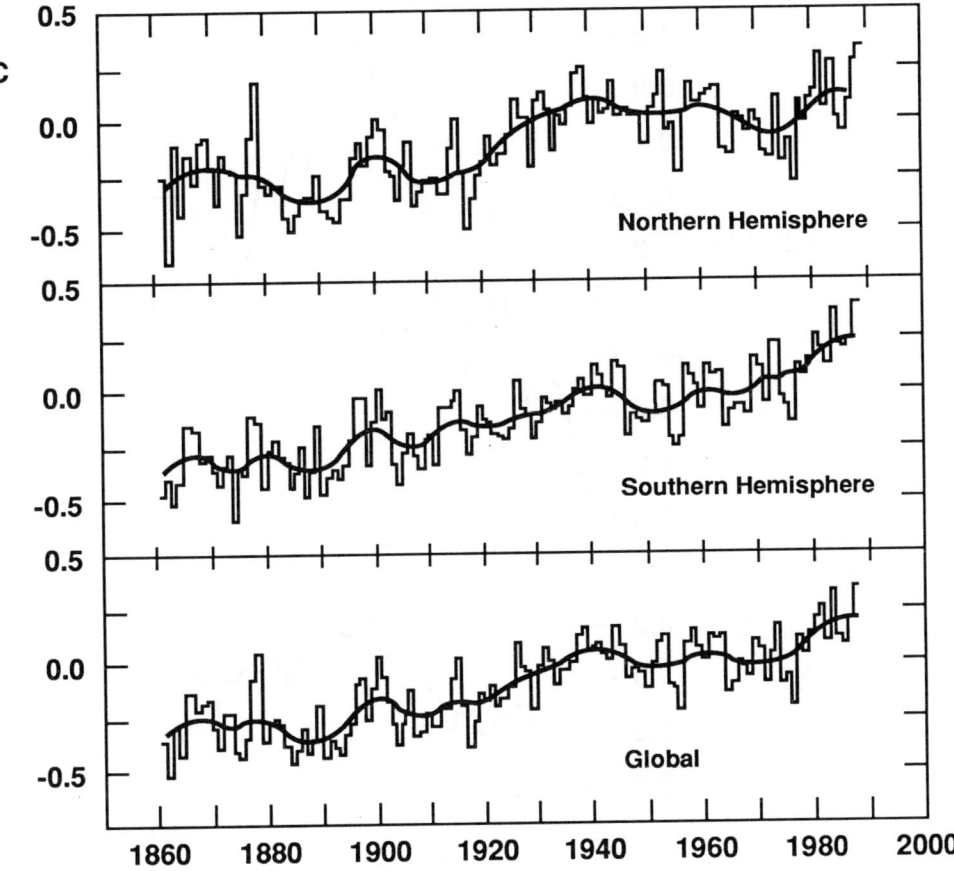

Figure 1

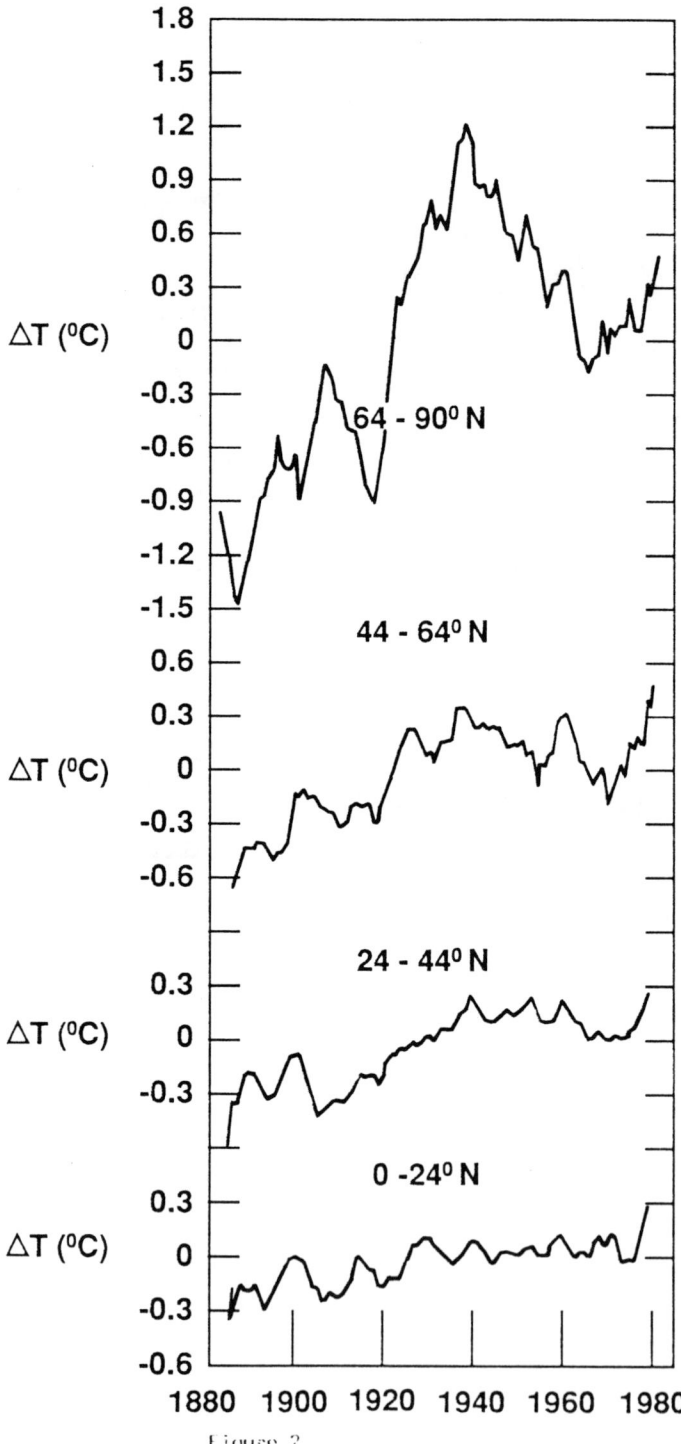

Figure 2

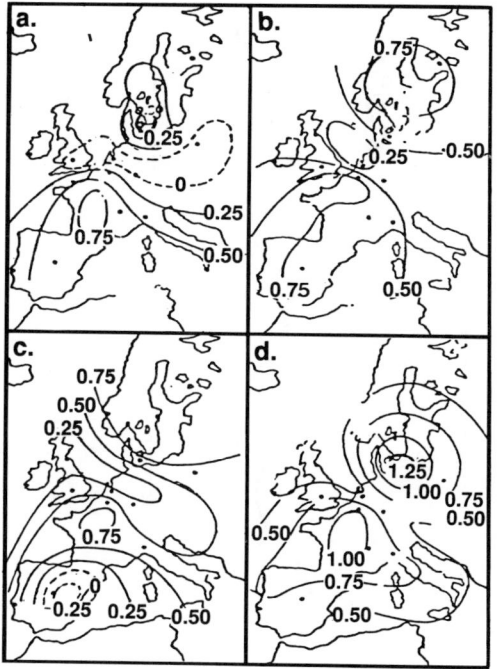

Figure 3

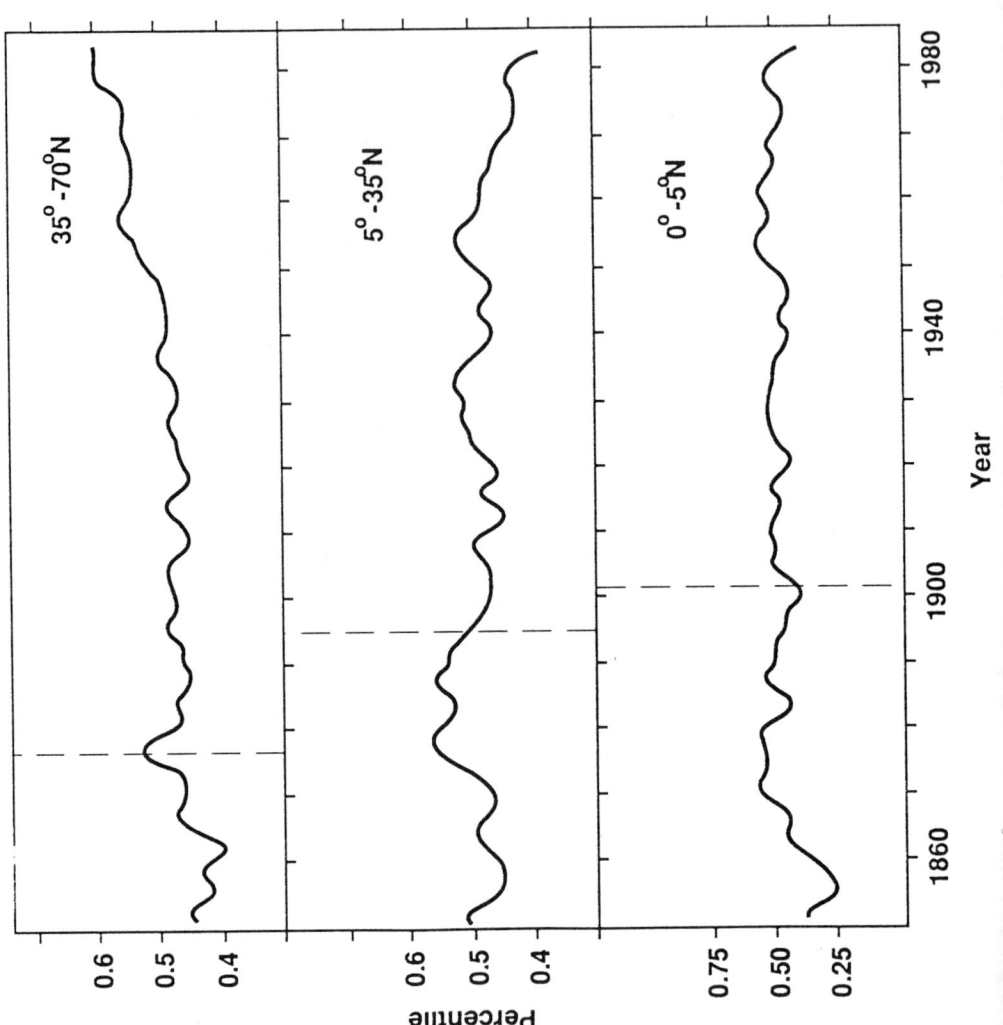

Figure 4

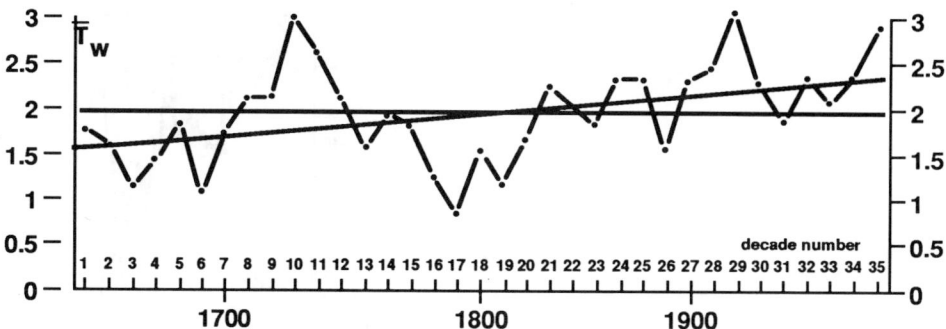

Figure 5

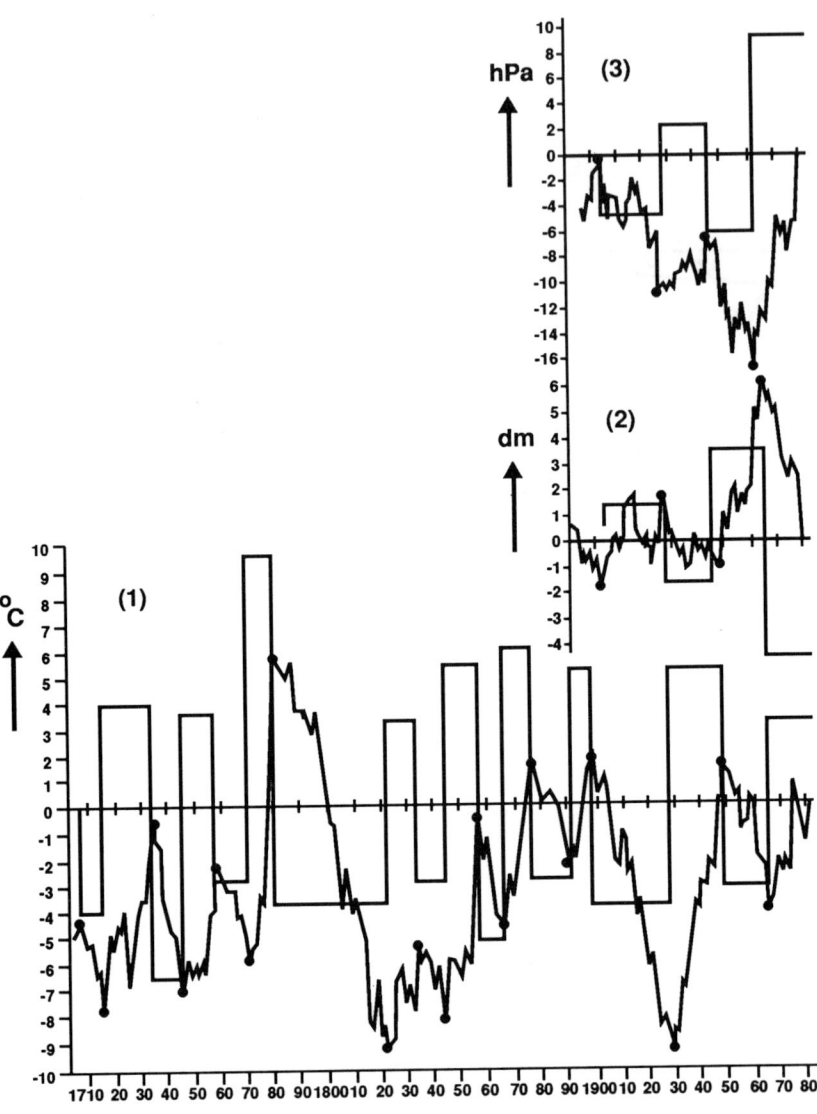

Figure 6

ALBEDO CHANGES AND SATELLITE OBSERVATIONS OF THE EARTH

M.Eckardt
Institut f. Meteorologie, Freie Universität Berlin

Summary
More than 70% of solar energy input to the Earth's climate
system is absorbed at the surface. The surface albedo as the
ratio of reflected to incident radiation determines the
amount of solar energy for conversion into heat. The reflection of solar radiation from many surfaces is found to be a
strong function of wavelength and depends upon the state of
the surface. Changes in surface albedo are caused by
variable snow cover and ice extent and by surface changes
due to anthropogenically induced desertification, deforestation and greenhouse warming. These albedo changes provide an
important source of variability within the climate system.
Satellite observations offer the potential of deriving and
monitoring surface and planetary albedos. An overview of
methods for determining albedos from satellite data is given
and discussed with respect to accuracy, correction methods
and use in general circulation models. Satellite-derived
albedos are the basis for computation of the vegetation
index (NDVI). In order to validate satellite observations a
number of field experiments have been conducted with special
emphasis to measurements of surface albedos and radiation
budget at the surface.

1. INTRODUCTION

The climatic system of the Earth consists of the components atmosphere, hydrosphere, cryosphere, biosphere and lithosphere with
the atmosphere as central component. The most important interactions occur between the atmosphere and the oceans as well as
between the atmosphere and the biosphere. The components have
quite different physical characteristics and respond to imposed
changes with different time scales. On a global annual average
the incoming energy flux density from the sun of 342 W/m² inter-
acts with the Earth-atmosphere system in a complex way. The solar
flux reaching the top of the atmosphere is different from that
reaching the surface due to scattering and absorption within the
atmosphere. Main absorbers are water vapor and ozone, reflections
are caused by dust, air (Rayleigh scattering) and clouds. One
part reaching the ground is backscattered from the sky after one
or more reflections from the surface. Thus the solar radiation
reaching the ground consists of direct radiation plus diffuse
radiation. Calculations of radiative transfer through the atmosphere are limited by incomplete informations of the distribution
and optical properties of atmospheric constituents. Both surface
and atmospheric properties are strongly wavelength-dependent.

About 75% of the incident solar radiation reaches the surface under a clear sky and overhead sun. The surface albedo is defined as ratio of reflected to incident solar radiation. Modifications of the surface albedo cause significant changes of the planetary albedo, that is the fraction of solar radiation that is reflected from the Earth-atmosphere system. Small changes in the energy balance of the Earth-atmosphere system therefore may be caused by large-scale changes in the surface albedo. Although land surfaces contribute only to less than one third to the total surface, their characteristics are of major importance, because they provide much of the time and space variability of the surface forcing of the atmosphere. Growing interest is addressed to human activities which lead to deforestation, desertification and global changes of temperature. Predictions of possible effects of greenhouse warming need simulation with climate models or general circulation models (GCM), which exhibit significant sensitivity to surface albedo changes. Therefore detailed and reliable data of the surface albedo are neccessary, they can be compiled mainly in two ways: satellite data composite and geographical land type classification including selection of typical locations. Satellite informations are restricted to clear sky conditions and must be corrected for several effects, but they provide the only data in inhospitable areas. Within the International Satellite Land Surface Climatology Project (ISLSCP) first attempts to validate satellite data from land surfaces are made.

Fig.1: A portion of NOAA 9 AVHRR image with two adjacent orbits covering parts of Europe, Mediterranean and northern Africa, 23 July 1985, Channel 2 (0.7-1.1 µm)

2. SURFACE ALBEDO

Land surface covers 29% of the total surface and in the Northern
Hemisphere it covers 39%. The importance of the land surface in
forcing the atmosphere comes from surface topography, thermal
contrast with the oceans, roughness and surface albedo. The
latter is highly variable in space and time and is dependent on:
  -wavelength
  -sun elevation (or angle of incident radiation)
  -ratio between direct and diffuse solar radiation
  -nature of the surface.
The figure 1 shows different reflectivities over land and water
surfaces (including a weak sunglint pattern around Sicily) under
clear skies. At the intersection of two adjacent orbits, which
were taken at 14:30 local time, the effects of solar and satellite geometry are evident. Figure 2 shows measurements of the
spectral albedo over different vegetated and nonvegetated surfaces, which have been made during field experiments in Europe
and Africa. In the following chapters the surface albedo for the
most important surfaces and their dependence on the above
mentioned effects is described.

2.1 SOIL ALBEDO

Soil albedos vary widely and depend on type of soil, colour,
structure (roughness), cultivation and moisture content. The
range of surface albedo is from 0.1 for organic black soil to
more than 0.5 for white sands. Dry soils have an albedo that is

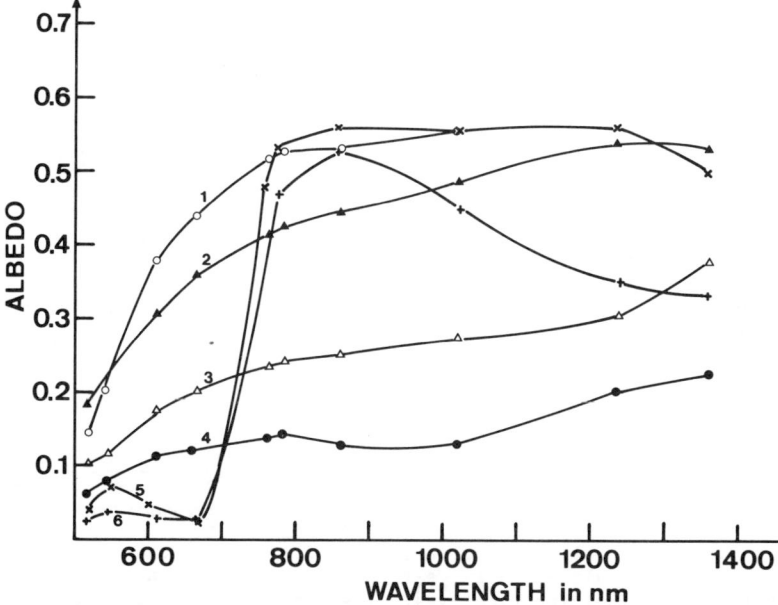

Fig.2: Spectral albedo over various surfaces: 1)yellow dune
2)bright sand, 3)dark sand, 4)laterit (all March 1988 Niger),
5)grassland (September, FRG), 6) wheat (June, FRG)

about 1.8 times higher than that of moist soils. The most drastic
albedo decrease appears when soil moisture is increasing up to
15-20%, a further supply of moisture does not further change the
albedo significantly. As can be seen from Fig.2 the albedo of
soils is increasing with wavelength, bright sands and dunes
have a steeper increase in the visible part of the spectrum.

2.2 SNOW ALBEDO

Snow-covered surfaces show a general decrease in albedo between
800 and 1000 nm, which is highly dependent upon the amount of
liquid water present. The highest values are achieved for fresh
dry snow throughout the 400-1000 nm range with more than 0.8, the
mean annual albedo for the Antarctic for instance is 0.86 with
small seasonal variations. The albedo varies also with density,
thickness and contamination of snow. Underlying vegetative cover
has also an influence on albedo with the tendency of decreasing
it, when vegetation (trees or other obstacles) are not completely
covered by snow. Snow-covered farmlands have significantly higher
albedo than forested areas. An angular dependence of reflectivities is not obvious.

2.3 ICE ALBEDO

Sea ice albedo shows a variability due to melting, snow cover and
development stage. Melting causes a reduction of albedo, most

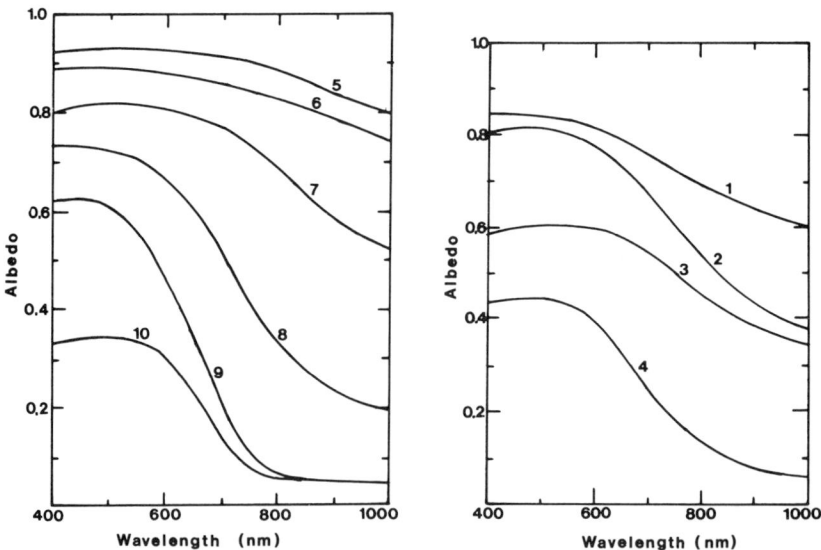

Fig.3:Spectral albedos over snow and ice: 1)frozen multi-year
white ice; 2)melting multi-year white ice; 3)melting first-year
white ice; 4)melting first-year blue ice; 5)dry snow; 6) wet new
snow over multi-year white ice; 7)melting old snow; 8)partially
refrozen melt pond with 3cm of ice; 9)early-season melt pond with
white bottom on multi-year ice; 10)old melt pond (30cm in depth)
on multi-year ice (after Grenfell and Maykut 1977)

126

pronounced at longer wavelengths. Lower salinity and higher air bubble density within the ice make multi-year ice less absorptive and leads to a higher albedo. Beyond 900 nm a rapid drop of albedo is observed. Furthermore increasing contamination of ice reduces the albedo. The spectral behavior of ice albedos is used for sea ice detection. Fig. 5 gives an example of the difference of AVHRR channel 1(VIS)-channel2(NIR), where white areas have the highest differences.

Fig.4: Portion of AVHRR image Central Europe with partly snow cover, 3 March 1986, Ch.2, 13.21 UTC

Fig.5: Sea ice in the Greenland Sea, AVHRR Ch1 - Ch2, 22 July 1977

2.4 VEGETATION ALBEDO

About 20% of the Earth's surface is covered by vegetation (vegetation cover more than 50%), but assuming a mean leaf area index of 3 a total area of leaf material exposed to the air would be in the order of $6.12*10^8$ km² (after Verstraete,Dickinson (24)). This very great surface affects the atmosphere by radiative interactions, water balance, energy balance and momentum absorption. The albedo of vegetation shows a sharp increase around 700 nm from 5-15% up to 30-60% in the near infrared (see also Fig.1). This depends on the type of the plant, season (development stage) and solar zenith angle. The variation of spectral albedo over a barley field with daytime is shown in Fig.6. The albedo of a canopy is lower than the leaf albedo which is caused by light trapping within the canopy. The height of a canopy also influences its albedo.

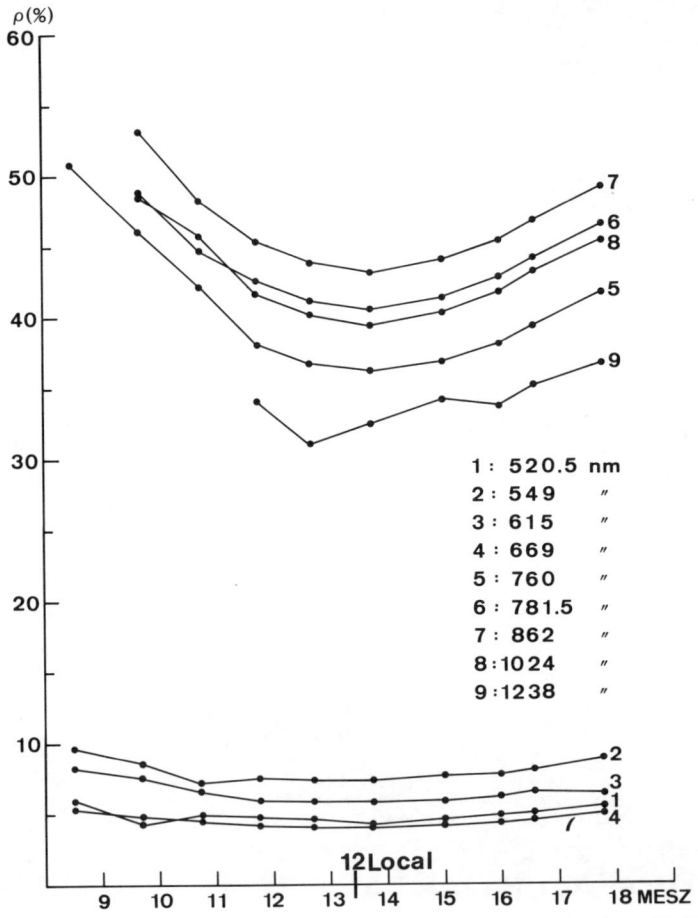

Fig.6: Daily variation of spectral albedo over a barley field 13 June 1988, northern Germany

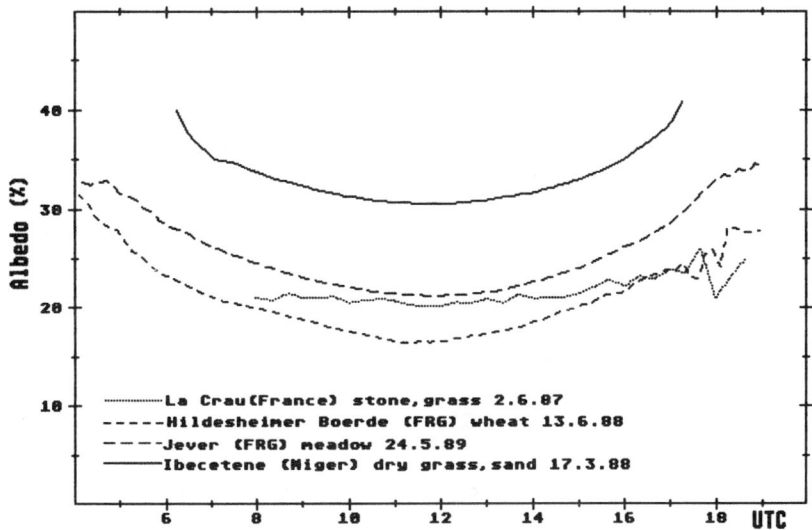

Fig.7: Daily variation of wavelength-averaged albedos
(0.3-3.0 µm) for various surfaces

The above figure shows the broadband albedos measured in 2m above
the surface. The site in Ibecetene (Niger) was covered by 50% of
dry Cram Cram grass and 50% light sand. The site in La Crau was
covered by small stone and very short and not very dense grass,
the other with over dense vegetation.

2.5 OCEAN ALBEDO
The sea surface albedo depends on sun elevation, roughness,
surface contamination and characteristics of the water basin, but
is generally low. The most important factor is the sea state in
combination with solar zenith angle. The ocean albedo increases
with increasing roughness for sun elevation >34°, whereas at sun
elevation <34° the pattern reverses (Ref.13). Albedos for water
surfaces exhibit very little wavelength-dependence.

3. SATELLITE OBSERVATIONS
Since atmospheric scattering has the tendency of increasing
observed clear-sky system albedo compared with the true surface
albedo, observations of surface albedo should be made close to
the surface. On the other hand measurements at a height of 2m do
not characterize adequately the reflectivity of extended areas
which is needed for climate models. Observations from aircraft,
balloons or satellites have to consider the transfer function of
the atmosphere and new problems arise with shading or mixing of
different surface types, radiances have to be interpreted in
terms of albedo of greater areas. For global coverage only
satellite data are convenient when albedos, surface temperatures
or other parameters should be evaluated. The physical basis for
estimation of surface albedo from satellite observations is the

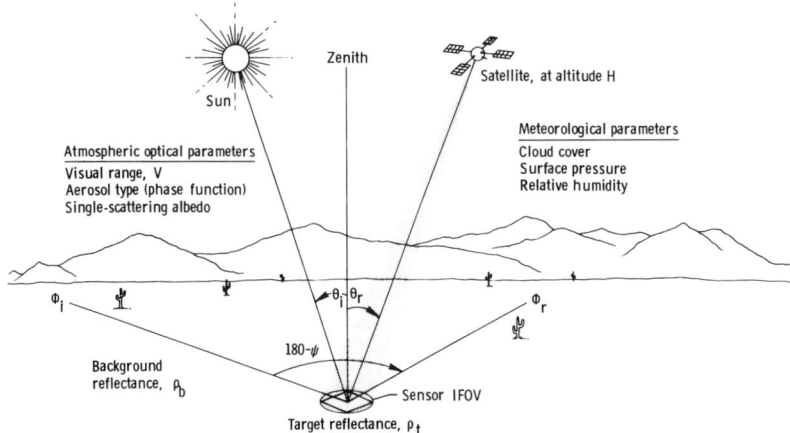

Fig.8: Parameters affecting apparent reflectance (from Bowker et al. 1985)

fact that reflected clear-sky solar radiation at top of the atmosphere depends mainly on the surface albedo. One of the major problems of deriving surface quantities from space is the presence of clouds. In order to transform satellite observed reflected radiance to surface albedo, the following procedures must be applied:

a. normalization to obtain top-of-the atmosphere bidirectional reflectance. The conversion to albedo requires knowledge of instantaneous solar zenith, viewing zenith and viewing azimuth angles. Nine-day groundtrack repeat cycles lead to errors of about ±10%, which could be avoided by computing monthly means.

b. correction of atmospheric scattering and absorption. The correction of atmospheric effects can be accomplished by radiative transfer models applicable for narrowband and broadband measurements. Different values for water vapor content, total aerosol optical depth and ozone have been used for simulations. There are limitations for global data sets, because optical properties and vertical profiles of aerosol are not available for many parts of the globe. Some authors have developed linear relationships between surface and planetary albedo (Preuß and Geleyn 1980, Ohring and Chen 1985, Koepke and Kriebel 1987 and Koepke 1989)for this purpose. Some other authors tried to calibrate satellite radiances with in-situ measurements of surface albedo and derived surface albedos from geostationary satellites or Landsat for special regions, where atmospheric conditions should not change (Pinty et al 1985).

c.correction for anistotropy of the atmosphere and the surface. Most methods of deriving surface albedos neglect this effect and take the surface as Lambertian reflector. But the anisotropy is of growing importance with increasing wavelength, since optical thickness decreases and spectral reflectivity of the surface increases, this is especially of importance for vegetation. However, over mixed surface types the anisotropic

reflective properties of singler elements may be masked. For low
solar zenith angles the shadowing effects are small and the azi-
muthal anisotropy is only slight.

      d.conversion of narrowband to broadband observations. In-
vestigations were made by Laszlo et al (14) for the AVHRR
channels and different surface types. The conversion into
broadband albedos, which are wanted, is best when using an
increasing number of narrowband channels. The relationship is
also affected by the solar zenith angle during observation,
change in atmospheric constituents and variation in surface
types.
Furthermore there exist uncertainties in satellite data due to
navigation, sampling and daytime of observation as well as sensor
design and degradation, satellite instability and deficiencies in
the optical system. All the effects mentioned above do not allow
monitoring daily variations of the surface albedo.
The following satellite sensors have been mainly used so far for
deriving surface albedos on a regional scale:

Radiometer	Spectral range	Nadir resolution	Local time
Scanning Radiometer(SR) ITOS 1 to NOAA 5	0.5-0.7µm	4 km	15/09
Advanced Very High Resolution Radiometer (AVHRR) TIROS-N to NOAA 11	0.55-0.68 µm 0.725-1.1 µm	1.1 km 1.1 km	15/13
Operational Linescan system (OLS) DMSP series	0.4-1.1 µm	2.8 km	12
Visible Infrared Spin Scan Radiometer (VISSR) SMS, GOES series	0.55-0.75 µm	0.9 km	geostat.
Visible and Infrared Radiometer Meteosat Series	0.4-1.1 µm	2.5 km	geostat.
Earth Radiation Budget (ERB) Nimbus 7	wide-angle channels		12
Earth Radiation Budget Experiment (ERBE) NOAA 9-10	"		14
Multispectral Scanner (MSS) Landsat series	0.5-0.6 µm 0.6-0.7 µm 0.7-0.8 µm 0.8-1.1 µm	80 m	9:30

4. VEGETATION INDEX AND FIELD EXPERIMENTS

Narrowband albedos from the AVHRR channnels 1 and 2 are used to compute a Normalized Difference Vegetation Index (NDVI). As was mentioned earlier, vegetation canopies show a rapid increase in albedo around 700 nm. The spectral response functions of the two AVHRR channels allow to derive information on the greenness. The difference of Channel 2 - Channel 1 is divided by their sum, this leads to some extent to the elimination of geometrical and atmospheric effects. For operational purposes, in a first step it is assumed that atmospheric effects can be neglected and that the surface is a Lambertian reflector. Taking only the maximum NDVI-value during a period of 7 - 10 days cancels out in most cases the variable cloudiness and effects due to the sun-sensor-target geometry.

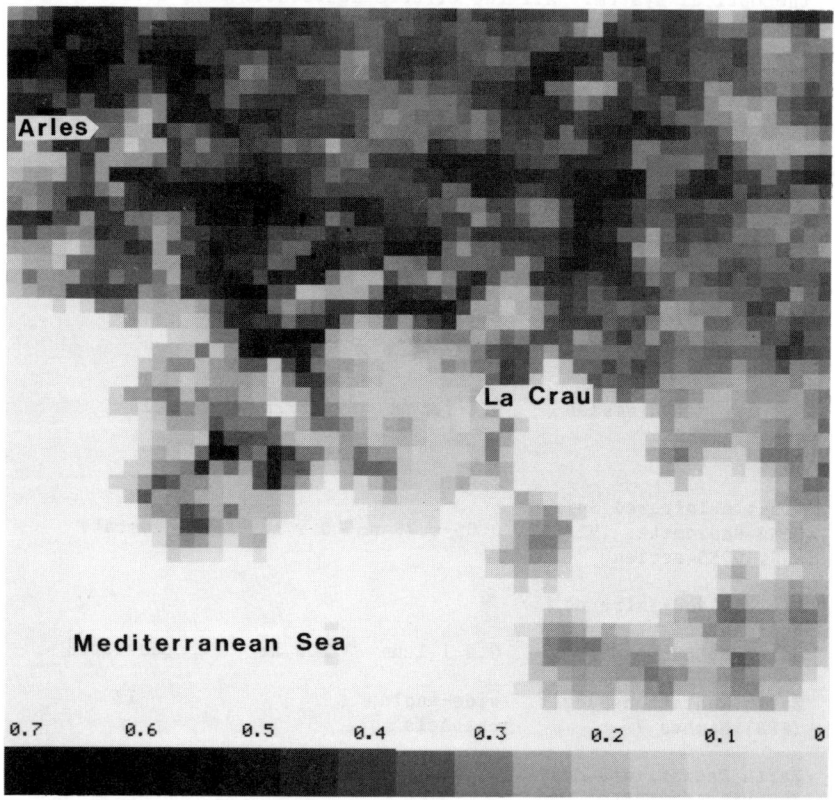

Fig.9: NDVI image of La Crau (southern France) and surroundings 2 June 1987

By constructing a NDVI data set covering the vegetation period in Europe for several years changes can be observed and related to land use and differences in seasonal vegetation development which in turn is coupled to atmopsheric conditions. These data sets are of great value for monitoring vegetation and surface changes, but

132

more work is neccessary for correcting the NDVI with respect to atmospheric and anisotropy factors. Since the NDVI is only an "index" and no physical quantity, it is neccessary to derive area averaged surface quantities that determine evaporation and heat fluxes. The goal of field experiments is to validate the methods applied for the evaluation of satellite measurements. These quantities (spectral surface reflectance, surface temperature and vegetation index) should be interpreted by experiments in terms of spectral/total albedo, net solar radiation flux, diurnal temperature amplitude, leaf area index and/or biomass production. Besides direct measurements soil moisture and heat fluxes can be estimated by SVAT models and combined to satellite data.

5. ALBEDO CHANGES AND CLIMATE MODELING

The sensitivity of existing climate and general circulation models to the specification of the surface albedo is quite different and there exists an increasing need for a reliable, global surface albedo data set. The diversity in albedo data makes parameterization in climate models very difficult. There have been only few attempts to produce global surface albedo data and the difficulties in deriving such an inventory from satellite observations are still considerable. Within the existing global data sets middle and high latitude values are less consistent. The removal of clouds in satellite data could be done by minimum-brightness maps, but increasing the time period also increases the problems which arise from surface variability (moisture, snow and vegetation). Snow-free land albedos are treated in current GCM's in three ways: fixed value for land and ocean, land albedo as function of latitude and specified geographical distribution including several surface and vegetation types. The treatment of snow and ice is even more diverse due to divergent observational data. In any case the surface albedo should have an accuracy of ±5%. There are two preferred regions for which the climate model sensitivity to surface albedo changes has been established: deserts and the cryosphere. Charney (1975) first proposed a bio-geophysical feedback mechanism for extending drought regions in the Sahel. Increased albedo from removal of vegetation leads to a net radiative loss and produced general subsidence. This is the cause for decreased cloudiness and precipitation. This mechanism was tested and confirmed with the use of a GCM by Charney et al.(1977), Sud and Fennessy (1982), Laval and Picon (1986) and Yamazaki (1989), partly also for other desert regions. Otterman (1977) showed observational evidence that human activities, especially overgrazing can increase the surface albedo significantly. But the complexity of interactions between radiation, cloudiness and atmospheric and surface moisture content cannot be properly modeled. Thus we get more an indication of possible impacts. Modeling vegetation is also very complex and therefore the presence of a canopy was simulated by parameterization of evaporation. Drastic changes in surface evaporation have a relatively large sensitivity in the models, so inclusion of vegetation in the models cannot be avoided. From recent simulations it showed up that vegetation, which produces low albedo and high values of evapotranspiration must have a positive feedback on rainfall. With respect to Global Change it is interesting to simulate tropical deforestation effects on climate. The following effects are established: increase in surface albedo, decrease in surface roughness length and an alteration of soil properties. But model parameterization schemes do not incorporate all these changes. Only few climate model experiments have been performed so far on this subject with prescribed changes of surface albedo (increases in clearing areas with grassland), surface hydrology and roughness length. From these a reduction in rainfall, evapo-transpiration and cloudiness resulted, but no significant tempe-rature changes occurred and hemispheric or global-scale changes were not discerned.

The magnitude of the snow/ice-albedo feedback is determined by the difference between cryospheric albedos and the background snow/ice-free albedos. The area covered by snow and ice is therefore important in the global climatic regions. For snow-covered land areas albedo varies considerably as a function of type and density of vegetation, Robinson et al.(1986) found that surface albedo of snow-covered deforested regions in the U.S. is about twice as high as when naturally vegetated. Yamazaki (1989) carried out GCM studies over an annual cycle and found that snow albedo is an important factor in controlling northern summer climate. The ice-albedo feedback is also believed to be signifi-cant for climatic perturbation on a large number of time scales. Surface albedo over open water is about 8-12% at low sun eleva-tion and increases over sea ice up to 60-80%, depending on snow cover, melt ponds and ice structure. The ice extent has great influences on the surface heat fluxes and therefore should be exactly parameterized in models. Up to now coupled cryosphere-at-mosphere models differ in the specification of ice albedos and shortwave surface energy fluxes. Sudden decreases in sea ice albedo by 30-40% are observed in the Arctic after melting of snow and formation of water puddles, in the Antarctic over great areas only first-year ice is observed leading to somewhat different conditions. Observational data from satellites in the marginal ice zones are needed. As an example Fig.10 shows a 25-year record of sea ice extent in the European sector of the Arctic, derived from high resolution satellite images. For the observed period a decrease in ice extent of 10-20% has been found.

In conclusion it can be said that an agreed and consistent set of surface albedo data is of urgent need for the climate modelers with which standard simulations can be made. Most models use climatological values of surface albedo, since satellite-derived albedos on a global scale are not yet available due to the shown difficulties. Some promising efforts have been done for specific regions which achieve the desired accuracy. Albedo chan-ges are simulated in climate models, but there exist only regio-nal observations and much more work should be done in establi-shing agreed long-term data sets especially in regions where albedo-climate feedbacks are important.

REFERENCES

(1) BARKER,H.D.and DAVIES,J.A. (1989): Surface albedo estimates from Nimbus-7 ERB data and a two-stream approximation of the radiative transfer equation. J.Cli. 2(5) 409-418.
(2) BOWKER,D. et al.(1985):Spectral reflectances of natural tar-gets for use in remote sensing studies. NASA Ref.Publ. 1139.

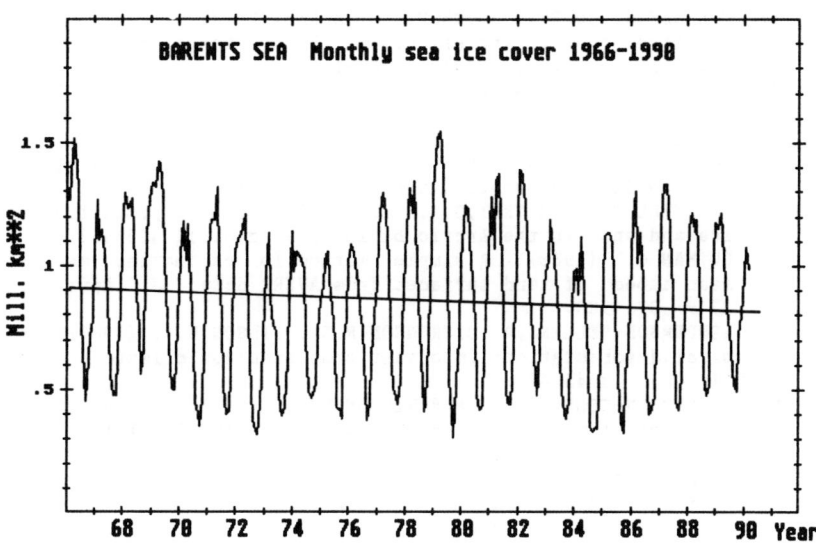

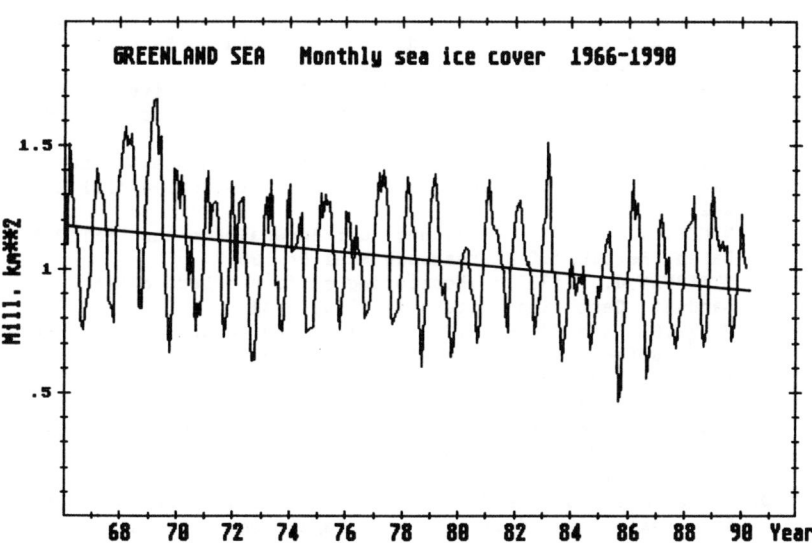

Fig.10: Monthly variation of sea ice extent for the European sector of the Arctic. a) Barents Sea 15° E-60° E, b) Greenland Sea 50° W-15° E.

(3)   CHARNEY,J.,QUIRK,W.J.,CHOW,S.H.,KORNFIELD,J.   (1977): A com-
      parative study of the effects of albedo change on drought in
      the Sahel. J.Atm.Sci.34(9) 1366-1386.
(4)   CHEN,T.S.and OHRING,G.   (1984): On the relationship between
      clear-sky planetary and surface albedos. J.Atm.Sci. 41(1)
      156-158.
(5)   COUREL,M.F., KANDEL,R.S., RASOOL,I. (1984): Surface albedo
      and the Sahel drought. Nature 307 528-531.
(6)   GRENFELL,T.C., MAYKUT,G.A. (1977): The optical properties of
      ice and snow in the Arctic basin. J.Glaciol.18 445-463.
(7)   GUTMAN,G.   (1988): A simple method for estimating monthly
      mean albedo of land surfaces from AVHRR data. J. Appl.Met.
      27(9) 973-988.
(8)   HENDERSON-SELLERS,A.,DICKINSON,R.E., WILSON,M.F. (1988):Tro-
      pical deforestation: important processes for climate models.
      Cli.Cha. 13 43-67.
(9)   HENDERSON-SELLERS,A. and WILSON,M.F. (1983): Surface albedo
      data for climate modeling. Rev.Geoph.Sp.Phys.21(8) 1743-1778
(10) HUMMEL,J.R.and RECK,R. (1979): A global surface albedo model
      J. Appl.Met. 18(3) 239-253.
(11) KOEPKE,P.and KRIEBEL,T.(1987): Improvements in the shortwave
      cloud-free radiation budget accuracy, part I: numerical stu-
      dy including surface anisotropy, part II: experimental study
      including mixed surface albedos. J.Cli.Appl.Met. 26(3) 374-
      409.
(12) KOEPKE,P. (1989): Removal of atmospheric effects from AVHRR
      albedos. J.Appl.Met.28 1341-1348.
(13) KONDRATYEV,K.Ya. et al.(1982): The shortwave albedo and the
      surface emissivity. in: Eagleson(ed.) Land surface processes
      in atmospheric circulation models 463-514.
(14) LASZLO, JACOBOWITZ and GRUBER,A. (1988): Relative merits of
      narrowband channels for estimating broadband albedos.
      J.Oc.Atm.Tech. 5(6) 757-773.
(15) LAVAL,K. and PICON,L. (1986): Effect of a change of the sur-
      face albedo of the Sahel on climate. J.Atm.Sci. 43(21) 2418-
      2429.
(16) NACKE,G.   (1989): Ableitung der Bodenalbedo aus METEOSAT-
      Daten. Ber.Inst.f.Meereskunde Kiel Nr.190
(17) OTTERMAN,J. (1977): Monitoring surface albedo change with
      Landsat. Geoph.Res.Letters 4(10) 441-444.
(18) PINTY,B. and SZEJWACH,G. (1985): A new technique for infer-
      ring surface albedo from satellite observations. J.Cli.Appl.
      Met. 24(8) 741-750.
(19) PINTY,B. and TANRE,D. (1987): The relation between incident
      and double-way transmittances: an application for the esti-
      mate of surface albedos from satellites over the African
      Sahel. J.Cli.Appl.Met. 26(8) 892-896.
(20) PREUSS,H.J. and GELEYN,J.F. (1980): Surface albedos derived
      from satellite data and their impact on forecast models.
      Arch.Met.Geoph.Biocl. A29 345-356.
(21) ROBINSON,D., MATTHEWS,E., KUKLA,G. (1986): Human-induced
      changes in winter surface albedo. Arch.Met.Geoph.Biocl. A34
      427-434.

(22) ROWNTREE,P.R. and SANGSTER,A.B. (1986): Remote sensing needs in climate model experiments with hydrological and albedo changes in the Sahel. ISLSCP Conf. Rome ESA SP-248 175-183.

(23) SUD,Y.C. and FENNESSY,M. (1982):A study of the influence of surface albedo on July circulation in semi-arid regions using the GLAS GCM. J.Cli. 2 105-125.

(24) VERSTRAETE,M.M. and DICKINSON,R.E. (1986): Modeling surface processes in atmospheric general circulation models. Annales Geophysicae 4 (4) 357-364.

(25) VULIS,I.L. and CESS,R.D. (1989): Interaction of surface and planetary directional albedos for vegetated regions. J.Climate 2(9) 986-996.

(26) YAMAZAKI,K. (1989) : A study of the impact of soil moisture and surface albedo changes on global climate using the MRI-GCM I. J.Met.Soc. Japan 67(1) 123-146.

THE PHYSICS AND DYNAMICS OF THE CLIMATE SYSTEM:

SIMULATION OF CLIMATE

John F B Mitchell
Dynamical Climatology Branch
Meteorological Office, Bracknell

Summary

The physical basis of numerical modelling of climate is outlined, using atmospheric models as an example. The use and verification of climate models is described, illustrated with data from simulations made with Meteorological Office models.

1. INTRODUCTION

It is well known that the climate of the earth has been very different from present, as for example during the major Ice Ages. Awareness of the possibility that man can change the atmospheric environment has been heightened in the last few years by publicity concerning "Acid rain", "The ozone hole" and the "Greenhouse effect". The advent of electronic computers has made it possible to develop 3-dimensional models of the atmosphere and oceans which have been used to investigate these phenomena. In these two lectures, we will consider the simulation of global climate, past, present and future.

Climate may be regarded as the statistical distribution of temperature, rainfall and other such parameters over a period of time at a given location. Thus, for example, we are not only concerned with the long-term mean monthly temperatures, but also the variability from day to day, and from year to year. In assessing the impacts of changes in climate, the changes in the frequency of extreme events such as frosts, gales, droughts and heatwaves are as important as the changes in mean parameters. As the climate varies over all timescales (Figure 1), the length of the averaging period will vary from 30 years or so for contemporary climate (for example, Schutz and Gates, 1971), through millennia when variations in orbital parameters are being considered, to geological timescales.

In order to simulate climate it is evident from the definition just given that we must be able to model temporal and spatial variability. Thus we are compelled to use 3-dimensional general circulation models (GCMs) of the atmosphere and oceans. Such models are expensive to run and it is often difficult to analyse their behaviour. Hence models of lower dimension have been used to investigate a wider range of parameter changes than is possible with a full model, or identify the dominant mechanisms in a GCM experiment.

Climate models, unlike numerical weather prediction models, cannot be validated against a large number of test cases. Although climate has varied in the past, the causes of these variations are not generally well known and our knowledge of past climates themselves is imprecise because of uncertainties in the interpretation of palaeoclimatic data. Thus, the usual scientific cycle of observation, making an hypothesis, validating the hypothesis against further independent observations and so on is difficult to achieve. It is therefore particularly important that climate models should be firmly based on physical principles and validated against whatever data there is available. These two topics are discussed further in the remainder of this paper.

2. ATMOSPHERIC MODELS AND THEIR PHYSICAL BASIS

In this section, the physical basis of climate models is discussed using atmospheric models as an example. A similar approach is taken in modelling the oceans, sea-ice and the land surface, but these components will not be discussed further. The purpose of this section is to outline the principles used in constructing a climate model as opposed to providing a comprehensive review of current climate models. Examples will be taken predominantly from the Meteorological Office climate model (Wilson and Mitchell, 1987a, henceforth referred to as WM; Slingo et al, 1988).

2.1 Dynamics and physical parametrizations

In a numerical model, the atmosphere is represented by values of wind, temperature and humidity which are held at various levels of grid points. For example, the model used by WM has 11 levels in the vertical and horizontal spacing of 5° latitude by 7.5° longitude (Figure 2) typical of current climate models. At any time t the state of the atmosphere is described by about 10^5 variables, and the values at some future time $t+\Delta t$ can be determined using the prognostic equations described below. Δt is of order 20 minutes. The Met. Office climate model (WM) requires 4.5 hours of elapsed time on Cyber 205 for each simulated year.

The prognostic equations are based on the laws of classical physics, and the local rates of change $\frac{\Delta Y_i}{\Delta t}$ at each location and level $\underline{X}$ are given by

$$\frac{\Delta Y_i}{\Delta t} = F_i(\underline{Y},\underline{X},t) \qquad Y_i = u,v,T,q \tag{1}$$

where u and v are the east-west and north-south components of wind, T is temperature, q is humidity, F_i is a function of all the variables $\underline{Y}$, and their spatial gradients at that location. Equation (9.2) may be expanded.

$$\frac{\Delta Y_i}{\Delta t} = \underline{\nabla} \cdot \underline{V} \; Y_i + Z + D + N \tag{2}$$
$$\qquad\qquad (a) \qquad (b) \; (c) \; (d)$$

Where the term

- (a) is the change in Y_i due to transport by the flow
- (b) represents the effects of pressure gradients and the earth's rotation in the wind equation, and of adiabatic vertical motion in the temperature equation

(c) are smoothing (diffusive) terms
(d) are the sources and sinks of momentum (i=u,v), heat (i=T) and
moisture (i=q).

The left-hand side and the terms (a) and (b) are often referred to as the
large scale dynamics. The equations for u and v may be transformed into
2 equations for divergence and the vertical component of vorticity, and so
may be solved using spectral techniques (see, for example, Schlesinger
1988): otherwise they are solved using finite difference techniques. In
solving the equations, care must be taken that there are no spurious sinks
of heat, moisture or momentum. This is particularly important for models
used in climate change experiments, since the sizes of the perturbations
typically considered are small (4 Wm^{-2} in the case of doubling carbon
dioxide concentrations (Lal and Ramanathan, 1984)). The diffusive terms
(c) are required to prevent energy accumulating in the smallest scales
resolved by the model grid and producing unmeteorological solutions. These
terms can be regarded as representing small-scale motions not resolved by
the grid.

The final term in 2 covers all the small-scale processes not incorporated
in (c), including boundary and surface processes, radiation and convection.
For some of these phenomena, notably radiation, there exists an accepted
theoretical basis, and accuracy is limited by the availability of computer
resources and the necessity of using grid-box values of parameters which in
reality may vary substantially within a grid box. For other processes,
there may be no accepted underlying theory, or the theory may be valid only
for limited idealized conditions. The subgrid scale parametrizations may
be derived on the basis of one or more of the following:

a. Simplified forms of the exact equation (if known)
b. Numerical experiments on a much finer mesh than feasible in the
GCM.
c. Observations from the atmosphere
d. Laboratory experiments
e. Sensitivity studies made with general circulation models.

To illustrate the diversity of these approaches, the derivation of the
radiative and convective parametrizations are considered in more detail.
In principle, the equations of radiative transfer are well known (see for
example Paltridge and Platt, 1976) and detailed measurements of the absorption of radiation as a function of wavelength have been made for the
principal atmospheric absorbers. Thus the accuracy of radiative fluxes
under clear skies is limited only by computational effort. Since the
explicit integration of the full equations for the vertical flux of
long-wave radiation F(z)

$$F(z) = \int_0^\infty d\nu \int^\infty \frac{d\mu}{\mu^3} \int_z^\infty dz' \Pi\, B(\nu,z') \frac{d}{dz'}, \exp\left\{-\mu \int\!\!\int \sum_i S_i(z'')f_i(\nu,z'')dz''\right\}$$

(from Rodgers, 1977) includes a double integration over the vertical
co-ordinate z, integrations over all angles represented by μ and all
frequencies ν and a summation over all spectral lines i, much effort has
been devoted to simplifying the equations and devising efficient
mathematical methods of solution. The Met. Office model currently uses an

emissivity approximation which involves a single integration over the vertical co-ordinate and a summation over 7 broad bands (Slingo and Wilderspin, 1986). The radiative properties of water clouds can be derived theoretically provided the droplet-size distribution is known. Ice clouds are comprised of non-spherical particles which may vary in shape and orientation, making it difficult to estimate the radiative properties theoretically. The major uncertainties in radiative effects of ice and water cloud arise in determining cloud properties (height, thickness, extent, water content and droplet (particle) characteristics) from grid scale model variables.

Much of the vertical transport of heat, moisture and momentum is the result of vertical motions on scales much smaller than a model grid box such as cumulus convection. Unlike radiative transfer, there is no generally accepted underlying theoretical basis for parametrizing convective motions. However, some general principles can be applied. If the atmosphere becomes statically unstable, the vertical profile will adjust towards the dry adiabat (or moist adiabat if saturation occurs), conserving moist static energy, $C_p T + Lq + gz$. This is the basis of adjustment schemes, such as the moist convective adjustment scheme of Manabe and Strickler (1964) which instantaneously removes convective instability, and the Betts-Miller scheme (Betts and Miller, 1984) which relaxes the vertical profile towards reference profiles defined from observational data. Other schemes (Arakawa and Schubert, 1974; Lyne and Rowntree, 1976, used in current Met. Office models) distinguish explicitly between small-scale ascent in convective towers and large-scale subsidence and warming of environmental air. A third approach taken by Kuo (1974) links convection directly to the large-scale convergence of moisture in a vertical column, including that due to surface evaporation, repartitioning the moisture supply between precipitation and moistening the column. In all but the moist convective adjustment scheme, there are explicit tunable parameters which may be chosen by testing the scheme against observed data such as those from the GARP Atlantic Tropical Experiment (GATE, Lyne et al, 1976) or by carrying out sensitivity experiments using the full model. A shortcoming of the former approach is that the large-scale convergence of heat and moisture is not well known, whereas in sensitivity experiments, errors in other parametrizations may degrade the performance of the convection scheme. Further guidance may be obtained from observations and numerical simulations of individual convective cells.

The factors guiding the development of various parametrizations used in a recent version of the Met. Office climate model (WM) are shown in Table 1. Note that apart from the snow free land reflectivities which are derived from observational data, adjustable parameters are prescribed with the same value over the whole globe, although separate values may be allocated to land and sea points. Hence there is a limit to the tuning possible in order to optimize the simulation of regional climate.

2.2   Using a climate model

Certain features, referred to here as boundary conditions, must be specified before a GCM can be integrated. For an atmosphere-only model, these include carbon dioxide and ozone concentrations, sea surface temperatures and sea-ice extents, orography and time of year. The normal procedure is then to integrate the model over the period of interest and look at the long-term statistics of the simulation. In the case of a climate change experiment, a second simulation is made changing only the relevant boundary conditions (for example carbon dioxide concentrations) and the long-term characteristics of the two simulations are compared. Note that even with fixed boundary conditions, the simulation will vary in time due to the inherent variability of climate. Hence it is necessary to perform statistical tests on the differences between the control and perturbed simulations in order to establish that the probability that such differences could have arisen by chance is small.

The length of simulation will depend on a variety of factors. First, the time taken for the model to adjust to the prescribed boundary conditions will vary from a week or so for prescribed sea surface temperature experiments to decades or even centuries in the case of coupled ocean-atmosphere models. Second, the length of time for which prescribed boundary conditions are valid will also vary, from a month or so for sea surface temperature anomalies to thousands of years in the case of orography. Third, the integrations must be sufficiently long (or numerous) to allow the statistics of the simulated climate to be defined. In practice, the length of a simulation is often limited by the computational resources available.

In the above approach, there are certain underlying assumptions whose validity may affect the relevance of the results to observed climate.

(a) A "mean" or quasi-equilibrium state exists in nature and in the model. As discussed in the introduction, the climate of the earth has varied considerably over a wide range of timescales. Thus in the real world at least, we do not know if statistically stationary climatic states exist. The problem of the stability of climatic states has been addressed using "simple" models (see for example Fraedrich 1978, Ghil and Childress, 1987) and will not be considered further here.

(b) The "mean" state is unique, and independent of the initial state. Lorenz (1975) suggested that the earth's climate may have more than one stable state, and that transitions between these states may be rare (climate is almost "intransitive"). It follows that the choice of initial conditions may determine which of the stable states evolve. For example, Wetherald and Manabe (1975) found that a simulation made with an idealized climate model which commenced with an ice covered surface remained ice-covered, whereas when the initial conditions were free of ice, only polar latitudes became ice-covered.

(c) The model converges, and converges to the correct solution. Although mathematical models of climate may be based on sound physical principles, this is not sufficient to guarantee that the model will produce the correct solution. We cannot check the convergence of such complex models using mathematical stability analyses, so the only way to check for convergence is to integrate the model and look at the results.

2.3   Verification

The validity of the model can be assessed on a variety of timescales against observational data. As climate models can be regarded as a development of numerical weather prediction models, one can specify initial conditions from meteorological analyses, perform short-range forecasts and compare the evolution of synoptic features in the model with those observed (for example, Carson and Cullen, 1976). A similar approach can be taken using long-range forecasts (Mansfield, 1986), though after a period of a week or so, the correspondence between individual simulated and observed features will be lost, and the forecasts of necessity become statistical rather than deterministic.

The traditional method of calculating atmospheric GCMs has been to compare the simulated means with climatological data (Manabe and Holloway, 1975; Mitchell, 1983). For example, by comparing a 4-year mean of simulated mean sea-level pressure from the version of the Met. Office GCM described by Slingo et al (1988) with observational data (Figure 3) one can see that the model reproduces all the main features of the observed circulation including the equatorial and mid-latitude pressure troughs, and the subtropical anticyclones which form over the ocean in summer and extend over land in winter. Note however that the simulated pressure ridge over the Rockies is weaker than observed, and the weak trough over Central Europe is much too strong in the model.

The variability in the model may also be assessed: Figure 4 shows the simulated and observed high pass filtered 500 mb heights which give an indication of the strength and position of the storm tracks. Again the agreement with observation is good, though the Atlantic storm track is weaker than observed, and penetrates too far south at the European end. One can assess variability on longer timescales including year to year in a similar manner.

The detailed evolution of the seasonal cycle can be assessed by comparing mean monthly grid point data with observations from suitable climatological stations (Reed, 1986). In Figure 5, the mean monthly midnight surface temperatures at a model grid point over eastern England are compared with observed 2 m minimum temperatures from Cambridge (Wilson and Mitchell, 1987b). Note the colder than observed spring temperatures. This is associated with more frequent than observed cold spells in the model (Figure 6) which are a result of too many or too strong outbreaks of easterly flow off the cold Eurasian continent (Reid, 1986).

The above approach may be used to establish that the mean climate and level of variability in the model are correct. In order to test that the structure of the model's response to specific perturbations is correct, one can examine the response of the model to observed changes in boundary conditions such as sea surface temperature anomalies, and compare the response with that observed. Palmer and Mansfield (1986) have examined the simulated and observed response to different sea surface anomalies in the tropical Pacific, and Folland et al (1986) have examined the relationship between sea surface temperatures and Sahel rainfall, using observations and numerical experiments. A weakness in this approach to model verifications is that in reality factors other than the changes in boundary conditions specified in the model may contribute to observed changes in the atmosphere, including inherent atmospheric variability.

The ability to simulate contemporary climate accurately is a necessary but not a sufficient condition for a model to produce reliable estimates of climate change. It is possible that in choosing values of disposable parameters which give an accurate simulation of present-day climate, the response of the model to changes in climate may be distorted. Thus, attempts have been made to simulate climates in the past, including the last glacial maximum 18,000 years before present (18 K bp) (Manabe and Broccoli, 1985) and the last northern hemisphere summer insolation maximum

(9 K bp), considered in the following chapter. Manabe and Broccoli compared the changes in surface temperature between 18 K bp and present from two different versions of the model with a palaeoclimatic estimate of the temperature change (Figure 7). However, the investigation was inconclusive because of both the small differences between the results from the two versions of the models and the uncertainties in the palaeoclimatic data.

3. OTHER COMPONENTS OF THE CLIMATE SYSTEM

The approaches used in developing atmospheric models may be applied to other components of the climate system including sea-ice, the oceans and continental ice-sheets. There is considerably less data from the ocean than from the atmosphere. In addition, synoptic eddies in the ocean have a typical length scale of 30 km as opposed to 1000 km typical of the atmosphere and so are not resolved by the current generation of global ocean GCMs. Thus ocean GCMs are even more difficult to validate than atmospheric models. Since the atmosphere, ocean and continental ice-sheets respond on timescales of days, centuries and thousands of years respectively, special care must be taken when representing the interaction between them. Schlesinger (1979) has listed several ways in which an atmospheric model may be coupled to an ocean model, and Bryan (1984) has tested methods for accelerating the convergence to equilibrium of ocean climate models. Alternatively, Hasselmann (1979) has suggested parametrizing the effect of quickly changing elements of a climate model (for example, the atmospheric component of a model which includes the deep ocean).

4. SUMMARY

There is a growing interest in the simulation of climate, both in order to predict future changes in climate due to man's activities, and to understand past climatic change. Climate models are based on the laws of classical physics, though processes which occur on scales smaller than the model grid (or timestep) must be approximated (parametrized). This may be done on the basis of theory, observations and laboratory and numerical experiments. Validation of models includes detailed verification of individual components of the model, and the comparison of results from the full model over a variety of timescales with observational data. Certain assumptions concerning the stability of climate and its dependence on specified boundary conditions are implied in the current use of GCMs in climate simulations.

5. ACKNOWLEDGEMENTS

This paper is an enlarged and updated version of a paper "The physical basis of climate modelling" which appeared in "Palaeoclimatic Analysis and Modelling" (Ed Ghazi) published by D Reidel, 1983.

6. REFERENCES

Arakawa, A. and Schubert, W. H. 1974 Interaction of a cumulus cloud ensemble with large-scale environment. Part I. J. Atmos. Sci., 31, 674-701.
Betts, A. K. and Miller, M. J. 1984 A new convective adjustment scheme. ECMWF Technical Report No 43, 65 pp.

Bryan, K. 1984 Accelerating the convergence to equilibrium of ocean-climate models. J. Phys. Oceanogr. 14 (4) 666-673.

Carson, D. J. and Cullen M. J. P. 1976 Intercomparison of short-range numerical forecsts using finite difference and finite element models from the UK Meteorological Office (Paper presented at the joint DMG/AMS International Conference on the Simulation of large-scale Atmospheric Processes, Hamburg, 30 August-4 September 1976). Met O 20 Tech. Note II/81, Meteorological Office, Bracknell.

Clarke, W. C. 1982 Carbon Dioxide Review: 1982 (Ed. W. C. Clarke) Clarendon Press, Oxford; Oxford University Press, New York. 469 pp.

Folland, C. K., Palmer, T. N. and Parker, D. E. 1986 Sahel Rainfall and Worldwide sea temperatures, 1901-85. Nature, 320, 602-607.

Fraedrich, K. 1978 Catastrophes and resilience of a zero-dimensional climate system with ice-albedo and greenhouse feedback. Quart. J. R. Met. Soc. 105, 147-167.

Ghil, M. and Childress, S. 1987 Topics in Geophysical Fluid Dynamics; Atmospheric Dynamics, Dynamo Theory, and Climate Dynamics. Applied Mathematical Sciences 60, pp 485, Springer Verlag, New York.

Hasselmann, K. 1979 On the problem of multiple time-scales in climate modelling. In "Man's Impact on Climate" (Eds. Backh, Pankrath and Kellogg). Developments in Atmospheric Science, 10, 43-55. Elsevier (Amsterdam, Oxford, New York).

Jones, P. D., Wigley, T. M. L., Folland, C. K., Parker, D. E., Angell, J. K, Lebedeff, S. and Hansen, J. E. 1988 Evidence for global warming in the past decade. Nature, 338, 790.

Kuo, H. L. 1974 Further studies of the parametrization of the influence of cumulus convection on large scale flow. J. Atmos. Sci., 31, 1232-1240.

Lal, M. and Ramanathan, V. 1984 Effects of moist convection and water vapour radiative processes on climate sensitivity. J. Atmos. Sci. 41, 2238-2249.

Lorenz, E. 1975 Climatic Predictability, WMO — GARP Publication Series No. 16, Geneva, 132-136.

Lyne, W. H. and Rowntree, P. R. 1976 Development of a convective parametrization using GATE data. Met O 20 Technical Note II/70, Meteorological Office, Bracknell.

Lyne, W. H., Rowntree, P. R., Temperton, C. and Walker, J. 1976 Numerical modelling using GATE data. Met. Mag. 105, 261-271.

Manabe, S. and Broccoli, A. O. 1985 A Comparison of Climate Model Sensitivity with data from the last glacial maximum. J. Atmos. Sci. 42, 2643-2651.

Manabe, S. and Holloway, J. L. Jnr. 1975 The Seasonal variation of the hydrologic cycle as simulated by a global model of the atmosphere. J. Geophys. Res., 80, 1617-1649.

Manabe, S. and Strickler, R. F. 1964 Thermal equilibrium of the atmosphere with a convective adjustment. J. Atmos. Sci. 21, 361-385.

Mansfield, D. A. 1986 The skill of dynamical long-range forecasts, including the effect of sea-surface temperature anomalies. Quart. J. R. Met. Soc., 112, 1145-1176.

Mitchell, J. F. B. 1983 The seasonal response of a general circulation model to changes in CO_2 and sea surface temperature. Q.J.R. Met. Soc., 109, 113-152.

Palmer, T. N. and Mansfield, D. A. 1986 Wintertime Circulation Anomalies During Past El Nino Events, using a High Resolution General Circulation Model II. Variability of the seasonal mean response. Quart J.R. Met. Soc., 112, 639-660.

Paltridge, G. W. and Platt, C. M. R. 1976 Radiative proceses in Meteorology and Climatology. Developments in Atmospheric Science, 5. Elsevier, New York, 318 pp.

Reed, D. N. 1986 "Simulation of temperature and precipitation over eastern England by an atmospheric general circulation model". Journal of Climatology 6, 233-253.

Roeckner, E., Schlese, U., Biercamp, J. and Loewe, P. 1987 Cloud optical depth feedbacks and climate modelling. Nature, 329, 138-139.

Rodgers, C. D. 1977 "Radiative Processes in the Atmosphere". In proceedings of ECMWF Seminars. "Parametrization of Physical Processes in the free atmosphere", 5-66.

Schlesinger, M. E. 1979 Comments on ocean-atmosphere coupling and discussion of the paper "A global ocean-atmosphere model with seasonal variation: Possible application to a study of climate sensitivity". Climate Research Institute, Report No. 39, Oregon State University.

Schlesinger, M. E. 1988 (Ed) Physically-based modelling and simulation of climate and climatic change. Vol. 1. NATO Advanced Study Institute Series. (Reidel, in Press).

Schutz, C. and Gates, W.L. 1971 Global Climatic data for surface, 800 mb, 400 mb (January), R-915-ARPA, RAND, Santa Monica.

Slingo, A. and Wilderspin, R. C. 1986 Development of a revised longwave radiation scheme for an atmospheric general circulation model. Q.J.R. Meteorol. Soc., 112, 371-386.

Slingo, A., Wilderspin, R. C. and Smith, R. N. B. 1988 The effect of improved physical parametrizations on simulations of cloudiness and the earth's radiation budget in the tropics. Submitted to Q.J.R. Meteorol. Soc. (Also Met O 20 DCTN 65).

Wetherald, R. T. and Manabe, S. 1975 The effects of changing the solar constant on the climate of a general circulation model. J. Atmos. Sci. 32, 2044-2059.

Wilson, C. A. and Mitchell, J. F. B. 1987a A Doubled CO_2 Climate Sensitivity experiment with a GCM including a simple ocean. J. Geophys. Res., 92, 13315-13343.

Wilson, C. A. and Mitchell, J. F. B. 1987b Simulated Climate and CO_2 — induced climate change over western Europe. Climate change, 10, 11-42.

7. Suggested Bibliography

Physically-based Modelling and Simulation of Climate and Climate Change
Vol.1 (1988) Schlesinger, M. E. (Ed), NATO Advanced Study Institute Series,
(Reidel, in Press). Excellent chapters by Bourke on spectral methods, and
Arakawa on finite differences, as well as other components of climate
models.

Numercial Weather Prediction and Dynamical Meteorology (1980)
Haltiner, G. J. and Williams, R. T. (John Wiley). Generally good on
numerical methods, but out-dated on physical parametrizations.

An Introduction to Three-Dimensional Climate Modelling (1986) Washington,
W. M. and Parkinson, C. L. There are not many books on climate modelling,
and this is the best to date.

The Global Climate (1984) Houghton, J. T. (Ed). (Cambridge University
Press) Contains a series of articles by different authors on various
aspects of the climate system.

Table 1   Choice of parameters in the atmospheric component of the model used by Wilson and Mitchell (1987)

	Parameter valid over globe	Tuning Experiments cf climatology	Theory	Detailed Models	Field Observations	Laboratory Experiments
Dynamics						
Diffusion Coefficient	X	X	(X)			
Time Smoothing	X	X				
Radiation						
Gaseous absorption	X		X	X	X	X
Cloud amounts	X	X			X	
Cloud properties	X	X			X	
Surface Reflectivity					X	
Convection						
Parcel size	X	X			X	
Entrainment and Detrainment	X	X	X		X	X
Evaporation	X	X	X		X	
Boundary Layer						
Drag coefficient	X		X		X	X
Surface roughness	X				X	
Other						
Soil moisture	X				X	
Heat flux through ice	X		X	X	X	X

FIGURE CAPTIONS

Figure: 1 (a) Estimated global mean surface temperature over the last 850,000 years (from Clarke, 1982)

      (b) Observed anomalies in global mean surface temperature 1901-1987 (from Jones et al, 1988).

2   The horizontal grid used by WM. This is typical of current climate models. The horizontal spacing is 5° latitude by 7.5° longitude.

3   Mean sea level pressure (contours every 4 mbs)

    (a) January, observed. (Schutz and Gates, 1971)

    (b) December to February, simulated using a high resolution version of the Met. Office GCM (see Slingo et al, 1988)

4   High pass filtered variance of 500 mb heights for the northern hemisphere for December to February. Contours every 1 mb.

    (a) From a high resolution version of the Met. Office GCM.

    (b) Observed, from Met. Office operational analyses, 1983-1986.

5   Monthly mean surface temperature over Eastern England. Solid line, climatological 2 m screen minimum temperatures at Cambridge; dashed line, simulated ground temperatures 00Z from a 5-layer atmospheric model. (Wilson and Mitchell, 1987b). The crosses are the observed mean monthly minimum.

6   Frequency distribution of surface temperature over central/eastern England (mean, annual and semi-annual cycle removed) at 0.5 K intervals.
Light line — From 3 years of Central England Temperature Series
Heavy line — From a 3-year simulation with a 5-layer atmospheric model (Reed, 1986).

7   Latitudinally averaged changes in surface temperature between 18 K bp and present. Crosses, sea surface temperature changes deduced from palaeoclimatic data. Solid and dashed lines, as simulated using a climate model with a mixed-layer ocean using prescribed and model generated cloud amounts respectively. (From Manabe and Broccoli, 1985).

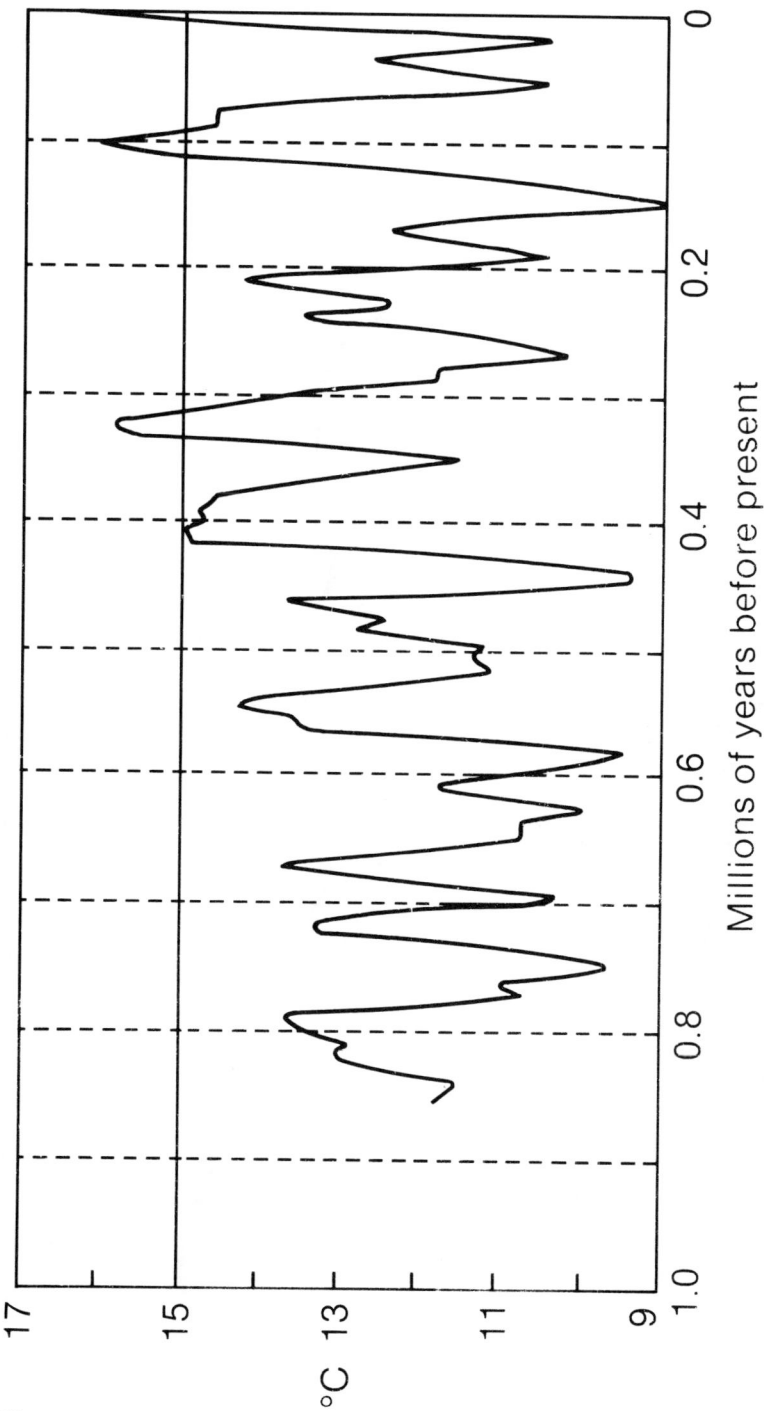

Figure 1a

Global Surface Air Temperature relative to 1950–79 average

(based on Jones *et al.* 1988)

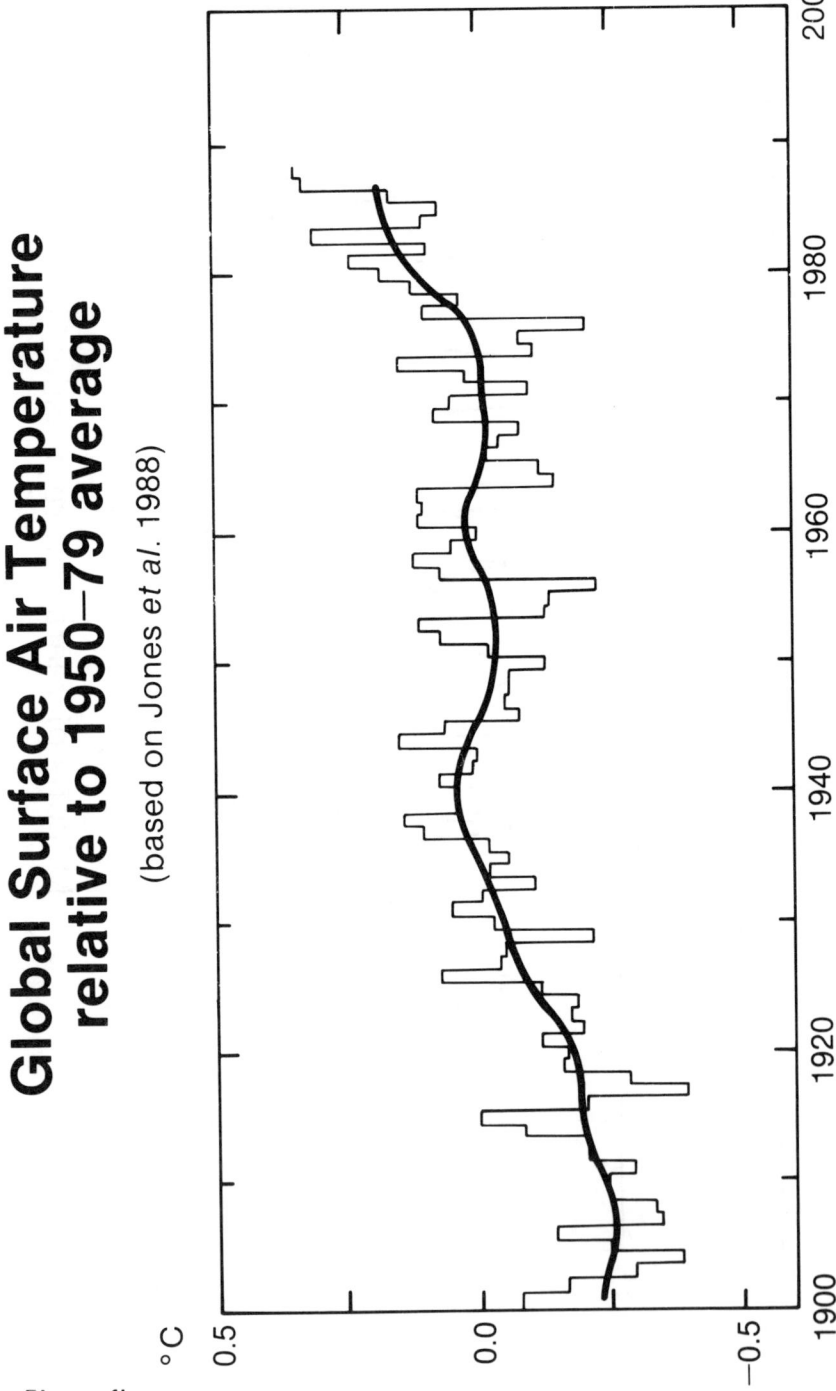

Figure 1b

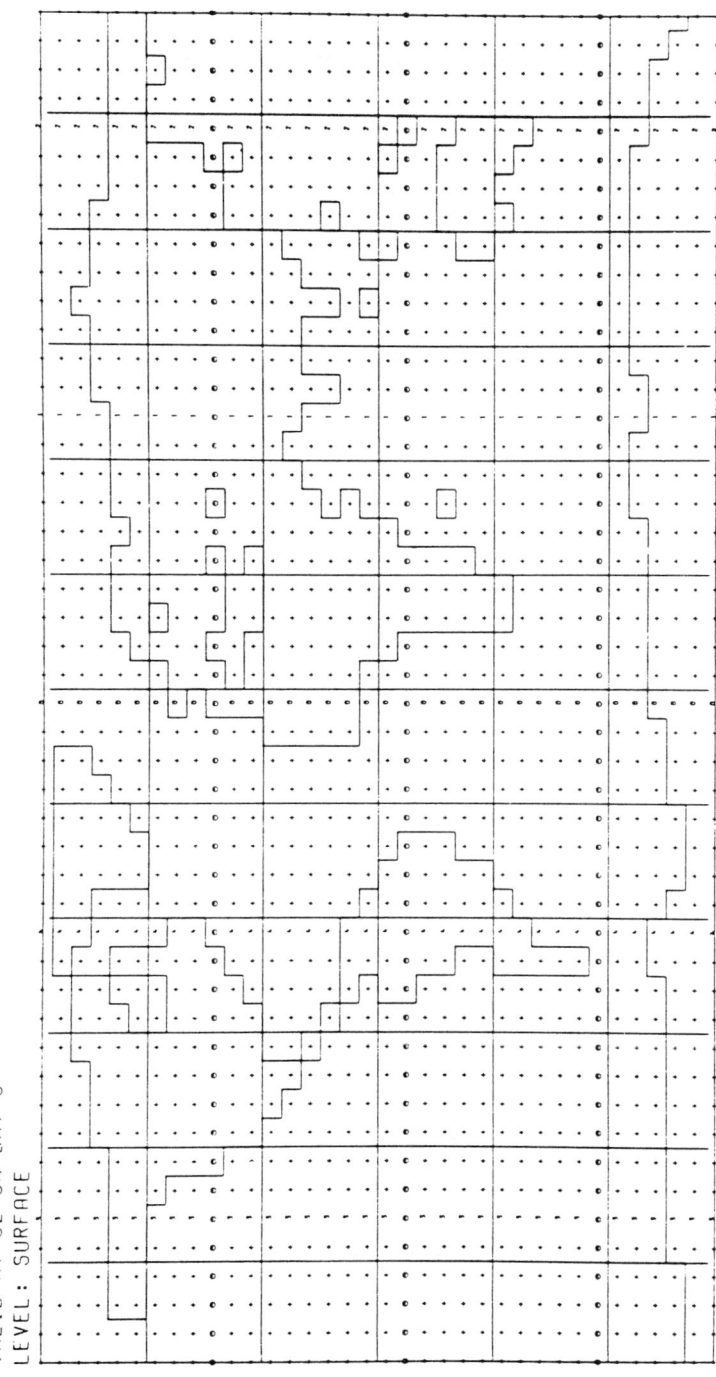

MODEL GRID COAST + GRID POINTS
EVERY 10TH ROW A SEQ OF 0 (5.0*7.5 GRID)
VALID AT 0Z ON DAY 0
LEVEL: SURFACE

Horizontal resolution of the 11-Layer Model when integrated on

the 5 x 7.5 degree latitude-longitude grid.

Figure 2

Mean-sea-level pressure (mb)

Observed (January)

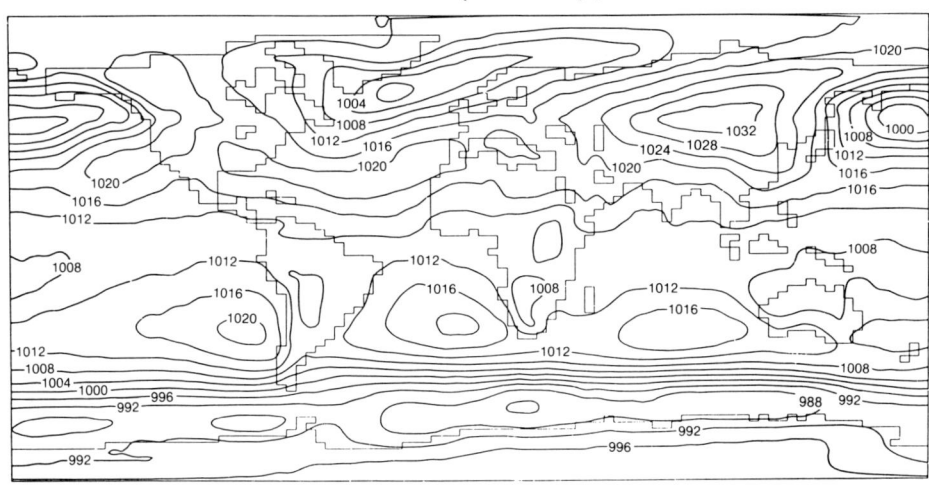

Figure 3a

Model (December to February)

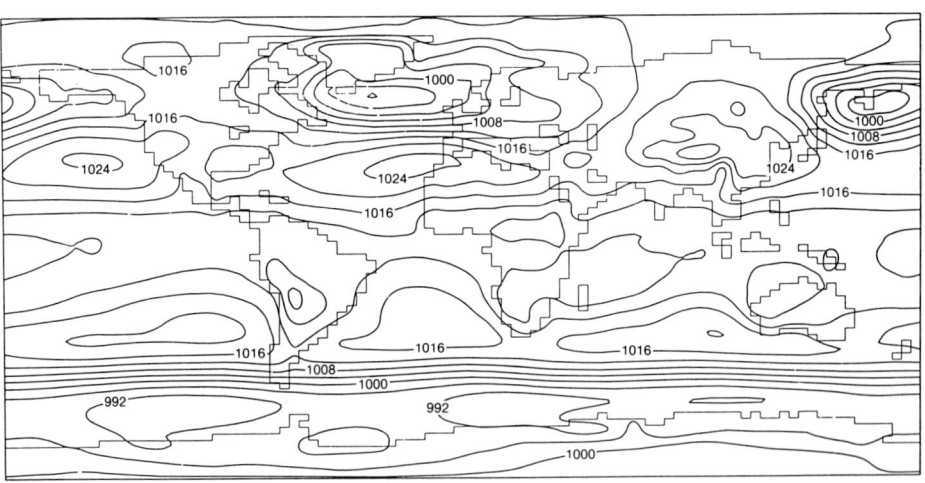

Figure 3b

High pass filtered variance of 500 mb heights
(December to February)

Observed

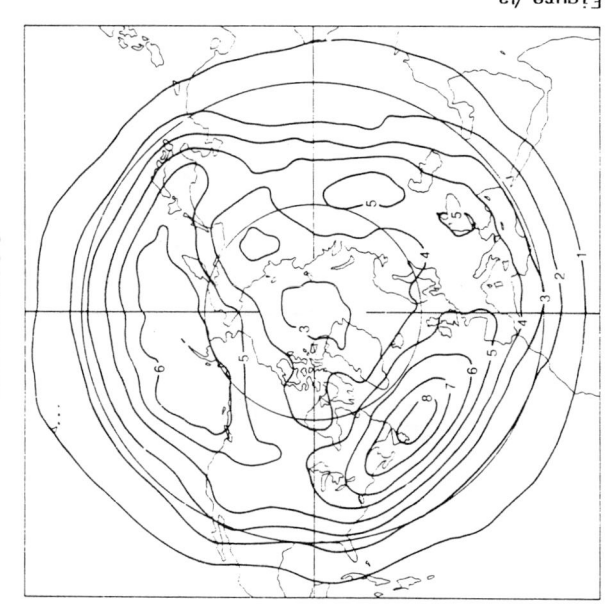

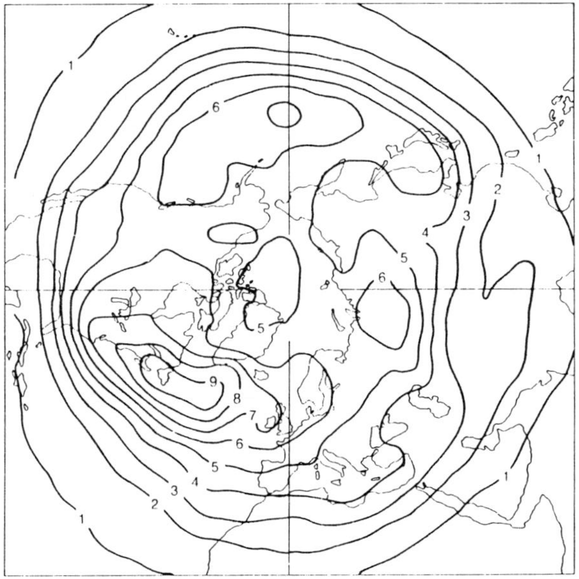

Figure 4a

Simulated

Figure 4b

155

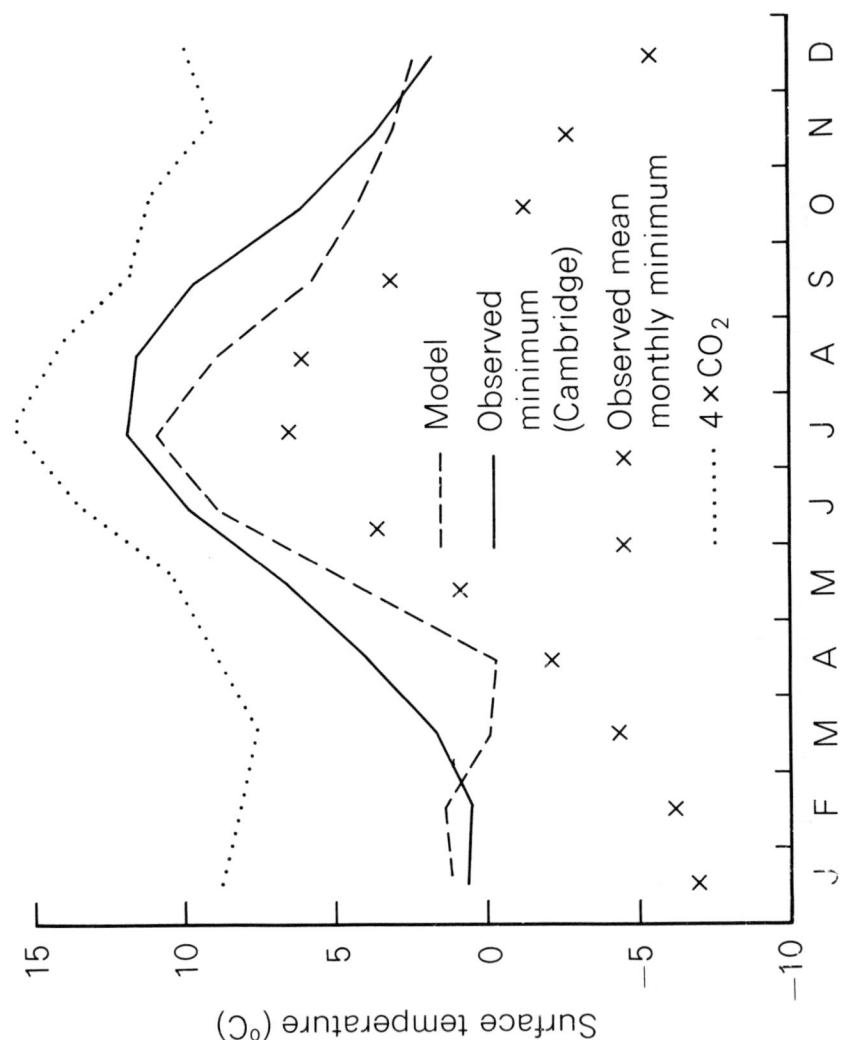

Figure 5

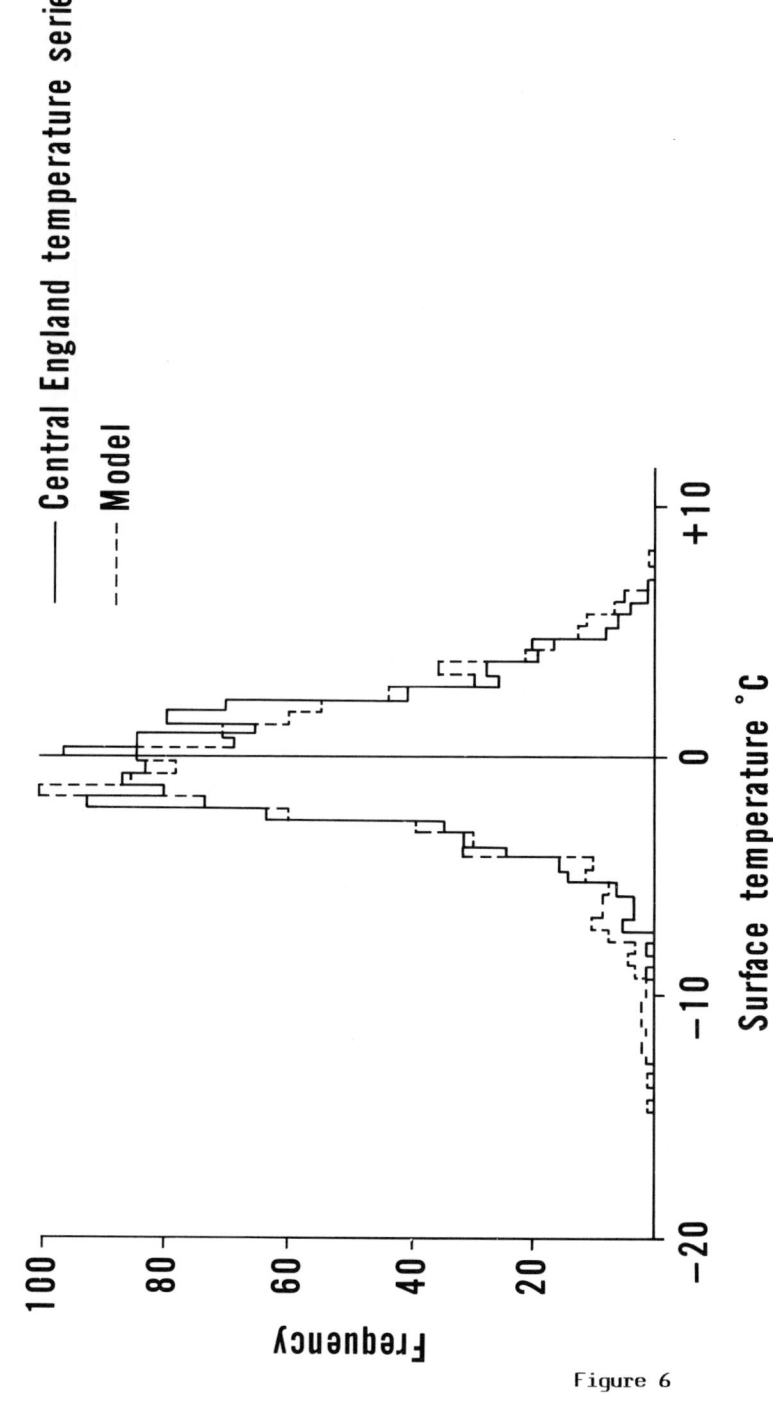

FREQUENCY DISTRIBUTION OF SURFACE TEMPERATURE

Central England temperature series

Model

Surface temperature °C

Frequency

Figure 6

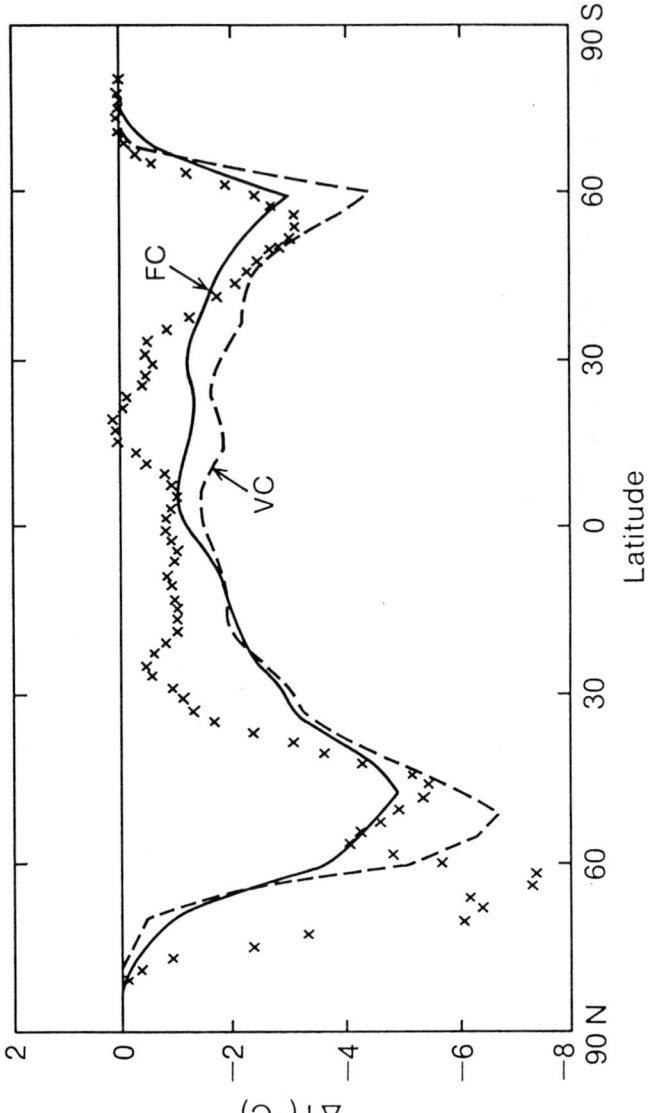

Figure 7

THE PHYSICS AND DYNAMICS OF THE CLIMATE SYSTEM

SIMULATION OF CLIMATE CHANGE

J F B Mitchell
Meteorological Office, Bracknell RG12 2SZ

Summary

The use of climate models to underfund and predict climate change is described, using examples from numerical studies of the effects of enhanced atmospheric CO_2.

1. INTRODUCTION, WHY MODEL CLIMATE?

The increases in atmospheric "Greenhouse" gases since 1860 have a radiative effect equivalent to a 40% increase in carbon dioxide concentrations, and by the middle of the next century, are expected to be equivalent to a doubling of carbon dioxide concentration. Simulations with detailed climate models indicate that this would produce a warming of 2 to 5 K in global mean surface temperature at equilibrium, with accompanying changes in precipitation, sea level and other parameters. The observed increase of 0.5 K since 1900 is consistent with the lower range of the estimated potential increase, allowing for a possible slowing of the global mean warming due to the ocean's large thermal inertia. Thus, man-made climate change is not just a possibility, it may already be occurring. Consequently, there is an ever pressing need to predict the likely changes in climate due to increases in trace gases in order to guide future energy policies and to minimize the possible climatic impacts. As discussed in Chapter 9, detailed 3-dimensional models of climate are the most promising method of providing the detailed information required for climatic impact assessment.

There is also considerable interest as to why the earth's climate has changed in the past (see Figure 1a). Climate models provide a means of testing theories of climate change. Conversely, periods from which reliable and wide-spread palaeoclimatic data are available may be used as a test of the fidelity of climate models, especially if the factors producing the climatic change are well known.

The rest of this paper is arranged as follows:

2. The "Greenhouse" effect.
3. The principal gases, past, present and future.
4. Climate feedbacks in CO_2 experiments.
5. Equilibrium climate change due to increased CO_2.
6. Modelling the transient response to increases in trace gases.
7. Uncertainties in the simulation and detection of the climatic effect of increased trace gases.
8. Appeals to the past; simulations for 9000 years before present (9 K bp).

9. Concluding remarks.
10. Acknowledgements.

2. THE "GREENHOUSE" EFFECT

The globally averaged radiative heat balance of the earth-atmosphere system may be approximated by

$$(S_0/4)(1-\alpha) = \sigma T_e^4 \tag{1}$$

where S_0 is the solar "constant", α is the fraction of solar radiation reflected by the earth and atmosphere, σ is Stefan's constant and T_e is the mean or effective radiative temperature of the system. T_e is the radiative equilibrium temperature that the earth's surface would reach if the atmosphere was transparent to longwave (thermal or infra-red) radiation. With the current value of α (0.3), T_e = 255 K, some 33 K below the current observed global mean surface temperature.

Now consider a single layer atmosphere, longwave emissivity ϵ and shortwave albedo α in radiative equilibrium with the surface (Figure 1). For convenience, it will be assumed that the atmosphere does not absorb solar radiation. The surface temperature T_* is given by

$$T_*^4 = T_e^4/(1-\epsilon/2) \tag{2}$$

and the atmospheric temperature T is given by

$$T^4 = \frac{T_*^4}{2} \tag{3}$$

Thus, gases which absorb and emit longwave radiation (greenhouse gases) warm the surface and cool the atmosphere. As an exercise, the reader can show that in a two-level atmosphere in radiative equilibrium, temperature decreases with height.

In practice, the troposphere is not generally in radiative equilibrium, but is heated convectively from the surface. The global mean lapse rate, about 6 K/km, is slightly less than the dry adiabatic lapse rate due to latent heat released in regions of saturated ascent. The radiative equilibrium temperature of 255 K corresponds to a height of 5.5 km (point a on the solid profile in Figure 2). An increase in the concentration of greenhouse gases raises the effective emitting level, (point b in Figure 2) and reduces the effective emitting temperature T_e and outgoing radiance σT_e^4. The system warms until the temperature at the new emitting level rises to the original value T_e (point c).

The stratosphere will cool if the concentration of greenhouse gases there is increased. The stratosphere is not convectively coupled to the surface but is in radiative equilibrium; an increase in longwave absorbers will produce enhanced emission to space which at equilibrium is compensated by a fall in emitting temperature.

3.   THE PRINCIPAL ABSORBERS, PAST, PRESENT AND FUTURE

The concentration and radiative characteristics of the climatically most important longwave absorbers in the contemporary atmosphere are listed in Table 1, based on Dickinson and Cicerone (1986) and Ramanathan et al (1985). The greenhouse heating is defined as the change in net downward flux at the tropopause due to the presence of each gas. Since increasing the concentration of greenhouse gases produces a slight increase in the downward flux from the stratosphere, the change in net flux at the tropopause rather than at the top of the atmosphere is considered (Schneider, 1975). Note that the troposphere and surface are strongly coupled and hence can be regarded as a single system. Water vapour is by far the most important "Greenhouse" gas, followed by carbon dioxide. Although the concentrations of chlorofluorocarbons (CFCs) CFC11 and CFC12 are 10^{-6} that of carbon dioxide, their contribution to radiative heating is only a factor 10^3 less. These gases have absorption bands which are both very strong and lie in the atmospheric "window" from about 833 cm^{-1} to 1250 cm^{-1}, where there is little absorption by other gases.

Estimates of the pre-industrial (1860) concentrations of the main trace gases (other than water vapour) are shown in Table 2. The concentrations of carbon dioxide, methane and nitrous oxide are deduced from ice-core sampling. CFCs are produced solely by industrial activity and so their pre-industrial concentration is zero. Gases other than CO_2 produce 30% of the radiative perturbation to date, and the total effect is already equivalent to half that due to doubling carbon dioxide concentrations. There is considerable uncertainty in the estimated concentrations for 2035; the first three gases in Table 2 undergo natural cycles which are not completely understood, and for all the gases the industrial contribution is dependent on the world economy and the effectiveness of recent agreements to limit production such as those for carbon dioxide and CFCs. (The assumptions made to derive the concentrations in 2035 are listed in Mitchell (1989) and references therein.) If the estimates in Table 2 are not greatly in error, the radiative effect of the increase in all the gases since 1860 will be equivalent to a doubling of CO_2 before 2035, and this would occur even if CFCs do not increase any further.

The effect of alternative scenarios can be estimated from Table 2 by noting that the radiative effect of carbon dioxide varies approximately logarithmically with concentration, (Augustsson and Ramanathan, 1977), and that of methane and nitrous oxide as the square root of their concentration. The effect of CFC11 and CFC12 vary linearly with concentration because of their low concentrations and the fact that they absorb at wavelengths where there is little absorption by other gases.

4.   CLIMATE FEEDBACKS IN CARBON DIOXIDE EXPERIMENTS

The radiative perturbation due to enhancing trace gases can (in principle) be calculated to any desired accuracy. Doubling carbon dioxide amounts increases the heating of the tropopause and surface by 4 Wm^{-2} (Lal and Ramanathan, 1984), which, assuming the current climate system behaves as a black body at mean temperatures near 255 K, would produce a warming of 1.1 K. However, the atmospheric response to changes in radiative heating is complex involving many processes which are not well understood. These may amplify the perturbation (positive feedbacks) or damp it (negative feedbacks).

Most equilibrium experiments to determine the climatic effects of increases in trace gases have assumed a doubling or quadrupling of carbon dioxide concentrations. The vertical profile of the radiative perturbation due to the other gases in Table 2 is qualitatively similar to that due to increasing carbon dioxide, (except near the tropopause where the minor gases produce a slight warming as opposed to a cooling (Ramanathan et al (1985)), so experiments with increased carbon dioxide only are probably adequate to represent the effect of all the gases. Here and in the next section, we will consider the equilibrium response to doubling carbon dioxide concentrations. First, we consider the main climate feedbacks which have been identified in numerical experiments.

A warmer atmosphere will hold more water vapour, and hence enhance the radiative perturbation due to carbon dioxide. Higher temperatures reduce the extent of highly reflective snow and ice, and so increase the absorption of solar radiation. In most simulations with increased carbon dioxide, cloud amount decreases, reducing the amount of insolation reflected back to space, and the mean cloud height increases, further enhancing the "Greenhouse" effect. It is also likely that cloud radiative properties may change. An increase in atmospheric moisture would tend to increase cloud water content and hence the reflectivity of cloud, producing a negative feedback. Thick clouds behave as black bodies at infra-red wavelengths, but thin high clouds may have an emissivity considerably less than unity. An increase in cloud water would tend to increase the longwave emissivity of high cloud and enhance the "greenhouse" effect, adding a further positive feedback.

The strength of individual feedbacks simulated in GCM experiments has been quantified using a simple zero-dimensional energy balance model

$$C_* \frac{d\Delta T}{dt} + \lambda \Delta T = \Delta Q \tag{4}$$

where ΔQ is the radiative perturbation, ΔT is the change in surface temperature, C_* is the effective heat capacity of the system, and λ is defined as the feedback parameter. At equilibrium, $\Delta T_{eq} = \Delta Q / \lambda$. Thus the smaller λ, the greater ΔT_{eq}, implying a larger positive feedback. Assuming the feedbacks are linear and independent,

$$\lambda = \sum \lambda_i \; ; \; \lambda_i = \Delta Q_i / \Delta T_{eq} \tag{5}$$

where i denotes the individual feedbacks, and ΔQ_i is the radiative perturbation due to the process i alone. For example, if i denotes changes in water vapour, then ΔQ_i is the radiative perturbation due to the increase in water vapour between the perturbed and control simulations, in the absence of changes in other parameters. Typical values for λ_i from GCM experiments are shown in Table 3. The water vapour feedback is easily the strongest, and there is good agreement between different models. The ice-snow albedo feedback is small, though it should be noted that the representation of sea-ice in current climate models is simplistic. The strength of the cloud feedback varies from model to model. Estimates of the effect of changes in cloud radiative properties range from a strong negative feedback (Somerville and Remer, 1984) to a weak positive feedback (Roeckner et al, 1987). The representation of clouds and their radiative properties are a source of major uncertainty in climate change experiments.

Other feedbacks have been analysed using one-dimensional radiative-convective models (for example, Schlesinger 1985). However, the reader is warned that the different modelling groups estimate the feedbacks in their experiments in different ways. The main value of feedback analysis is in identifying the processes which most affect climate sensitivity, and in helping to identify the reasons for discrepancies between models.

5. SIMULATED EQUILIBRIUM CLIMATE CHANGE DUE TO INCREASED CARBON DIOXIDE

The increases in carbon dioxide and other trace gases have been occurring on timescales of decades to centuries. It is generally assumed that one can ignore changes in the continental ice sheets on this timescale, but not changes in the ocean or sea-ice. Coupled ocean-atmosphere models are at present prohibitively expensive to run to equilibrium, so over the last decade atmospheric models coupled to a simple oceanic mixed layer of prescribed depth have been used to model the effects of increased CO_2. In order to allow for the advection of heat by the oceans, some models have included a prescribed oceanic heat flux C such that the mixed layer temperature T is given by

$$\rho h C_p \frac{dT}{dt} = S + C \qquad (6)$$

where ρ, h and C_p are respectively the density, depth and specific heat of the well mixed layer and S is the net downward heat flux from the atmosphere. C may be derived by prescribing the sea (surface) temperature T from climatology in the atmospheric model and integrating through several annual cycles to determine S as a function of time of year and location. Knowing S, h and dT/dt, one can derive C as a function of time and space, and prescribe it in equation 6 to determine the evolution of T in the coupled atmosphere mixed layer model.

In the remainder of this section, the effect on simulated climate of doubling carbon dioxide will be outlined using examples primarily from Wilson and Mitchell (1987a), henceforth referred to as WM. Schlesinger and Mitchell (1987) have reviewed three other comparable modelling studies; a fifth study is described by Schlesinger and Zhao (1987). Where appropriate, discrepancies between models are discussed.

Increasing carbon dioxide concentrations leads to a warming of the troposphere and a cooling of the stratosphere (Figure 3). In the tropics, the tropospheric warming increases with height, since in low latitudes the lapse rate adjusts towards the moist adiabatic which decreases with increasing temperature. The increase in warming with height is more pronounced in models with penetrative convection schemes as opposed to those with moist convective adjustment. Penetrative schemes are more efficient in transporting moisture upwards, and hence in warming upper levels.

In high latitudes in winter, the low-level inversion confines the warming near the surface. The reduction of snow and sea-ice leads to considerable enhancement of the warming in high latitudes in the winter hemisphere. Note that the removal of sea-ice in summer leads to increased absorption of solar radiation and storage of heat in the mixed layer, delaying the onset of freezing in autumn and early winter (Manabe and Stouffer (1980)). This leads to a pronounced warming, or more precisely, a reduced cooling relative to the control. In contrast, over sea-ice in summer, temperatures generally rise to the melting point of ice and are maintained there in both the control and anomaly simulation. Even where sea-ice is temporally removed in summer, the thermal inertia of the oceanic mixed layer substantially reduces the magnitude of the warming.

The influence of sea-ice is seen more clearly in the geographical distributions of seasonal mean changes in surface temperature (Figure 4). During December to February the maximum warming occurs in the Arctic, Hudson Bay and near Kamchatka where sea-ice is thinner or removed altogether in the $2xCO_2$ simulation, and the minimum occurs over Antarctic sea-ice in the Ross and Weddell Seas. In models in which sea-ice extents are excessive in the simulation of present day climate, the high latitude warming is exaggerated and the latitude of the maximum warming may be displaced equatorwards. The warming is generally less over the ocean than over land. Over land, evaporation may be limited by the dryness of the surface so that a greater proportion of the increase in radiative heating is used to raise surface temperature as opposed to increasing evaporation. In middle latitudes in winter, removal of snow in the $2xCO_2$ simulation also enhances the surface warming.

The warming is a minimum over the tropical oceans. The saturation vapour pressure for water increases exponentially with temperature, so assuming the relative humidity is approximately constant, potential evaporation also increases non-linearly with temperature. As a result more of the increase in radiative heating of the surface is converted to latent heat and less to raising surface temperatures and sensible heat as one goes from cold to warm regions. Increased evaporative cooling at the surface is linked to increased warming due to latent heat aloft, consistent with the reduction in lapse rate in the tropics noted above.

The changes in cloud are qualitatively similar from model to model (Figure 5). In general, there is an increase in high cloud and a reduction in upper and middle tropospheric cloud which in some models is most pronounced in the tropics and mid-latitudes. The changes in low cloud vary considerably from model to model, though increases in low cloud are generally restricted to middle and high latitudes. Thus in low latitudes, where insolation is greatest, there is a reduction in planetary albedo (high cloud is generally assigned a low albedo) leading to enhanced absorption of solar radiation, and in the extratropics, there are substantial increases in high cloud tending to reduce longwave cooling to space.

The warming of the atmosphere leads to an increase in specific humidity (41% in WM) and evaporation and precipitation (15% in WM). Precipitation does not increase everywhere, (Figure 6) and the spatial scale of the change is smaller than for temperature (compare Figure 4). Whereas there is substantial agreement between the temperature changes simulated by different models, there is little agreement between the regional details of the changes in precipitation, especially in the tropics. All models produce increased precipitation in high latitudes, particularly in winter. The increase in water vapour produces an increase in the transport of moisture into high latitudes, giving enhanced moisture convergence and precipitation. Most models simulate areas of decreased precipitation outside the intertropical convergence zone (ITCZ) and the subtropics, but there is little spatial correlation from model to model. In general, precipitation in the ITCZ is enhanced.

Changes in the variability as well as the mean climate are important in the assessment of the impacts of climatic change. Wilson and Mitchell (1987b) investigated changes in the variability over Western Europe in a

4xCO$_2$ sensitivity experiment. They found, for example, statistically significant reductions in the frequency and intensity of summer precipitation over Southern Italy (Figure 7).

There is a growing concern that increases in trace gases will lead to drier conditions in the northern and mid-latitude continents in summer. Even if mean summer precipitation increases (Figure 5) the soil may still become drier, as found by WM. Three out of five recent studies found a drying in mid-latitudes (Fig 8), whereas one model (Washington and Meehl, 1984) produced a wetter land surface on enhancing CO$_2$.

Mitchell and Warrilow (1987) have investigated the importance of land surface parametrizations on determining the simulated hydrological response to increased CO$_2$. In the standard formulation, snowmelt is used to augment soil moisture, or is run off if the ground is saturated. In the control simulation, snowmelt saturates the ground in late spring and the surface then dries rapidly through the summer (Figure 9a). On enhancing CO$_2$, precipitation increases in winter, raising soil moisture levels, but snowmelt and the associated summer drying occur earlier but from the same (saturated) level as in the control. The experiment was repeated, changing only the treatment of snowmelt such that all snowmelt is lost as run-off over frozen ground. The loss of snowmelt leads to a much drier control run (Figure 9b) and much less pronounced drying during summer. On enhancing CO$_2$, the enhanced precipitation in winter now enters in the soil, and though the summer drying starts earlier, it starts from a much higher level, and so the soil remains wetter for most of the summer. The model used by Washington and Meehl (1984) produces an excessively dry surface in the control simulation, apparently as a result of using too low a land surface albedo, and so behaves in a similar manner to the second version of the model described above.

Equilibrium simulations to date using models with a full dynamical representation of the ocean have used idealized geography and annual mean insolation (for example, Spelman and Manabe (1984); Manabe and Bryan (1985)). These studies suggest that the inclusion of the deep ocean does not substantially alter the equilibrium response. A faithful representation of the ocean is however essential if one is to model the time dependent response of climate to increases in greenhouse gases. This is considered in the following section.

6. MODELLING THE TRANSIENT RESPONSE TO INCREASES IN TRACE GASES

Although the concentration of "greenhouse" gases has and will continue to increase gradually, the temperature response of the atmosphere is likely to be slowed down by the large thermal inertia of the ocean. If the oceans are assumed to have a constant heat capacity C$_*$ (equation 4), then the evolution of the temperature change ΔT due to an instantaneous increase in heating ΔQ at t=0 is given by

$$\Delta T = \frac{\Delta Q}{\lambda} (1 - \exp(-\lambda t/C_*)) \qquad (7)$$

The time τ taken to reach (1-1/e) of the equilibrium response is C_*/λ. Assuming $\lambda = 1$ Wm^{-2} K^{-1} and that the warming is mixed uniformly through the deep ocean, τ is about 500 years. On the other hand, if the warming is confined to the observed seasonal ocean mixed layer which has a depth of order 100 m, then τ is of order 10 years. These two estimates are probably upper and lower limits respectively, and the actual timescale for the oceans depends on the efficiency with which heat is mixed into the deep ocean and so varies between different parts of the ocean.

Rough estimates of the effect of oceanic inertia have been made using one-dimensional diffusive models of the ocean. The evolution of global mean temperatures assuming a doubling of effective carbon dioxide concentrations between 1850 and 2050 and vertical diffusion coefficients of 2.25 m^2 sec^{-1} are shown in Figure 10. Two cases are shown corresponding to an equilibrium increase of 4 K and 1.1 K due to doubling CO_2. For each case, the solid curve denotes the equilibrium response, and the dashed curves, the response of the diffusive ocean.

The diffusive model gives a very crude representation of vertical mixing in the ocean, and at equilibrium produces a temperature anomaly which is uniform with depth. Transient experiments have been carried out using coupled ocean atmosphere models in which carbon dioxide concentrations have been instantaneously doubled. The degree to which the warming penetrates the ocean varies with latitude, as can be seen in Figure 10 which shows the warming of an annually averaged ocean-atmosphere model with idealized geography 25 years after quadrupling carbon dioxide concentrations. Also shown in Figure 10 is the fraction of the equilibrium response achieved after 25 years, and the equilibrium response itself. Estimates of the response time τ derived from simple models at GCMs range from 10 to 100 years (Schlesinger, 1986) and possible reasons for the wide range of estimates have been investigated using simple diffusion models (Wigley and Schlesinger, 1985).

A numerical study carried out by Bryan et al (1988) indicates that the response in high latitudes of the southern hemisphere may be substantially slower than elsewhere because of the larger fractional ocean coverage and the existence of a meridional cell upwelling unmodified water from great depth. Hansen et al (1988) have estimated the time dependent response from 1958 to 2020, using a high, medium and low scenario for future increases in trace gases. In their model, the ocean is represented by a seasonally varying mixed layer. The mixed layer depth and advection of heat are assumed to remain unchanged as the concentration of trace gases is increased, and heat transfer into the deep ocean is represented by simple diffusion, the coefficient of diffusion being specified geographically on the basis of observations. As a result of their study, Hansen et al (1988) claim that the climatic effects should be clearly discernible in the 1990s. In contrast to Bryan et al (1988), they find that high southern latitudes is one of the regions where the warming should be detected earliest.

7.   Uncertainties in the simulation and detection of the climatic effect of increases in trace gases.

    (a)   There are uncertainties in the pre-industrial concentration of carbon dioxide, and hence in the change in radiative forcing since then.
    (b)   The size of the equilibrium response to a given increase in gases is known only within a factor of 2 or so (Table 3), due mainly to uncertainties in the modelling of cloud and cloud radiative properties. These problems are currently an area of intensive research. The nature of the regional response, particularly in the

hydrological cycle, varies from model to model. These discrepancies
may be reduced by using higher horizontal resolution and improved
physical paramatrizations.

(c) There is much to learn concerning the vertical mixing of heat in
the oceans and its representation in numerical models. It is not
known whether or not one has to resolve oceanic eddies in order to
produce a reliable simulation of the changes in oceanic heat
transport. Current global ocean GCMs have a horizontal resolution of
300-500 km. These require large horizontal diffusion coefficients
which maintain numerical stability but also smooth out the more
intense ocean currents.

(d) The climatic effects of enhanced trace gas concentrations have to
be detected against the background of the natural variability of the
atmosphere and ocean. The instrumental temperature record exhibits
variability on timescales up to several decades. Some indication of
the inherent variability of climate on the decadal timescales may be
obtained from long (of order 100 years) simulations of coupled ocean
atmosphere models Hence there remains some uncertainty in the
observed warming.

(e) There remains the possibility that other factors are masking (or
enhancing) the effects of increases in "greenhouse" gases. There has
been much speculation concerning the possibility of changes in the
solar "constant". Accurate measurements of the solar constant have
only been available for the last decade or so, and these indicate a
decrease of 0.1% between 1980 and 1985 (Willson et al, 1986). (The
climatic effect of a 2% increase in solar constant is thought to be
approximately equivalent to that due to a doubling of carbon dioxide
concentrations, Wetherald and Manabe, 1975.) Variations in the
concentration and composition of stratospheric aerosol resulting from
volcanic eruptions have also been cited as a cause of climatic change,
but the indices used to categorize volcanic dust do not allow a
quantitative estimate of the associated radiative effects.

8. APPEALS TO THE PAST; A SIMULATION FOR 9000 BEFORE PRESENT (9 Kbp)

In order to try and limit some of the uncertainties discussed in the
previous section, attempts have been made to model past climates and
compare the simulated changes with the changes from the present deduced
from palaeoclimatic data. Models have also been used to help understand
past climates (for example Kutzbach and Guetter, 1986). The two periods
chosen most frequently are the last glacial maximum (18 Kbp) and part of
the Holocene around 9 Kbp. During the latter period, perihelion almost
coincided with the summer as opposed to the winter solstice, enhancing
insolation during the northern summer, especially in high northern
latitudes. Here, a brief summary of some recent simulations for 9 Kbp
(Mitchell et al, 1988) will be given. This period is of interest because

(a) The changes in boundary conditions are simple; the changes in
orbital parameters are well known, CO_2 concentrations are believed to
have been similar to present, and the ice sheet over North America was
relatively small in extent.

(b) There is a large amount of palaeoclimatic data available from
that period.

In the initial experiment, only the orbital parameters were altered. The enhanced insolation during the boreal summer produces an increase in surface temperature over the northern mid-latitude continents (Figure 12a). This enhances the land-sea contrast, lowering surface pressures over the continents (Figure 12b) and enhancing the monsoon precipitation over Venezuela, East Africa and Southern Asia (Figure 12c). The land surface becomes drier in mid-latitudes (Figure 13a) and wetter over most of the tropics. This is consistent with estimates of lake levels for the period (Figure 13b) which are lower than average in mid-latitudes and above average in the tropics. Although the broad pattern of simulated changes is in agreement with palaeoclimatic data, there are discrepancies, as over West Africa, where the model produces a drier surface at 9 Kbp, whereas the lake level data suggest that it was wetter then. This shortcoming has been the subject of further investigation.

9. CONCLUDING REMARKS
General circulation models have been used to investigate the climatic effects of a wide variety of other phenomena including sea surface temperature and sea-ice anomalies, changes in land surface characteristics due to desertification or deforestation, "nuclear winter" and waste heat from energy parks. Experiments have also been carried out to determine the influence of topography, land-sea contrasts and so forth on the general circulation of the atmosphere.

Although the use of climate models has become more frequent, each experiment should still be analysed carefully before one attempts to interpret the results. In doing so, one must bear in mind the limitations of the model and how such limitations will distort the simulated response. One can attempt to do this by isolating the mechanisms producing the changes in climate in the model, and assessing the physical realism of each step. Where there is some uncertainty concerning a parametrization on which the model's response is dependent, the experiment can be repeated using alternative forms of that parametrization to determine whether or not the uncertainties affect the results. One also has to consider whether or not all the physical processes relevant to the experiment have been included in the model. In short, one should always treat results from numerical models with a certain amount of cynicism until they are supported by a critical assessment of the model used and the simulated response.

10. ACKNOWLEDGEMENTS
The development, running and assessment of climate models is not a task that can be carried out by an individual. I am indebted to members of the Dynamical Climatology Branch of the Meteorological Office over the last

10 years for their help and support. I am also grateful to colleagues at other modelling centres who have provided additional information on their experiments. Peter Rowntree and Alayne Street-Perrott kindly provided Figures 10 and 13(b) respectively.

11. REFERENCES

Augustsson, T. and Ramanathan, V. 1977 A radiative — convective model study of the CO_2 climate problem. J. Atmos. Sci., 34, 448-51.

Bryan, K., Manabe, S. and Spelman, M.J. 1988 Interhemispheric Asymmetry in the transient response of a coupled ocean-atmosphere model to a CO_2 forcing. J. Phys. Oceanogr. 18, 851-867.

Dickinson, R.E. and Cicerone, R.J. 1986 Future global warming from atmospheric trace gases. Nature, 319, 109-115.

Hansen, J., Fung, I., Lacis, A., Rind, D., Lebedeff, S., Reudy, R. and Russell, G. 1988 Global climatic changes as forecast by Goddard Institute for Space Studies Three Dimensional Model. J. Geophys Res. (To appear).

Hansen, J. Lacis, A., Rind, D., Russell, L, Stone, P., Fung, I., Ruedy, R. and Lerner, J. 1984 Climate Sensitivity, Analysis of Feedback Mechanisms. In Climate Processes and Climate Sensitivity (ed J. Hansen and T Takahashi) Geophysical Monograph 29, 130-163. American Geophysical Union, Washington DC.

Kutzbach, J. E. and Guetter, P. J. 1986 The influence of changing orbital parameters and surface boundary conditions on climate simulations for the past 18,000 years. J. Atmos Sci. 43, 1726-1759.

Lal, M. and Ramanathan, V. 1984 The effects of moist convection and water vapour radiative processes on climate sensitivity. J. Atmos. Sci. 41, 2238-2249.

Manabe, S. and Bryan, K.Jnr. 1985 CO_2-Induced Change in a Coupled Ocean-Atmosphere Model and the Paleoclimatic Implications. J. Geophys. Res. 90, C11, 11,689-11,707.

Manabe, S. and Stouffer, R.J. 1980 Sensitivity of a global climate model to an increase in the CO_2 concentration in the atmosphere. J. Geophys. Res. 85, 5529-5554.

Mitchell, J.F.B. and Warrilow, D.A. 1987 Summer dryness in northern mid-latitudes due to increased CO_2. Nature 330, 238-240.

Mitchell, J.F.B., Grahame, N.S. and Needham, K.J. 1988 Climate simulations for 9000 years before present: Seasonal variations and the effect of the Laurentide ice sheet. J. Geophys Res. 93, 8283-8303.

Mitchell, J.F.B. 1989 The "greenhouse" effect and climate change. Submitted to Revs. of Geophysics.

Ramanathan, V., Cicerone, R.J., Singh, H.B. and Kiehl, J.T. 1985 Trace Gas and Trends and their potential role in climate change. J. Geophys. Res. 90, 5547-5566. J. Geophys. Res.

Roeckner, E., Schleses, U., Biercamp. J. and Loewe, P. 1987 Cloud optical depth feedbacks and climate modelling. Nature, 329, 138-139.

Schlesinger, M.E. 1985 Feedback analysis of results from energy balance and radiative-convective models. In "Projecting the climatic effects of increasing atmospheric carbon dioxide", edited by M.C. MacCracken and F.M. Luther, pp 280-319, US Department of Energy, Washington DC, 1985. (Available as NTIS, DOE (ER-2037 from Natl. Tech Inf.Serv., Springfield Va.)

Schlesinger, M.E. 1986 Equilibrium and transient climatic warming induced by increased atmospheric CO_2. Climate Dynamics 1, 35-51.

Schlesinger, M.E. and Mitchell, J.F.B. 1987 Climate model simulations of the equilibrium climatic response to increased carbon dioxide. Reviews of Geophysics 25, 760-798.

Schlesinger, M.E. and Zhao, Z. 1987 Seasonal climate changes induced by doubled CO_2 as simulated by the OSU atmospheric GCM/mixed layer model Oregon State University Climatic Institute Report 70, 73pp.

Schneider, S.H. 1975 On the Carbon Dioxide Climate Confusion. J.Atmc 32, 2060-2066.

Somerville, R.C.J. and Remer, L.A. 1984 Cloud Optical Thickness Feedbc the CO_2 Climate problem. J. Geophys. Res. 89, 9668-9672.

Spelman, M.J. and Manabe, S. 1984 Influence of Oceanic Heat Transport upon the sensitivity of a model climate. J. Geophys. Res. 89, 571-586.

Washington, W.M. and Meehl, G.A. 1984 Seasonal Cycle Experiment on the Climate Sensitivity Due to a Doubling of CO_2 with an Atmospheric General Circulation Model Coupled to a Simple Mixed Layer Ocean Model. J. Geophys. Res. 89, 9475-9503.

Wetherald, R.T. and Manabe, S. 1975 The effects of changing of the solar constant on the climate of a general circulation model. J. Atmos. Sci. 32, 2044-2059.

Wetherald, R.T. and Manabe, S. 1986 An investigation of cloud cover change in response to thermal forcing" Climate Change. 8, 5-24.

Wigley, T.M.L. and Schlesinger, M.E. 1985 Analytical solution for the effect of increasing CO_2 on global mean temperature. Nature 315, 649-652.

Willson, R.L., Hudson, H.S., Frohlich, C. and Brusa, R.W. 1986 Long term downward trend in solar radiance. Science, 234, 1114-1117.

Wilson, C.A. and Mitchell, J.F.B. 1987a A Doubled CO_2 Climate Sensitivity experiment with a GCM including a simple ocean. J. Geophys. Res, 92, 13315-13343.

Wilson, C.A. and Mitchell, J.F.B. 1987b Simulated climate and CO_2 induced climate change over Western Europe. Climatic Change, 10, 11-42.

12. BIBLIOGRAPHY

Physically-based modelling and simulation of climate and climate change
    Vol 2 (1988) Schlesinger, M.E. (Ed) NATO Advanced Study Institute Series (Reidel, in Press). This volume covers applications of climate models as opposed to their development.

The "Greenhouse" Effect, Climatic Change and Ecosystems: 1987
    Bolin, B., Doos, B.R., Jager, J. and Warrick, R.A. (Eds). SCOPE 29, Wiley, New York. Probably the best book to date covering the "greenhouse" effect and related topics.

Climate Model Simulations of the Equilibrium Response to Increased Carbon Dioxide: 1987
    Schlesinger, M.E. and Mitchell, J.F.B., Rev. of Geophysics, 25, 760-798. A detailed review of CO_2 simulations up to 1986.

Issues in Atmospheric and Oceanic Modelling, Part A. Climate Dynamics: 1985
    Manabe, S. (Ed). Advances in Geophysics Vol. 28, Academic Press (London). Chapters by Dickinson on Climatic sensitivity, Manabe and Wetherald on CO_2 and hydrology and Kutzbach on modeling of palaeoclimatics are of particular interest.

Palaeoclimatic Analysis and Modeling: 1985 Hecht, A.D. (Ed) John Wiley and Sons (New York). An excellent book describing both methods on analysis of palaeoclimatic data and some numerical experiments.

Table 1   Current concentrations and "greenhouse heating" due to trace gases. The main absorption bands and band strengths (a measure of the probability of a molecule absorbing a photon at the band wavelength) are shown for the less abundant gases.

Gas	Concentration (ppm)	Principal absorption bands		Greenhouse heating Wm^{-2}
		Position (cm^{-1})	Strength $cm^{-1}(atm\ cm)^{-1}$ STP	
Water Vapour	~3000			~100
Carbon Dioxide	345	667	(many bands)	~50
Methane	1.7	1306	185	1.7
Nitrous Oxide	0.30	1285	235	1.3
Ozone	$10-100 \times 10^{-3}$	1041	376	1.3
CFC11	0.22×10^{-3}	846 1085	1965 736	0.06
CFC12	0.38×10^{-3}	915 1095 1152	1568 1239 836	0.12

Table 2.   Past and projected greenhouse gas concentrations and associated changes in "greenhouse" heating ΔQ

Gas	Assumed 1860 Concentration	ΔQ 1860 to 1985	Estimated 2035 Concentration (ppm)	Estimated ΔQ 1985 to 2035 (Wm^{-2}) Range
Carbon Dioxide	275.	1.3	475	1.8
Methane	1.1	0.4	2.8	0.5
Nitrous Oxide	0.28	0.05	0.38	0.15
CFC11	0	0.06	1.6×10^{-3}	0.35
CFC12	0	<u>0.12</u>	2.8×10^{-3}	<u>0.69</u>
TOTAL		1.9		3.5

Table 3.   ANALYSIS OF FEEDBACKS AND TEMPERATURE
            CHANGES DUE TO DOUBLING CO_2, BASED ON
            TYPICAL VALUES FROM GCM EXPERIMENTS

FEEDBACK	STRENGTH ($Wm^{-2} K^{-1}$)	TEMP. CHANGE (Cumulative)
NONE	+3.7	1.1
WATER VAPOUR	-1.4	1.7
ICE/SNOW	-0.3	2.1
CLOUD AMOUNT	-0.9	4.0
CLOUD RADIATIVE PROPERTIES	-0.1 to 0.5	2.7 - 4.5

FIGURE CAPTIONS

1. In idealized single layer atmosphere, reflectivity α, longwave
    emissivity ε and temperature T in radiative equilibrium with the
    surface at temperature T_*.

2. The effect on vertical temperature profile of increasing atmospheric
    CO_2 concentrations (schematic). Increasing CO_2 raises the emitting
    level from a to b, and the profile then warms from b to c to restore
    the effective emitting temperature to its original value.

3. The height/latitude distribution of simulated temperature changes due
    to doubling CO_2 concentrations. Areas of decrease are stippled (a)
    Northern winter (December to February) (b) Northern Summer (June to
    August). From Wilson and Mitchell (1987a).

4. Simulated changes in surface temperature due to doubling CO_2
    concentrations. (a) Northern winter (b) Northern Summer.

5. The height/latitude distribution of simulated annual mean changes in
    cloud from three different models. Areas of decrease are stippled, and
    the units are percent of total cover. Top, from Wetherald and Manabe
    (1986); Centre, from Hansen et al (1984); bottom, from Washington and
    Meehl (1984).

6. Simulated changes in precipitation rate (mm/day) due to doubling CO_2
    concentrations. Areas of decrease are stippled (a) Northern winter (b)
    Northern Summer. From Wilson and Mitchell (1987a).

7. Simulated daily precipitation totals (mm) at a grid point over Southern
    Italy. Top panel, with quadripled CO_2 amounts and enhanced sea surface
    temperatures: bottom panel, with present CO_2 amounts and sea surface
    temperatures. From Wilson and Mitchell (1987b).

8.   Simulated changes in soil moisture (cm) due to doubling CO_2 amounts.
     (a) From Manabe and Wetherald (1987) (b) Hansen et al (1984)
     (c) Washington and Meehl (1984) (d) Schlesinger and Zhao (1987)
     (e) Wilson and Mitchell (1987a).

9.   Simulated monthly mean soil moisture averaged over land between 45° and
     60°N for $1xCO_2$ (solid line) and $2xCO_2$ (dashed) line. (a) Allowing
     snowmelt to augment soil moisture (b) Running off snowmelt if the
     ground is frozen. From Mitchell and Warrilow (1987).

10. Estimates of the time dependent response to doubling effective CO_2
     concentrations between 1850 and 2050, made using a one-dimensional
     diffusive ocean model with a vertical diffusion of 2.25 cm^2 sec^{-1}. The
     solid curves represent the instantaneous equilibrium values, the dashed
     curves the actual response allowing for oceanic thermal inertia. The
     upper curves allow for climate feedbacks, the lower curve neglect
     feedbacks. (Rowntree 1988, personal communication).

11. Height (depth)/latitude diagrams of the simulated changes in
     atmospheric (oceanic) temperatures due to quadrupling CO_2 amounts in an
     annually averaged coupled ocean-atmosphere model using idealized
     topography. Upper panel, the changes 25 years after instantaneously
     quadrupling CO_2 amounts. Middle panel, the percentage of the
     equilibrium response achieved after 25 years. Lower panel, the
     equilibrium response. From Spelman and Manabe (1984).

12. Simulated changes during northern summer due to change in the earth's
     orbital parameters to those appropriate to 9 Kbp. (a) Surface
     temperature (K), (b) Sea-level pressure (mb), (c) Precipitation rate
     (mm/day). Areas of decrease are stippled. From Mitchell et al
     (1988).

13. (a) Simulated changes in soil moisture (9 Kbp-present). Contours at 0, ±2, ±5 cm. Areas of decrease are stippled, and reductions greater than 2 cm are stippled heavily. From Mitchell et al (1988).
(b) Changes in lake level status between 9 Kbp and present. (Street-Perrott, 1988, personal communication.)

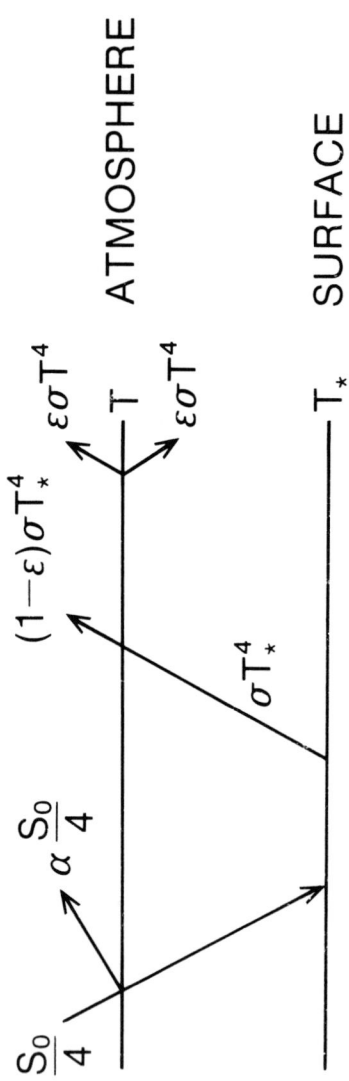

Figure 1

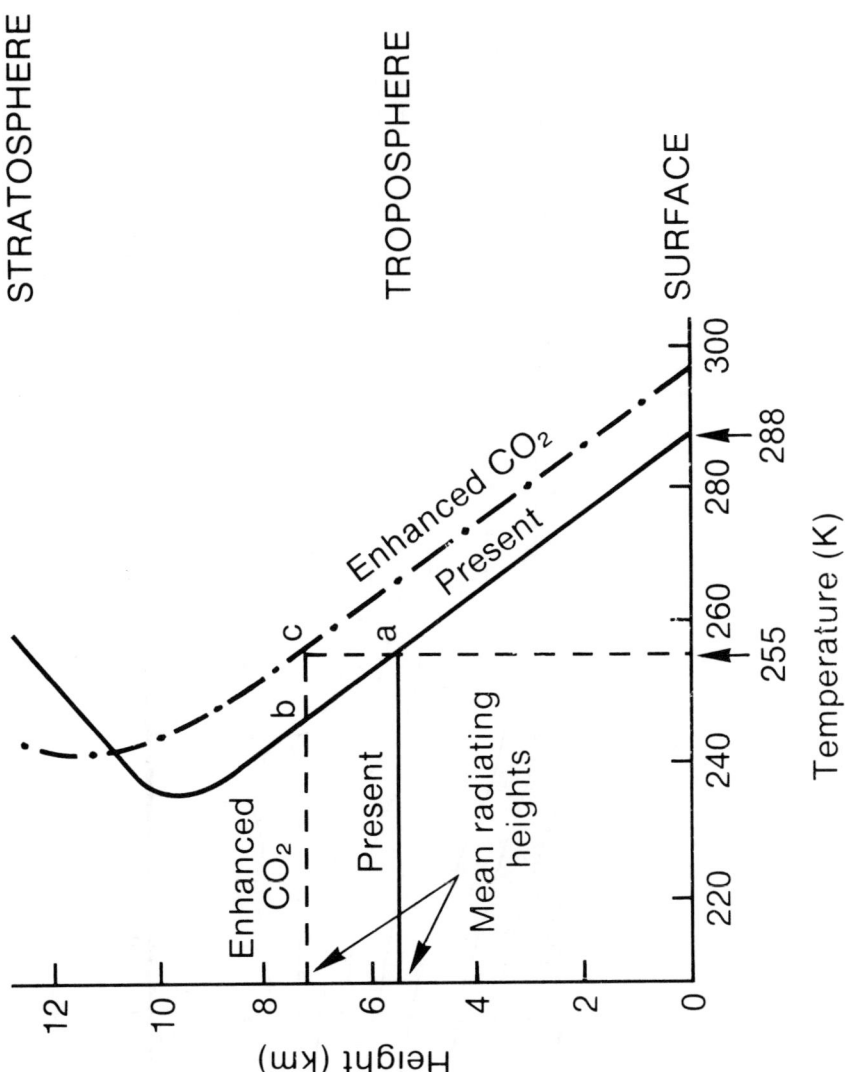

Figure 2

Figure 3a

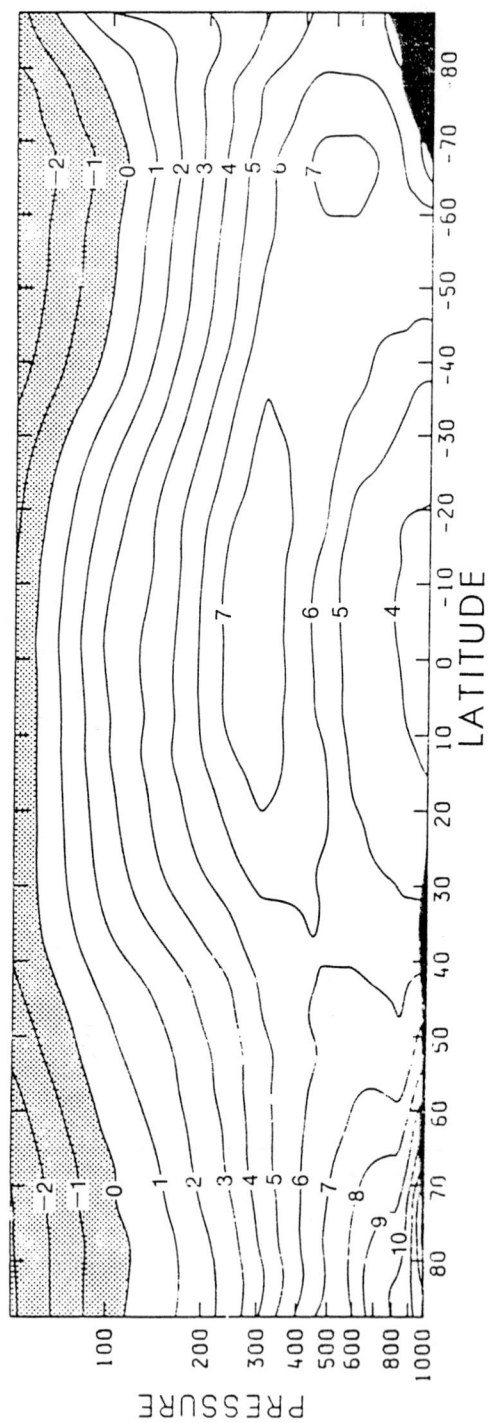

Figure 3b

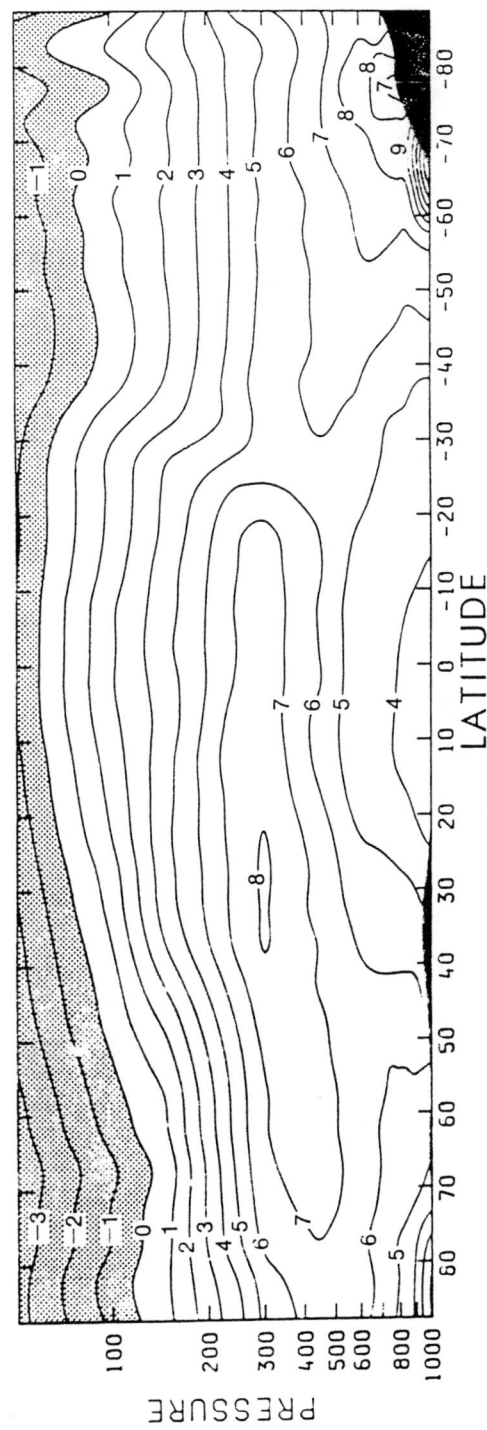

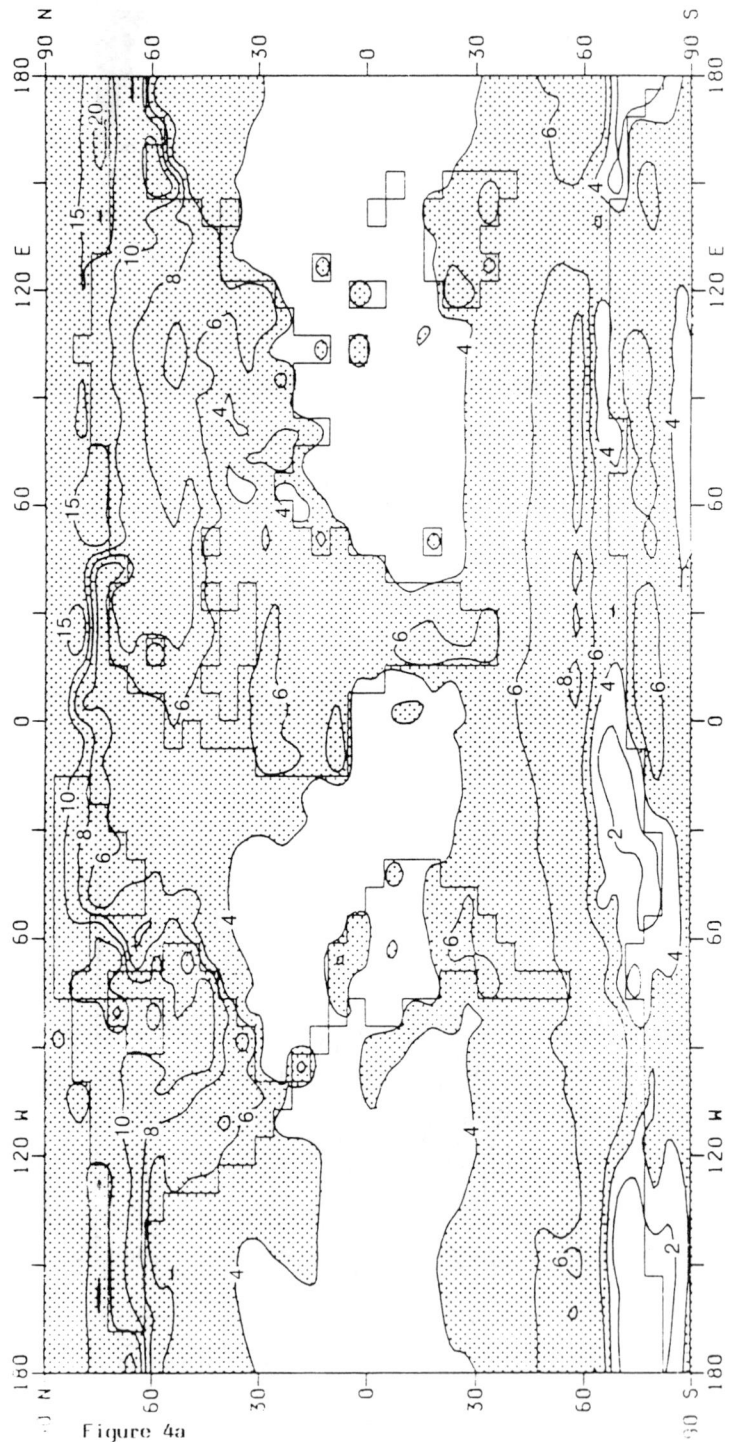

Figure 4a

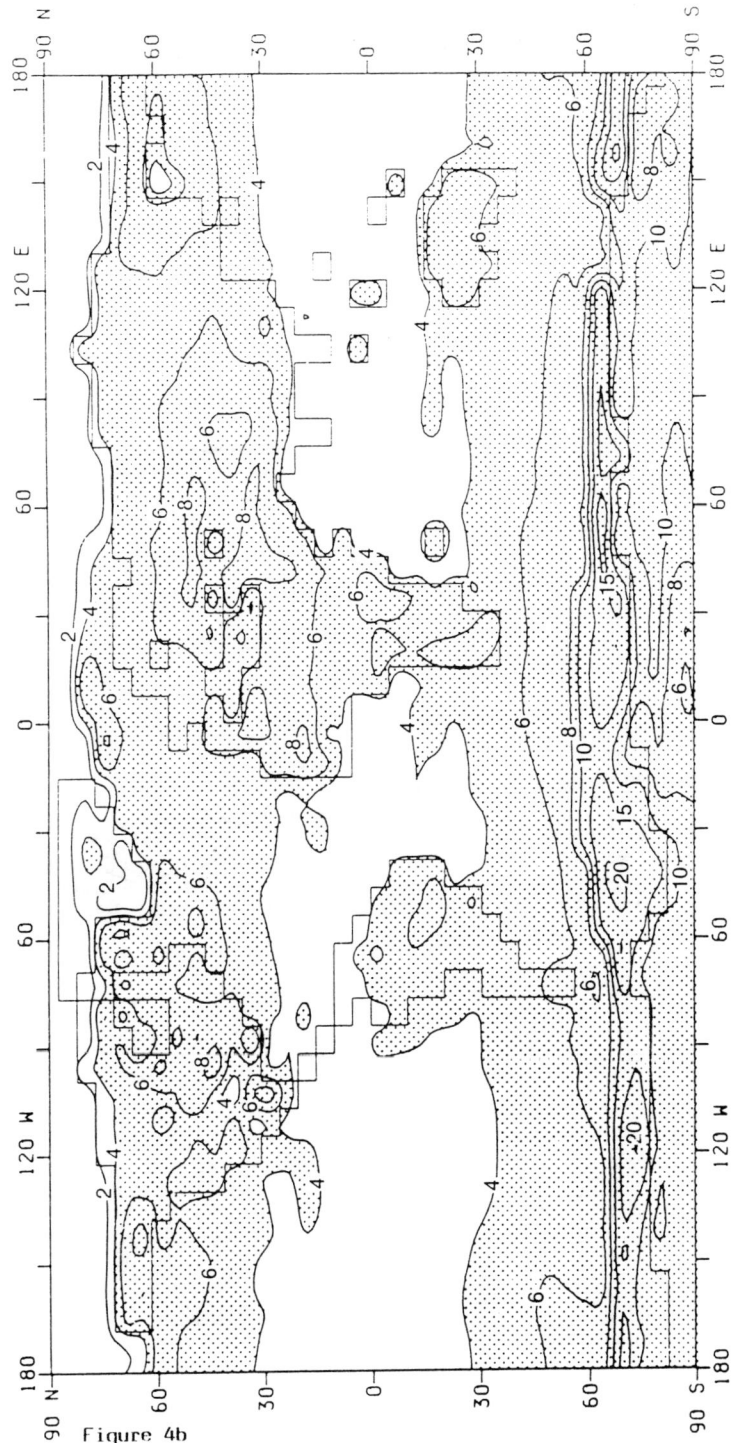

Figure 4b

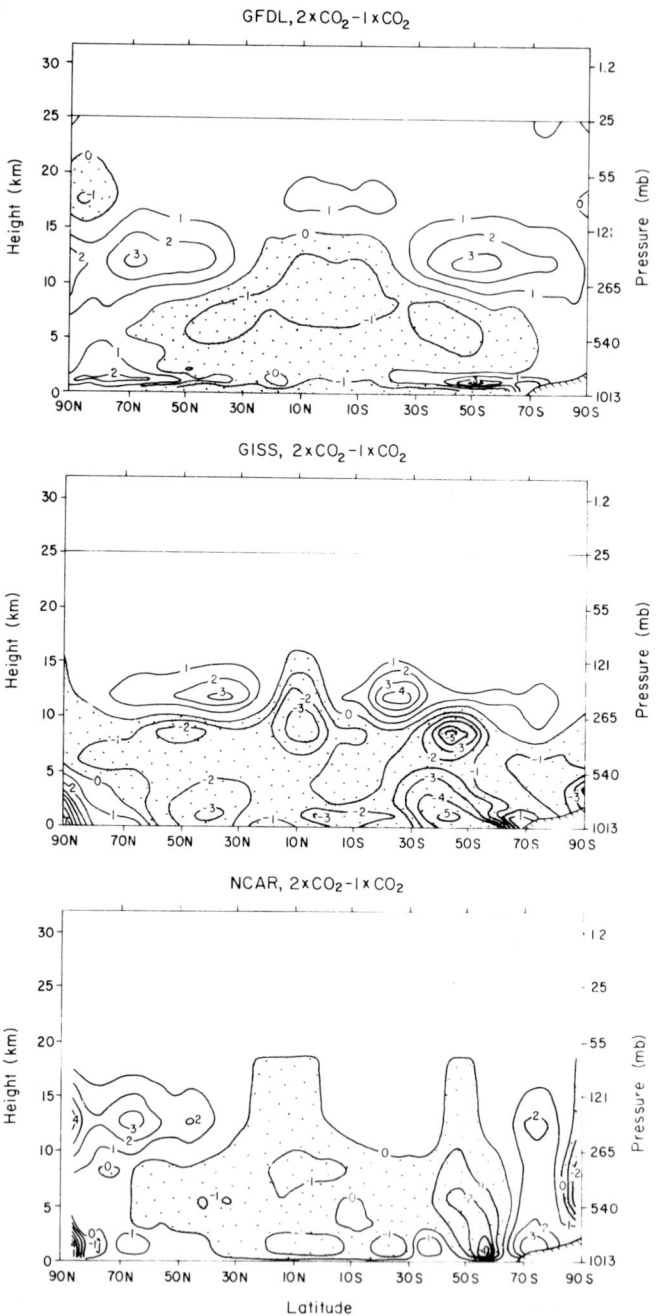

Figure 5
ZONAL MEAN CLOUDINESS DIFFERENCES FOR YEAR

GFDL, $2 \times CO_2 - 1 \times CO_2$

GISS, $2 \times CO_2 - 1 \times CO_2$

NCAR, $2 \times CO_2 - 1 \times CO_2$

Figure 6a

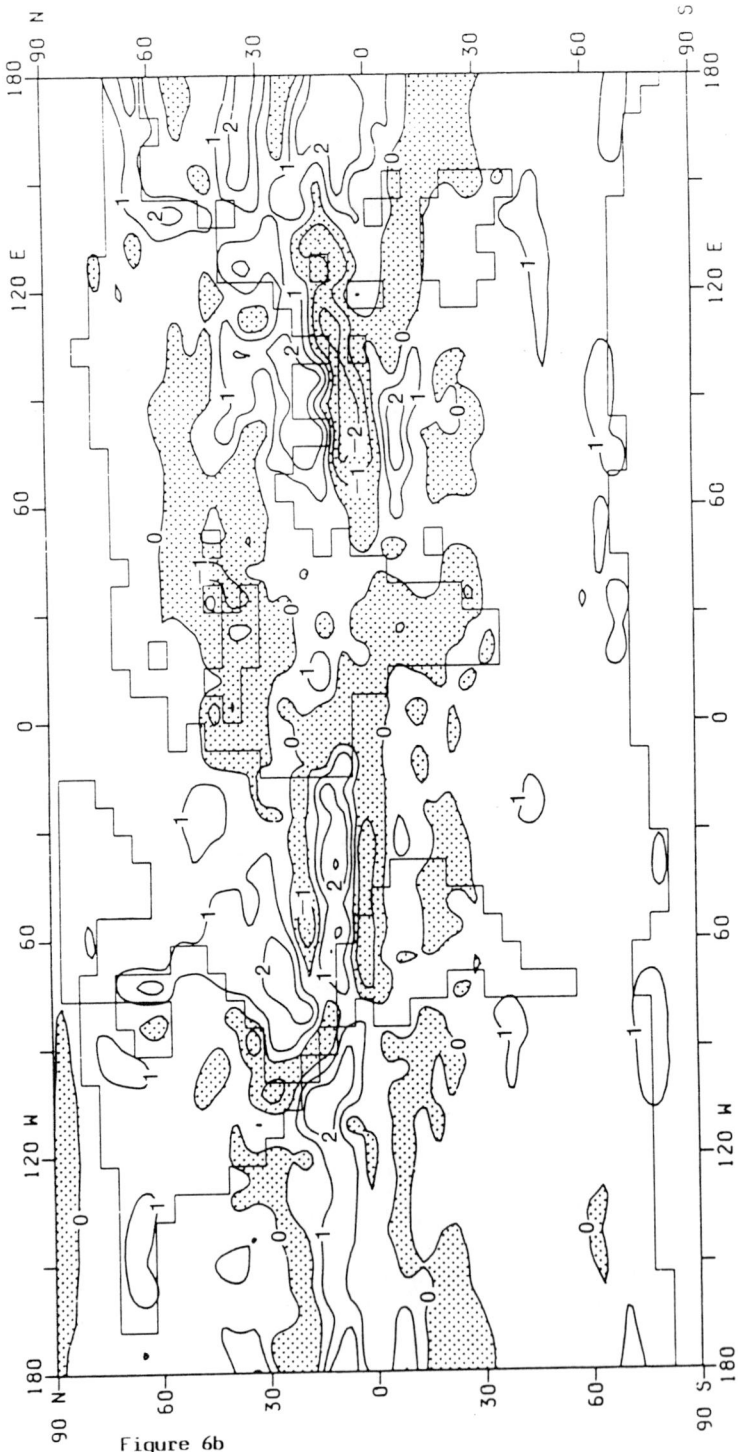

Figure 6b

Figure 7
PRECIPITATION SOUTHERN ITALY FOR CONTROL (TOP)
AND 4 X CO_2 (BOTTOM)

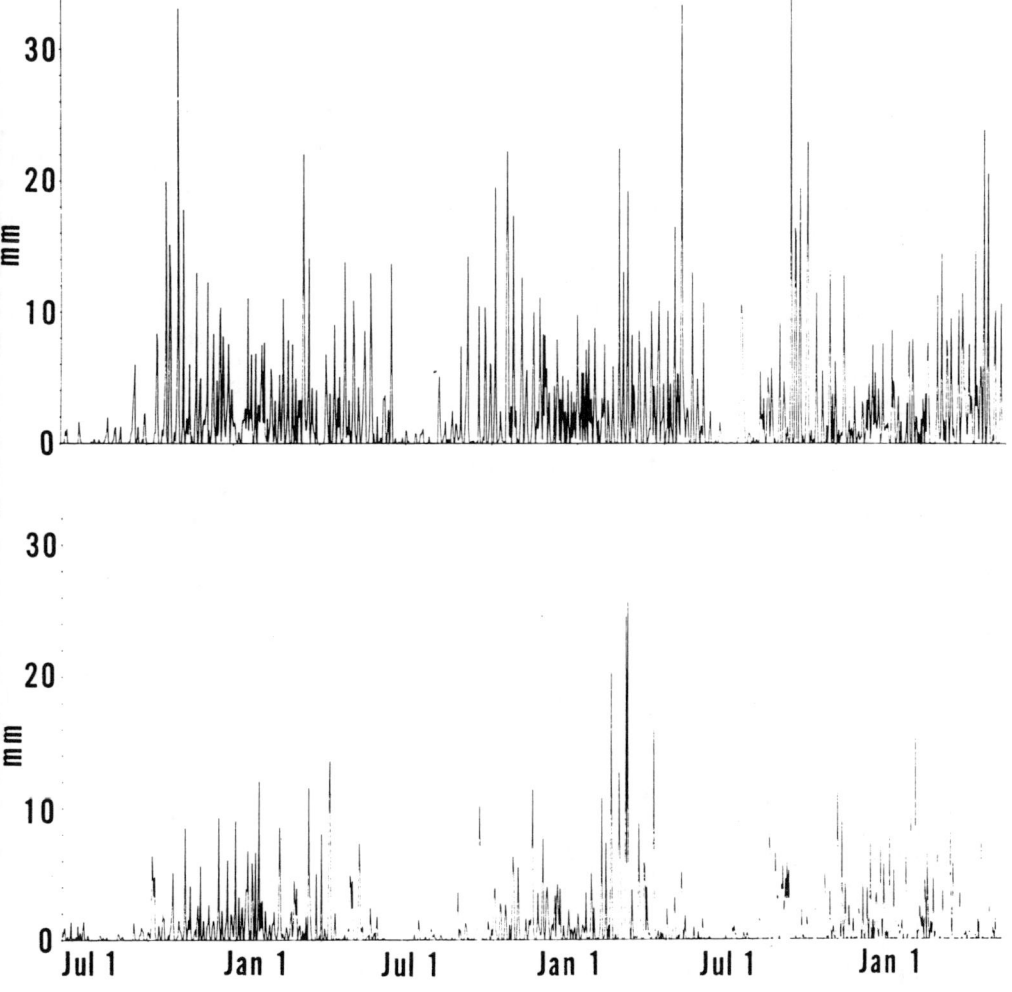

SOIL WATER DIFFERENCES FOR JJA

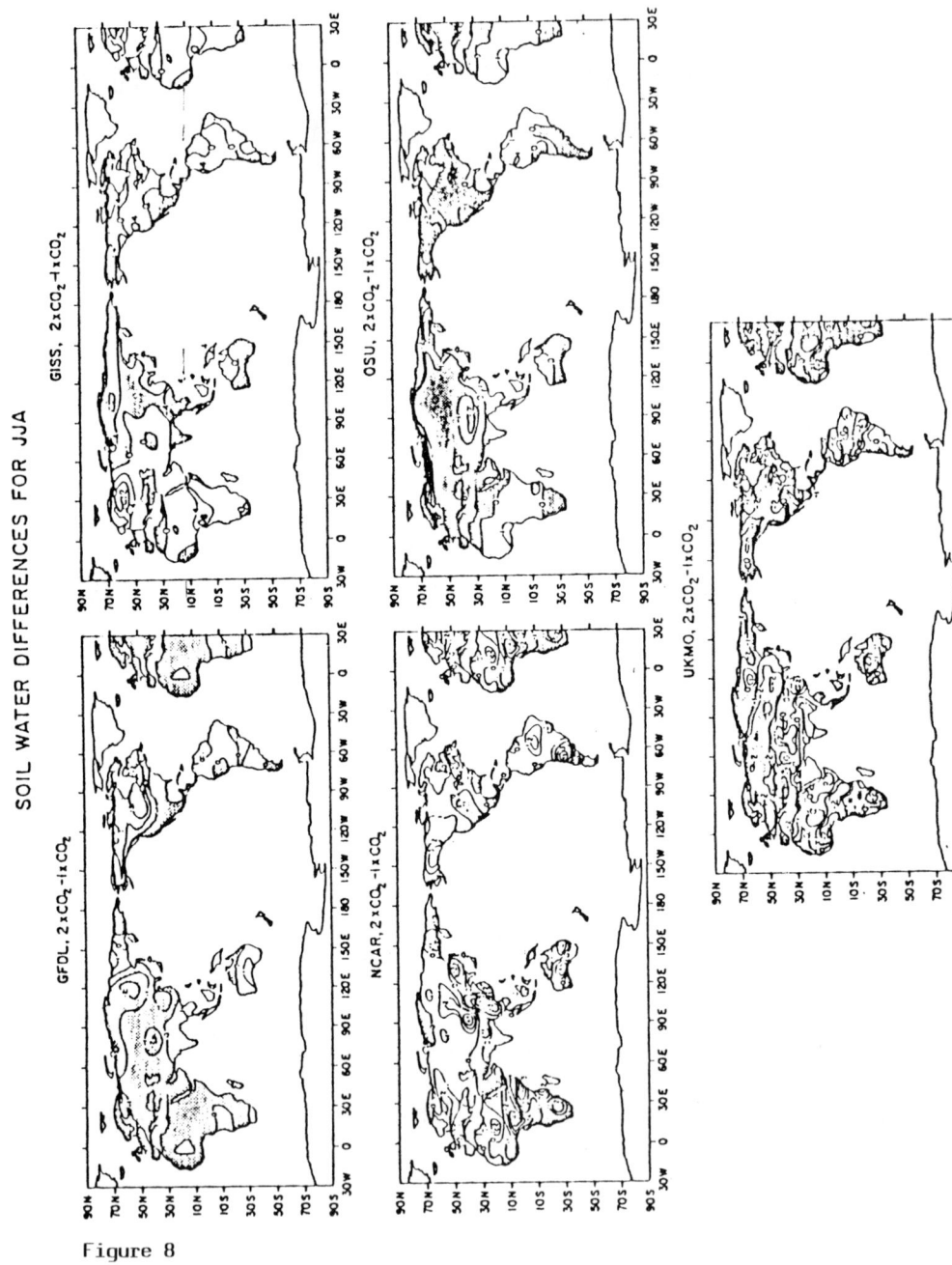

Figure 8

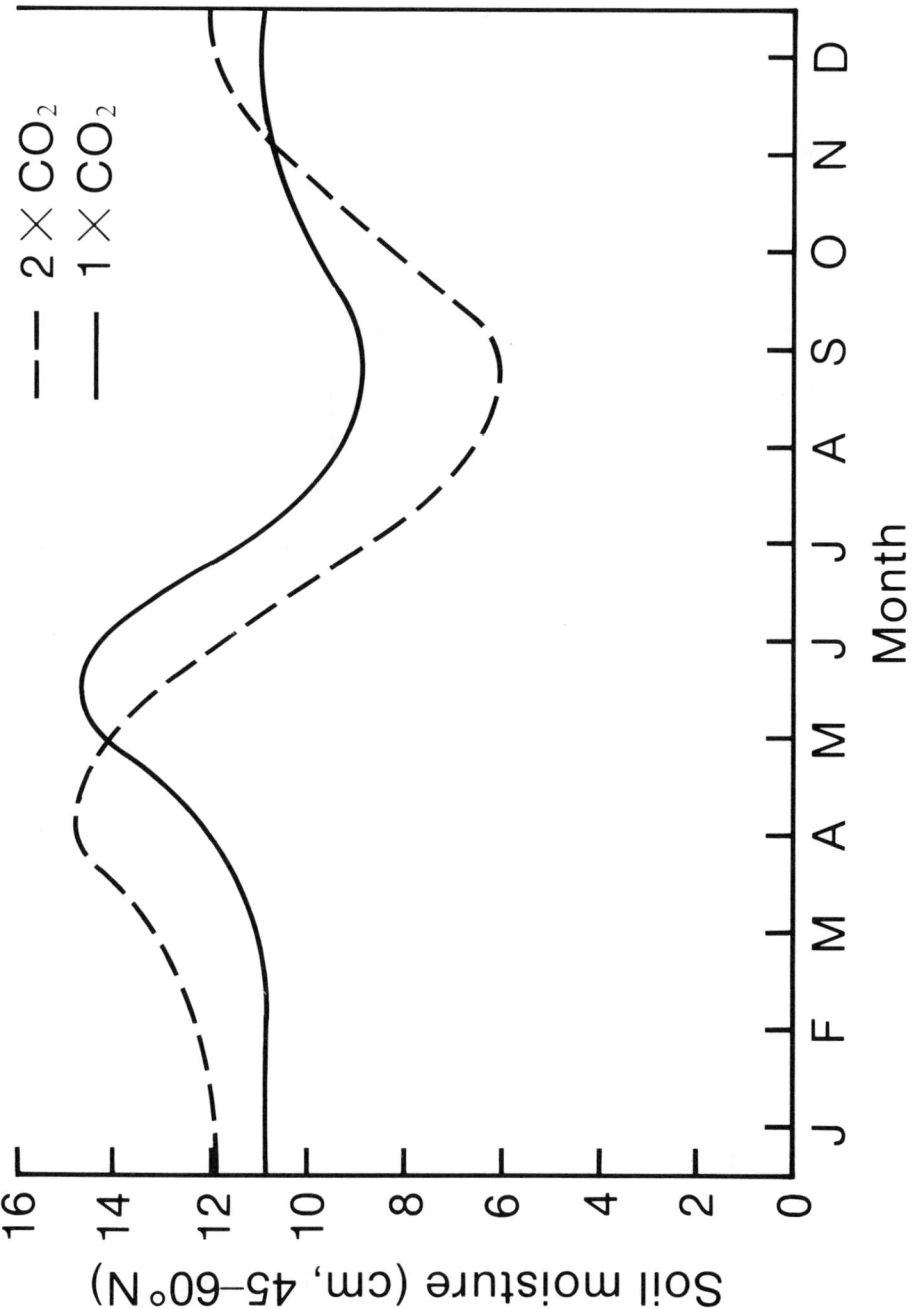

Figure 9a

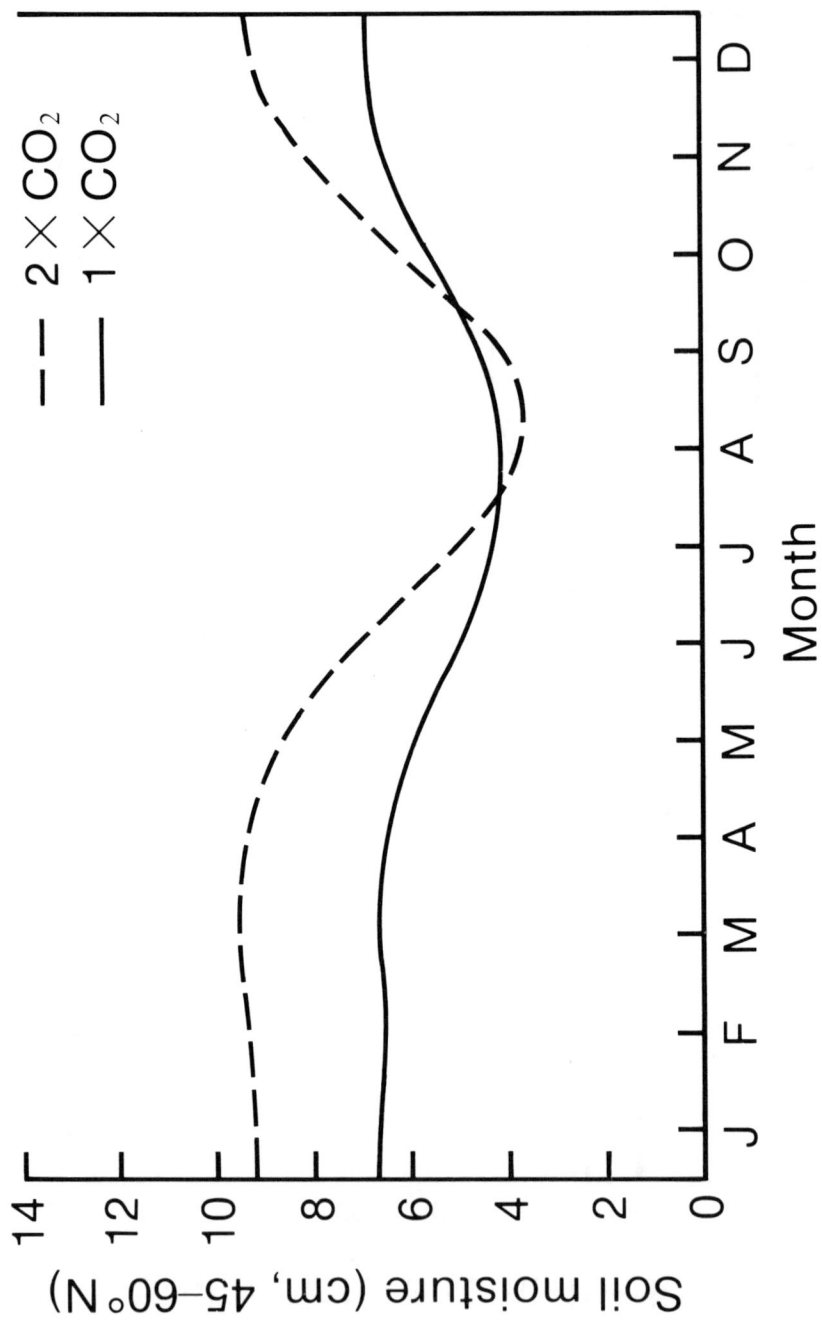

Figure 9b

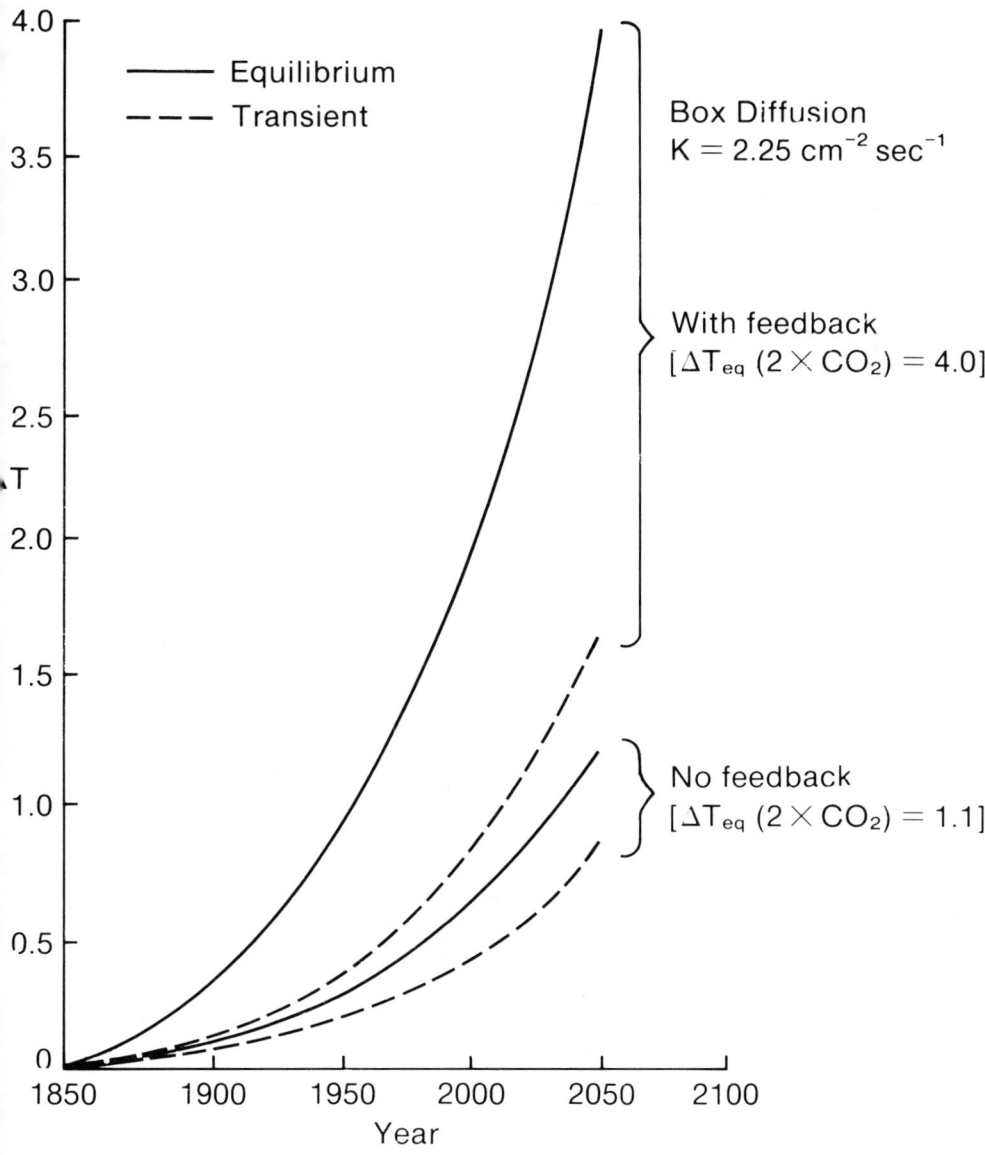

Figure 10

Figure 11a

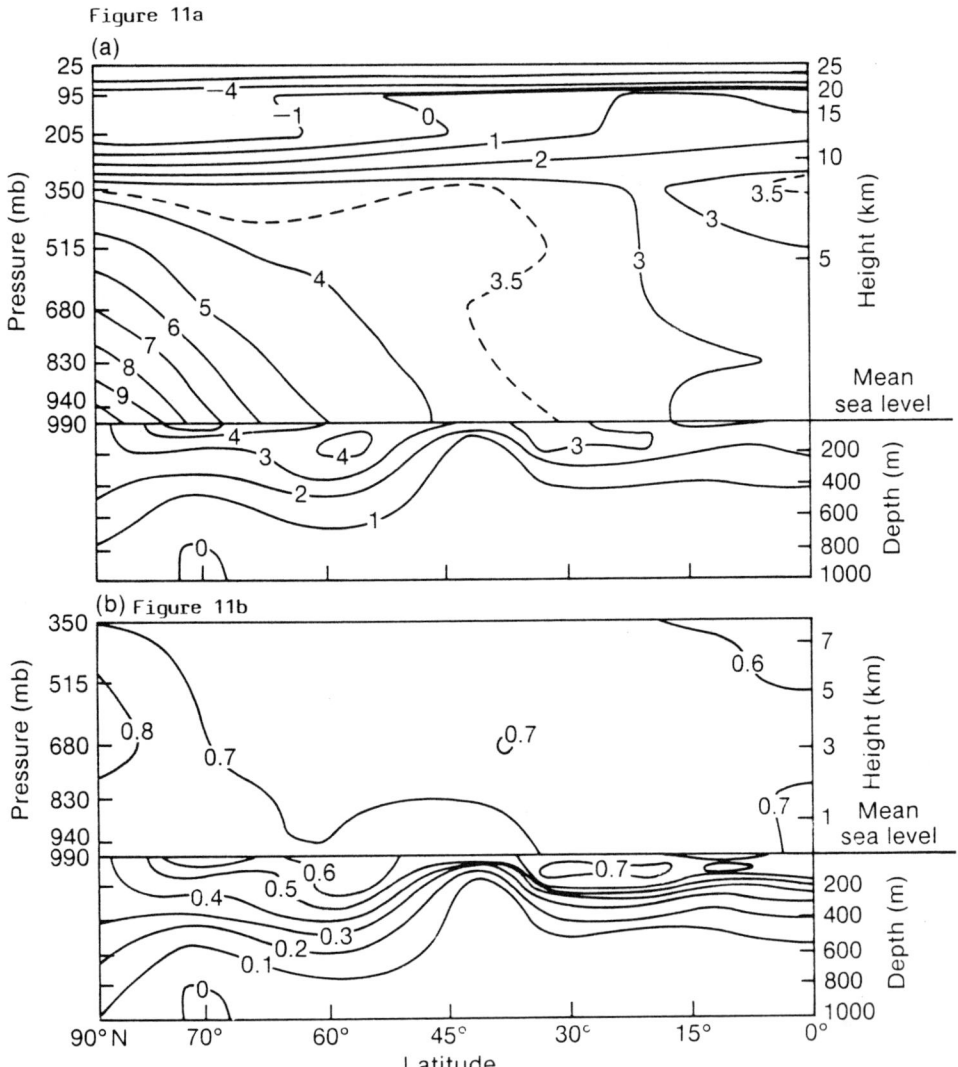

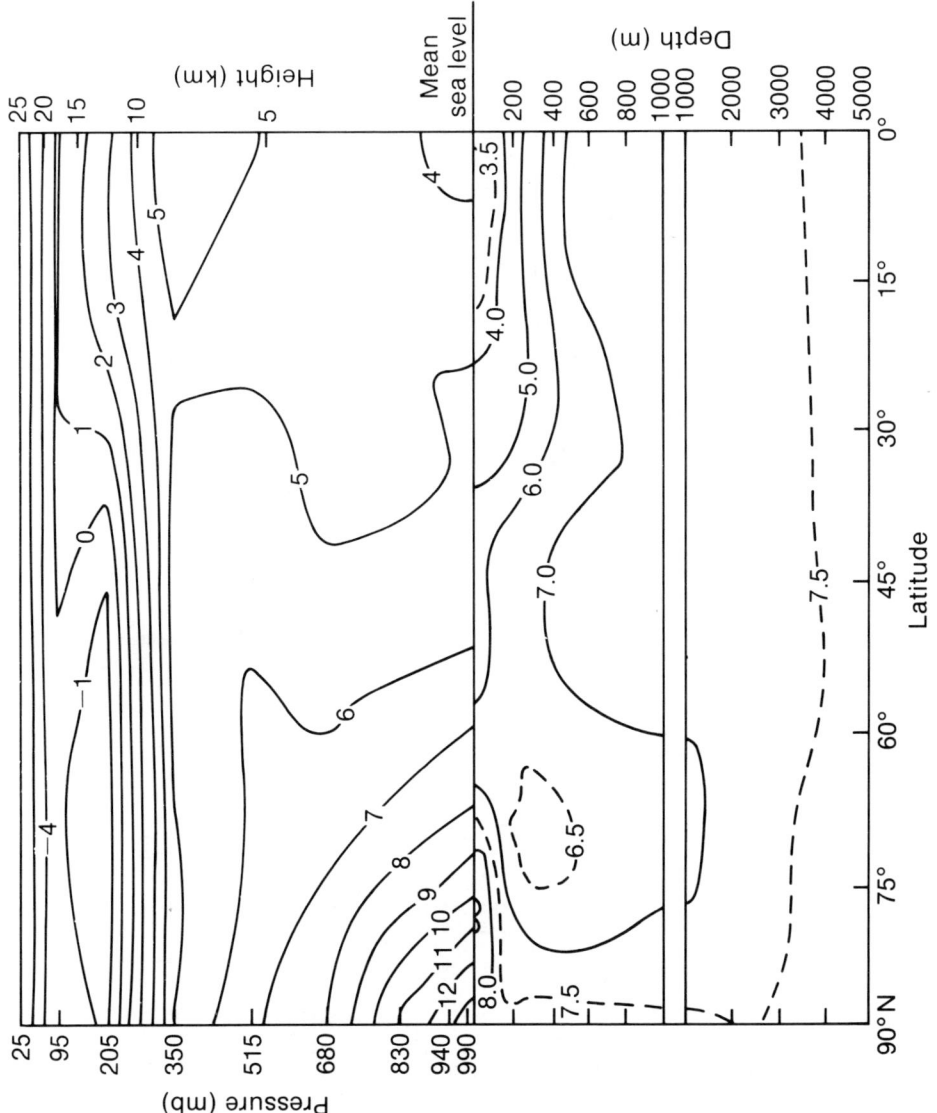

Figure 11c

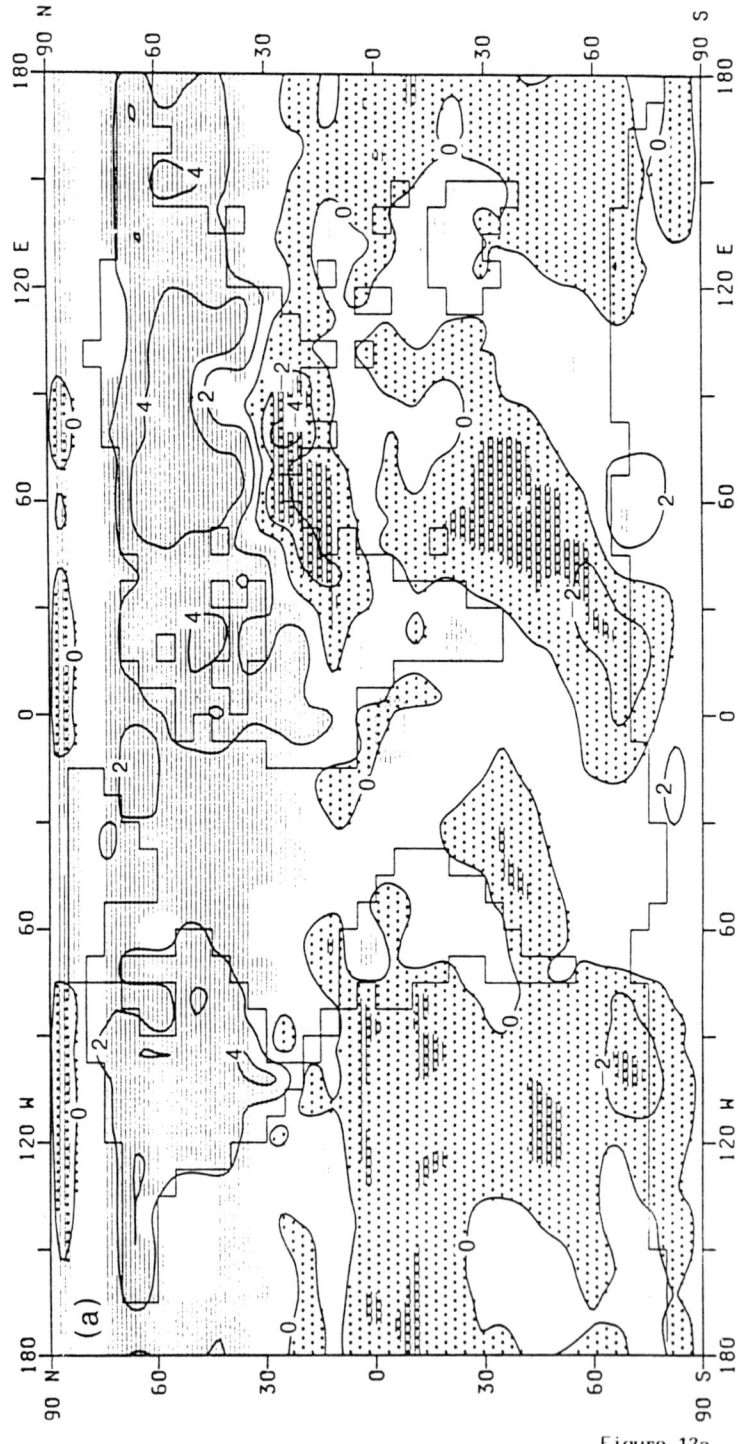

Figure 12a

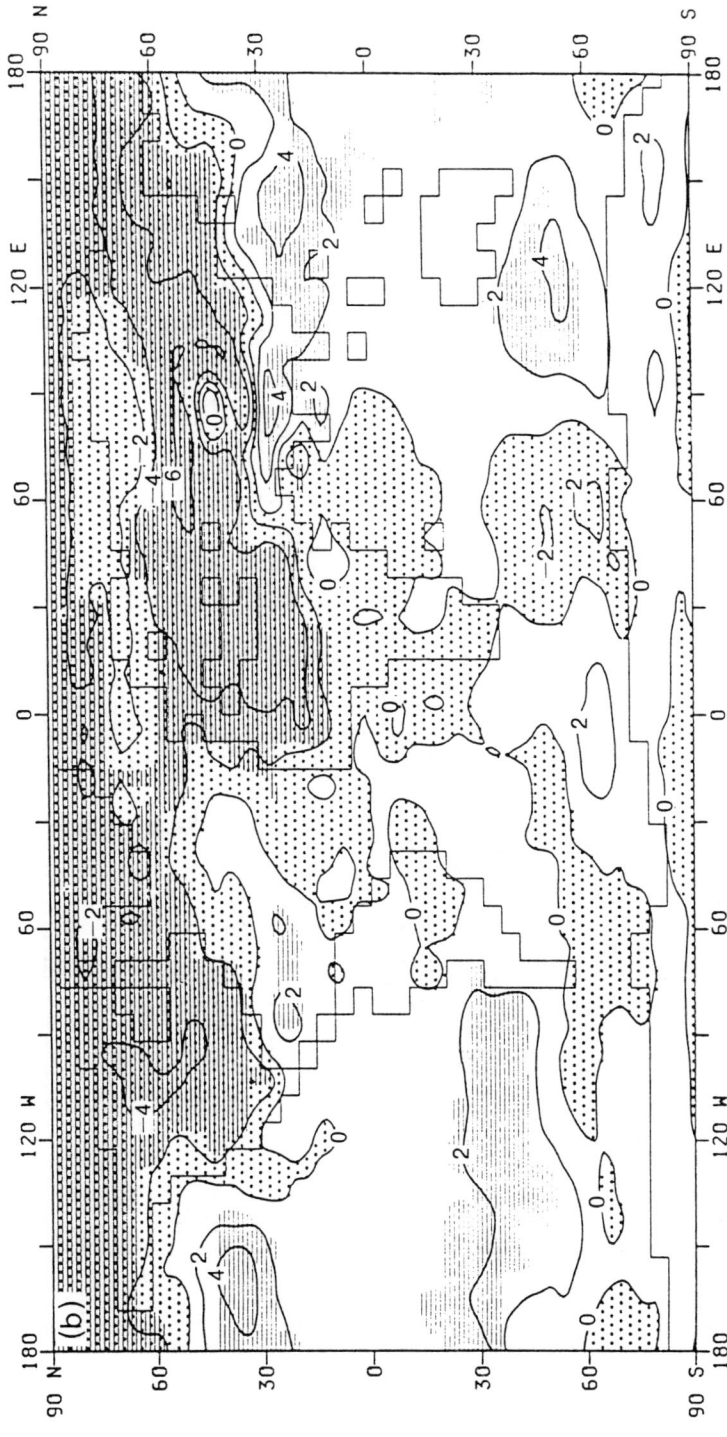

Figure 12b

193

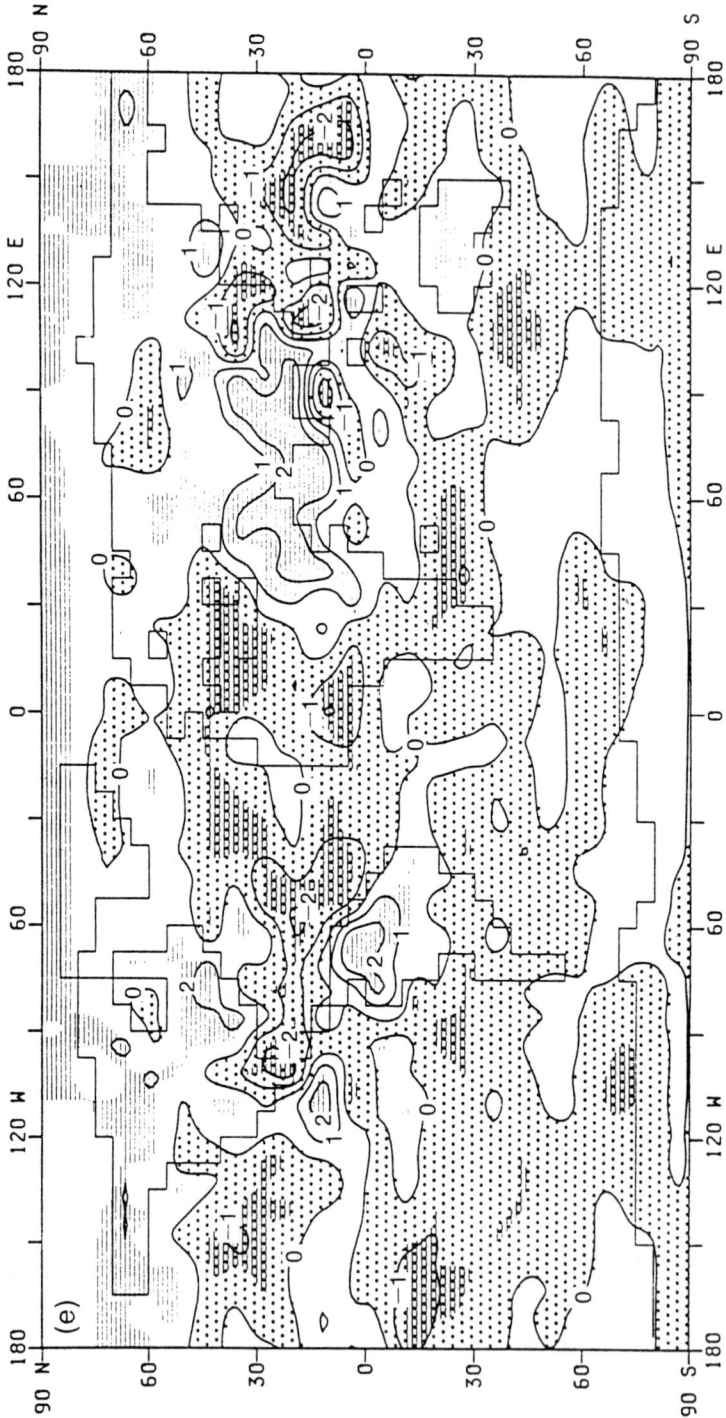

Figure 12c

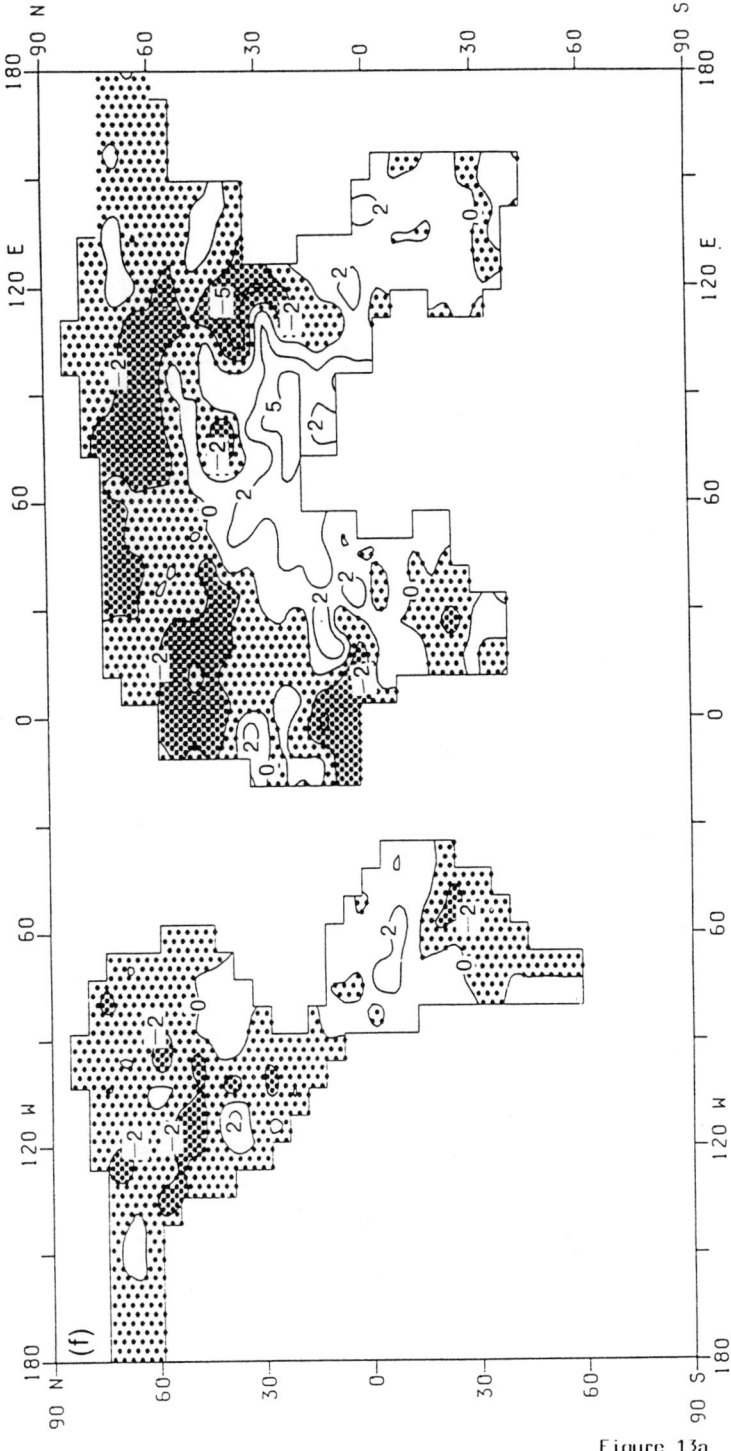

Figure 13a

Figure 13b

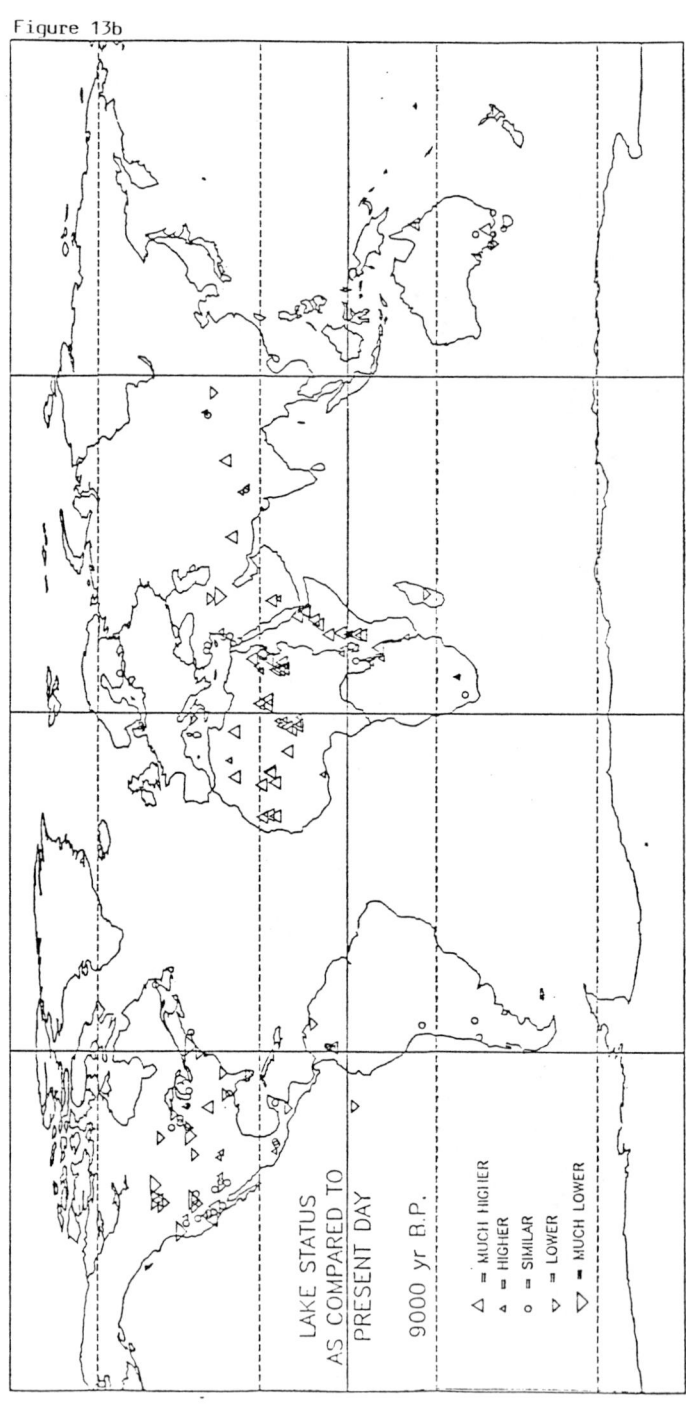

LAKE STATUS
AS COMPARED TO
PRESENT DAY

9000 yr B.P.

△ = MUCH HIGHER
▲ = HIGHER
o = SIMILAR
▽ = LOWER
▽ = MUCH LOWER

THE HYDROLOGICAL CYCLE

J.P. PEIXOTO
Director of the Geophysical Institute Infante D. Luís
the University of Lisbon

1. THE NATURE OF THE PROBLEM
1.1. The water cycle

The global hydrosphere is formed by various reservoirs interconnected by the transfers of water in any of its three phases. In decreasing order of water amount held in storage, the five reservoirs of the hydrosphere are: the world oceans, the ice masses and snow deposits, the terrestrial waters, the atmosphere and, finally, the biosphere.

Vast quantities of water are continuously on the move in the climatic system. Under the direct or indirect influence of solar energy, water evaporates from the oceans and the continents and is transpired by plants and animals into the atmosphere. Within the atmosphere the ascending moist air develops, by cooling, into visible moisture in the form of clouds (condensation). Thus, in the atmosphere, water is stored and transported in the condensed phase (liquid water and ice crystals) as clouds, or in the vapor phase (water vapor). Under the action of gravity it falls on the continents and oceans in the form of rain, snow, hail, or other forms of precipitation. The precipitated water then returns partly to the atmosphere, through evaporation and evapotranspiration, infiltrates into the ground, or runs off over or under the ground to the rivers and streams, which carry it back to the oceans and seas (see Figs. 1 and 2) and becomes ready to start a new journey.

This gigantic and complex system of distilation, pumping and transport of water substance in its many forms originates an unending circulation that constitutes the hydrological cycle. Basicaly it is a consequence of the conservation of water substance. Not all water of the climatic system takes part in this continuous cycle. Some remains for shorter or longer periods of time in the atmosphere, in the biosphere, as snow accumulations in the cryosphere, in the oceans, river valleys, reservoirs and lakes, or as connate waters chemically or physically bound to the lithosphere (soil and rocks), etc.

The hydrologic cycle has two major branches – the terrestrial and the atmospheric branches. The terrestrial branch consists of the inflow, outflow and storage of water in its various forms on and in the continents and in the oceans, while the atmospheric branch consists of the atmospheric transports of water, mainly in the vapor phase. The two branches of the hydrologic cycle join at the interface between the atmosphere and the earth's surface (including the ocean). Fig. 1 and 2 show schematically the distribution and the flow of water in the climatic system.

The loss or 'output' of water from the earth's surface, through, evaporation and evapotranspiration, is the 'input' of water for the atmospheric branch, whereas precipitation, the atmospheric 'output', may be regarded as a gain for the terrestrial branch of the hydrological cycle. Water is thus one of the crucial links between the various components of the climatic system. It is clear that the atmospheric and terrestrial branches of the hydrological cycle have to be taken as a whole, bringing together the sister disciplines of meteorology and hydrology.

As we know, the precipitation and evaporation over the globe are

197

unevenly distributed. However, on the average, there is an excess of precipitation over evaporation in the equatorial region, associated with the intertropical convergence zone, and a similar excess in midlatitudes associated with the pertubations along the polar front, whereas in the subtropical and polar regions evaporation exceeds precipitation. Thus the water vapor released, mainly over the subtropical oceans, is continuously transported both equatorward and poleward to maintain the moisture supply for the observed precipitation belts. This shows how fundamental and relevant is the role played by the atmosphere and its general circulation as a forcing factor in maintaining the hydrological cycle. This fact has been recognized by climatologists, hydrologists and glaciologists. However, quantitative studies of the gaseous hydrosphere and of its aerial runoff, the main component of the atmospheric branch, have only been possible in recent decades since the development of an adequate network of aerological stations. Traditionally, most of the research has been focused on the earth-bound branch of hydrological cycle. However, progress along this line has been kept back by the difficulties in obtaining reliable values of evaporation, change of water storage and precipitation. Estimates of evaporation, based on the diffusion theory or on the energy-budget approach, are uncertain and may differ widely.

Precipitation data over the oceans are still rather scarce and difficult to interpret, especially if they are taken on islands.

Water can exist in the climatic system in various phases, and the amounts of thermodynamic energy involved in the phase transitions, namely in the form of various latent heats are very large. For example, condensation and freezing release considerable amounts of heat, while evaporation and melting absorb equivalent amounts of heat. Evaporation occurs mainly at the surface of the oceans and continents leading to a transfer of latent heat into the atmosphere, that is released as sensible heat when the water vapor condenses. Furthermore, water vapor is an important selective absorber of solar and terrestrial radiation. These facts make the water a crucial factor in the energetics of the climatic system.

1.2. Water in the climatic system

The global amount of water substance in the climatic system does not vary appreciably, although some new water is being produced by volcanoes and hot springs. Also a certain amount of water vapor may be destroyed in the upper atmosphere through photodissociation by solar radiation. All these effects are negligible, so that the total amount of water on earth can be considered to be constant.

The main reservoirs of water on the earth, namely the oceans, cryosphere, terrestrial waters and atmosphere, are shown schematically in Fig. 2.. The interior of the earth also contains an appreciable amount of water, either dissolved or chemically combined with solid or molten rock, but there is no satisfactory estimate of the amount of water thus locked up.

As shown in Fig. 2. about $1,350 \times 10^{15} m^3$ of water (about 97% of all the water in the hydrosphere) is contained in the oceans; while $33.6 \times 10^{15} m^3$ (the remaining 24%) is found on and in the continents, mostly in the glaciers of the Artic and the Antarctic. The atmosphere contains $0.013 \times 10^{15} m^3$ or only one hundred-thousandth of the total water present on the earth. When we consider the total amount of water that falls on the earth's surface, it seems an astonishing fact that this can happen with such extremely small amount of the moisture stored in the atmosphere at a given instant. The influence of this small amount of water vapor on the climate and hydrological resources is far out of proportion to its mass.

The transfer rates between the various reservoirs are also shown in Fig. 2.. They are associated mainly with precipitation, evaporation and runoff.

The water on the continents is distributed in several reservoirs, namely in glaciers (25×10^{15} m^3), ground water (8.4×10^{15} m^3), lakes and rivers (0.2×10^{15} m^3), and in the living matter of the biosphere (0.0006×10^{15} m^3). The amount of water locked in the polar ice is impressively large, totaling some 1.8% of all the water in the hydrosphere. Of the total amount of underground water, vadose water (water present in soils) accounts for only 0.066×1^{15} m^3. The remainder is about evenly divided between reservoirs deeper than 800 m and reservoirs shallower than that level.

The residence time of water in the various reservoirs can be deduced from the ratio of the amount of water in the particular reservoir and the accumulation or depletion rate as presented in Fig. 2.. It is found to vary from about nine days for atmospheric water vapor to thousands of years for the polar ice and the oceans.

2. BALANCE REQUIREMENTS
2.1. Classic equation of hydrology
The water balance requirement for the terrestrial branch of the hydrological cycle leads to the so called classic equation of hydrology. Applying the principle of continuity to a specific region, the balance equation for the terrestrial branch may be written as

$$\bar{S} = P - (E + \bar{R}_o + R_u)$$

where S is the rate of storage of water; P the precipitation rate (in liquid and solid phase) E the evaporation rate (wich includes evapotranspiration over land and sublimation over snow and ice); R_o is surface runoff and R_u the subterranean runoff.

For a large land region, the net subterranean runoff is usually small so that the classic equation can be simplified to the form

$$\{\bar{S}\} = \{\overline{P-E}\} - \{\bar{R}_o\}$$

where ($-$) denotes a time average and { } a space average over the region of area A, $\{\bar{S}\}$ is the rate of change of total surface and subterranean water storage, $\{\overline{P-E}\}$ is the average rate of precipitation minus evaporation per unit area, and $\{\bar{R}_o\}$ is the average rate of runoff. For long periods of time and large areas $\{S\}$ tends to be small compared to the other terms, and the equation can be reduced to

$$\{\bar{E}\} = \{\bar{P}\} - \{\bar{R}_o\} \quad .$$

Traditionally, the quantity of major interest for the classical hydrology has been runoff wich can be measured fairly accurately through gauging the extensive networks of streams. Precipitation has been measured, very extensively, since it is the primary cause for runoff.

In an appendix to this lecture, we present a general description of the terrestrial branch of the hydrological cycle.

Measurements of evaporation and evapotranspiration and of the change in water storage are very difficult to make, although some semi-empirical estimates of these quantities are available for local areas. Since there is an increasing need for large-scale estimates of evapotranspiration and soil moisture storage for the planning of more efficient irrigation systems, for various water resource projects and, in general, for the optimum use of the available water, hydrologists are now beginning to study the atmospheric branch of the water cycle.

2.2. Balance equation for water vapor

The study of atmospheric branch of the hydrological cycle is based on the balance requirements of water vapor in the atmosphere. For a general discussion see Peixoto (1973) and Peixoto and Oort (1983).

The amount of water vapor, **W**, contained in a unit area column of air which extends from the earth's surface to the top of the atmosphere is given by the expression

$$W = \int_o^{p_o} q \, \frac{dp}{g}$$

where **q** is the specific humidity, **g** the acceleration of gravity and **p** the pressure.

It is called the precipitable water in the atmosphere. It represents the amount of liquid water that would result if all the water vapor in the unit column of the atmosphere was condensed. It is usually expressed in units of kg m^{-2} or mm.

The horizontal transport of water vapor is obtained by multiplying the wind vector velocity by the specific humidity (amount of water vapor per kg of moist air). The total horizontal transport vector, integrated along the vertical, is the "aerial runoff" $\vec{Q}$ defined as:

$$\vec{Q} = \int_o^{p_o} q \, \vec{v} \, \frac{dp}{g} = Q_\lambda \, \vec{i} + Q_\phi \vec{j}$$

where $\vec{v}$ is the wind vector, Q_λ the zonal component (paralel to the latitude circles), Q_ϕ the meridional component and $\vec{i}$ and $\vec{j}$ the unit vectors tangent to a paralel and to a meridian, respectively.

The main sources and sinks of water vapor in the atmosphere are primarily evaporation and condensation and, to a lesser extent (except near the earth's surface), diffusion from the surroundings. Thus, s(q) = e − c, where **e** is the rate of evaporation (plus sublimation) and **c** the rate of condensation per unit mass (in units of grams of water per kg moist air per sec). For the total atmosphere, integrating along the vertical, the previous difference becomes **E-P** at the earth's surface, since the evaporated water within the atmosphere is very small.

From the balance requirements it is easy to show that the excess of evaporation over precipitation **(E-P)**, at the earth's surface, is balanced by the local rate of change of water vapor storage $\partial W/\partial t$ and by the net inflow, or the net outflow, of water vapor, given by the divergence of $\vec{Q}$ and represented mathematically by div $\vec{Q}$ (Peixoto, 1973). Thus, we obtain a simplified general balance equation for the water vapor in the atmosphere:

$$\frac{\partial W}{\partial t} + \text{div} \, \vec{Q} = E-P$$

Averaging in time and in space over a region bounded by a conceptual vertical wall (for example, a river-drainage based or an interior sea) the previous equation can be written in the form:

$$\{\partial \overline{W}/\partial t\} + \{\text{div} \, \overline{\vec{Q}}\} = \{\overline{E-P}\}.$$

Using the Gauss theorem, this equation may be written in a form which is often more useful for regional studies:

$$\left\{\frac{\partial W}{\partial t}\right\} + \frac{1}{A} \oint (\vec{Q}.\vec{n}) \ dl = \{\overline{E-P}\}$$

where **n** denotes the outward normal unit vector to the boundary $[\mathcal{l}]$ of the region with an area A.

These equations describe the atmospheric branch of the hydrological cycle. Except in the case of severe storms and for short intervals of time, the rate of change of precipitable water, $\partial W/\partial t$, is very small compared with the other terms. For sufficiently long periods of time, this term can be disregarded, so that $\{\text{div } \vec{Q}\} = \{\overline{E-P}\}$. Thus the divergence of water vapor is found over those regions of the globe where evaporation exceeds precipitation, whereas convergence is found where precipitation is greater than evaporation.

The term **E-P** establishes the connection between the terrestrial and the atmospheric branches of the hydrological cycle. Elimination of $\overline{E-P}$ between these two equations yields

$$\{\overline{R}_o\} + \{\overline{S}\} = -\{\text{div } \vec{Q}\} - \{\partial \overline{W}/\partial t\}$$

which shows how the two branches of the hydrological cycle are linked together. This equation can be regarded as a general equation of hydrology which combines both branches of the hydrologic cycle. If, besides the aerological terms, $\{\overline{R}_o\}$ and $\{\overline{P}\}$ are known over a certain region, one can estimate the rate of change in ground water and the rate of evaporation.

Over long periods of time, such as a year, changes in storage in the land and in the atmosphere become small so that, for example, for a continent the surface and subsurface runoff have to be exactly balanced by the aerial "runoff" into that continent from the surrounding ocean areas. Furthermore, when the entire global atmosphere is considered over a long period of time, all transport and storage terms vanish, and we can conclude that the global-mean evaporation has to be equal to the global-mean precipitation (see Appendix).

3. ATMOSPHERIC BRANCH OF THE HYDROLOGICAL CYCLE
3.1. Data
The basic data necessary to evaluate the various fields, **W**, $\vec{Q}$ and div $\vec{Q}$ are the daily values of specific humidity and wind components at various levels. This was done using the rawinsonde data for the ten year period, May 1963 through April 1973 by the Geophysical Fluid Dynamics Laboratory/ /NOAA, Princeton University. The network of rawinsonde stations covered the entire globe from pole to pole. In addition to the mean annual fields for the entire 10-year period, statistics of higher orders were also evaluated (Oort, 1983).

3.2. Water vapor storage in the atmosphere
The spatial distribution of the annual mean precipitable water content, $\overline{W}$, is represented in Fig. 3. With only a few exceptions, the analysis shows a continuous decrease of precipitable water from the equatorial regions, where it attains the highest values, to the north and south poles.

The departures from zonal symmetry are associated with the physiography of the earth's surface, and are apparent in both hemispheres. As a general rule, the precipitable water is higher over the oceans than over the continents. The distribution over the Southern Hemisphere is practically zonal, since the ocean coverage exceeds by far that of the continents. As expected, the lowest values of W (5 kg m^{-2}) occur over subpolar and polar regions.

The precipitable water over the desert areas is considerably smaller than the corresponding zonal average, mainly due to strong subsidence. This effect is pronounced in the eastern portions of the large semi-permanent antyciclones of the subtropics. In addition, the effects of high terrain on the precipitable water distribution are illustrated by relatively dry areas (often W < 10 kg^{-2}) over the major mountain ranges, such as the Rockies, Himalayas, highlands of Ethiopia, and the Andes. The effects of topography and the land-sea contrast in the Southern Hemisphere are shown by the dipping of the 20 kg m^{-2} isoline towards lower latitudes.

The humidity decreases rapidly with height, almost following an exponential law. It also decreases with latitude. More than 50% of the water vapor is concentrated below the 850 mb surface (1500 m), while more than 90% is confined to the layer below 500 mb (5.600 m). The seasonal variations are more intense in the Nothern than in the Southern Hemisphere, as expected from the corresponding temperature variations.

Using the grid point values of the W maps, the zonally averaged storage of water in the atmosphere $[\bar{W}]$ can be evaluated. The results are shown in Fig. 4.

These profiles give the gross distribution of water vapor in the atmosphere. The seasonal profiles are also included. They show a maximum in the equatorial zone, with a slight seasonal migration into the summer hemisphere, and a monotonic decrease to polar latitudes with the steepest gradients in the subtropics.

The global water content in the atmosphere obtained from the $[\bar{W}]$ - profile, is on the order of 13.1x10^{15} kg which is equivalent to a uniform layer of about 2.5 cm of water covering the globe. Seasonal changes in the hemispheric water content are more pronounced in the Northern than in the Southern Hemisphere with summer minus winter values of 3.0 and 1.8x10^{15} kg (or 1.2 and 0.7 cm of water covering the hemisphere), respectively.

Assuming a mean annual precipitation value over the globe of 1.0 m, the ratio of the amount of water in the air and the precipitation rate leads to a residence time of water in the atmosphere of (0.025/1.00) yr or about 9 days. This indicates that the water vapor in the atmosphere is replenished about forty times a year.

3.3. The mean water vapor transport in the atmosphere

Our discussion of the role of the general circulation in the hydrological cycle commences with an analysis of the vertically integrated atmospheric moisture flow in terms of the $\vec{Q}$ vector field, the aerial runoff, as shown in Fig.5.

The $\vec{Q}$ maps provide a good indication of the prevailing movements of the main moist air masses in the atmosphere. The $\vec{Q}$ field is not uniform in intensity and direction. The intensity is largest over ocean areas (e.g. Pacific Atlantic, South Indian Ocean, etc.). The field configuration reflects the main characteristics of the general circulation in the lower part of the atmosphere, as expected. There is, in general, a transfer of water vapor from the oceans into the continents. Even when the vectors are mainly zonal, their intensities over oceans are larger, by far, than over the continents so that a net inland flow of water results. In some cases this inland transfer is evident from the vector direction (e.g. Gulf of Mexico, etc.). The non uniformity of $\vec{Q}$ field distribution leads immediately to the impossibility of accepting the evaporation-precipitation **in situ** theory.

Although the zonal (east-west) component exceeds in general the meridional (south-north) component, the $\vec{Q}$ field is far from being truly zonal. The meridional component, in certain aspects, is far more important for the water balance of the globe than the zonal component.

To illustrate, in a concise form, the main characteristics of the mean meridional flux and its importance for the water budget of the earth, meridional profiles of averaged values of the meridional flux for yearly and seasonal conditions are shown in Fig. 6. The inspection of the profiles shows that in midlatitudes, the meridional transports are poleward in both hemispheres, with maxima near 40° latitude and with small seasonal variations. In the tropical zone, the mean annual transports are positive south of the equator and negative to the north of it. There is evidence for a strong interaction between the two hemispheres as shown by the seasonal curves. In fact, the cross-equatorial flow in the Hadley cells changes direction with the seasons leading to a water vapor flux into the Northern Hemisphere during JJA of about 18.8×10^{8} kg s^{-1} and a flow into the Southern Hemisphere during DJF of about -13.6×10^{8} kg s^{-1}. For the year as a whole, there is a net influx into the Northern Hemisphere of 3.2×10^{8} kg s^{-1}. In other words, on an annual basis, the Southern Hemisphere supplies a considerable amount of water vapor to the Northern Hemisphere. The cross-equatorial flow of water vapor implies an annual excess of precipitation over evaporation in the Northern Hemisphere of 39 mm yr^{-1} and an excess of 58 mm for the three months of the JJA season. The water vapor exported by the Northern Hemisphere during the winter season corresponds to an excess of evaporation over precipitation in that hemisphere of about 42 mm for the three months of the DJF season.

Although the meridional flux of moisture over midlatitudes varies considerably with the seasons, it is predominantly poleward throughout the year. This transport is mainly accomplished by baroclinic lows associated with the polar front and by stationary eddies, such as subpolar lows and subtropical anticyclones, together with their transient pulsations. The largest variability during the year is associated with the movement and changes in strength of the Hadley cells. The lower branches of these cells, in both Hemispheres, are very effective in transporting moisture into the intertropical convergence zone (ITCZ).

3.4. Vertical transport of water vapor

The vertical transport of water vapor in the atmosphere plays an essential role in the hydrological cycle, since it links the terrestrial and atmospheric branches. It is also another requirement that results from the conservation of the water substance.

Water vapor is transported upward from the lowest levels in order to balance the losses in the higher atmosphere due to condensation and precipitation. Convection associated with vertical instability and cumulus activity play an important role in this transport particularly in the tropics and in summer over the continents when convective activity pumps large amounts of water vapor into mid atmosphere.

The belt of maximum upward transport over the equatorial region is, of course, associated with the ascending branches of the Hadley cells, whereas the upward flux in middle and high latitudes must be connected with the quasi-stationary low-pressure systems. The centers of maximum downward flux occur mainly in the eastern parts of the subtropical anticyclones over the oceans with the prevailing subsidence.

4. MEAN PLANETARY WATER BALANCE AND THE GENERAL CIRCULATION OF THE ATMOSPHERE

4.1. The mean divergence of water vapor

The importance of the divergence field of water vapor in the atmosphere depends on it relationship with $(\overline{E-P})$, as discussed earlier. Thus, divergence maps are of great interest for the study of the planetary water balance, since regions of mean positive divergence $(\overline{E-P} > 0)$ constitute main source regions of water vapor, whereas the regions of convergenge $(\overline{E-P} < 0)$ are sink regions of water vapor for the atmosphere. Using the grid point values of the meridional and zonal components of $\overline{Q}$ it is possible to evaluate the divergence of $\overline{Q}$ (div $\overline{Q}$). The resulting map for the mean yearly conditions is presented in Fig. 7.

Convergence (negative divergence) generally prevails over the equatorial and mid to high latitude zones, while divergence predominates in the subtropics. The convergence and divergence centers are, as a rule, more intense over the ocean than over land. The equatorial convergence of water vapor is associated with the **ITCZ**. Water vapor is carried towards the region of mean rising motion by the lower branches of the Hadley cell, leading to heavy precipitation confirming the observed distributions of P. The belt of convergence of water vapor consists of various centers located over the headwaters and drainage basins of large river systems, such as the Amazon in South America, The Ubangi, Congo, Senegal and Blue Nile in Africa and the Indus, Ganges, Mekong and Yangtze in Southeast Asia.

The subtropical belts of divergence coincide largely with the arid zones of the globe. They are associated with strong evaporation over the oceans and with subsidence that prevails over the large subtropical anticyclones. Their latitudinal oscillation follows the early movement of the subtropical antyciclones. The distribution of divergence over the oceans is easy to understand since there is always water available for evaporation and the prevailing ocean currents will advect the necessary fresh water to maintain equilibrium. However, the situation is more difficult to explain when divergence occurs over land. In this case, surface and underground flows from less arid regions must supply the water required to counterbalance the observed excess of evaporation over precipitation (Starr and Peixoto, 1958).

The middle to high latitude convergence in both hemispheres is mainly associated with the transient cyclonic lows that accompany the polar front. Over the polar regions, there are some indications of a slight divergence, especially in the vicinity of the north pole.

The evaporation is very large over the oceanic regions with strong divergence of water vapor and consequently with high salinity values. On the other hand, in regions of convergence, such as in equatorial latitudes, the excess of fresh water from rain will dilute the ocean water, leading to lower salinity values. As expected, the observed configuration of the salinity field (see Fig. 8.) reveals a high correlation with the mean annual divergence map of water vapor.

Zonally averaged profiles of the $[\text{div } \overline{Q} \, (\simeq \overline{E-P})]$ fields are shown in Fig. 9. The profiles synthesize the mean features of the water vapor divergence fields already discussed. In all cases, they indicate a strong convergence with an excess of mean precipitation over mean evaporation, $[\text{div } \overline{Q}] \simeq [\overline{E-P}] < 0$, over the equatorial region and strong divergence in the subtropics. The divergence intensifies during the winter season. As can be seen from climatological maps, precipitation has a minimum and evaporation a maximum in each of the subtropical zones.

It is instructive to compare the meridional transport $\overline{Q}_\phi$ and $[\text{div } \overline{Q}]$ profiles, since the latter profiles are the derivatives of the $\overline{Q}$ curves.

They show a large export of water vapor from the zones of divergence and a strong import of moisture into the belts of convergence.

4.2. The aerial runoff and the water balance

The role of the general circulation in the hydrological cycle was shown well through maps of the vertically-integrated atmospheric moisture flow, in terms of the $\vec{Q}$ vector field, the so-called aerial runoff. From the $\vec{Q}$ field it is possible to draw the corresponding streamlines (see Fig. 5).

In a hypothetical steady state, the streamlines would show the prevailing paths of water vapor in the atmosphere after its release from the various source regions at the earth's surface.

They provide a good indication of the prevailing movements of the main moist air masses in the atmosphere and the sites of their formation. They show, again, that the main sources of water vapor for the atmosphere are located over the subtropical oceans and that most of the water vapor, necessary for precipitation over the continents, comes from the oceans. In steady state conditions, this net inflow of moisture to the continents must be compensated by runoff of the rivers into the oceans. The air masses also receive some moisture over the continents due to evapotranspiration and evaporation from lakes, soil, etc. However, this last moisture supply constitutes only a small fraction of the water that falls locally, as precipitation over land. These results confirm that the **in situ** theory (i.e., local evaporation would provide the water vapor required for the local precipitation) cannot be accepted, as a general rule.

Since the water vapor transport occurs, mainly, in the lower troposphere, the realm of weather phenomena, it is clearly affected by the earth's topography. Indeed, the absence of large mountains along the Atlantic coast in Europe favors the deep penetration of moisture from the Atlantic Ocean into the Eurasian Continent and the Mediterranean region. On the other hand, the existence of the Rocky Mountains parallel to the west coast of North America does not allow moisture from the Pacific Ocean to penetrate deeply into the American continent. Most of the moisture falling as precipitation over North America seems to be supplied by water vapor originating over the warm waters of the Gulf of Mexico, with a deep northward intrusion of water vapor in all seasons. On the other hand, most of the moisture over South America comes from the Atlantic Ocean.

When the hydrosphere of the climatic system is considered as a whole, there must be, on the average, a compensating meridional flux of water substance in the terrestrial hydrosphere opposite to that wich is observed in the gaseous hydrosphere. Thus, balance considerations require a small net southward transport of water by ocean currents and rivers in midlatitudes of the Northern Hemisphere, as well as across the equator in the subequatorial zone. At other latitudes, the net meridional flux of liquid water must be northward. Preliminary results (Hellerman, 1981) indicate that there is a detectable meridional runoff by major rivers in the required direction, although it is much smaller than the net contribution by ocean flows.

In order to get some idea of three-dimensional circulation of water in the atmosphere, a streamline representation of the mean meridional transport for yearly conditions is also presented in Fig. 10.

Inspection of the cross-section confirms that the water vapor circulates in the lower troposphere, and that the maximum transports occur in the planetary boundary layer. The streamlines that begin at the earth's surface are due to the excess of evaporation over precipitation, while those that end at the earth's surface define the sink regions for atmospheric water. The yearly cross-section shows that there is an export of water vapor from the zones, between 12°N and 35°N, and 10°S and 35°S. In the Northern Hemisphere, a considerable portion of this water vapor (to the right of the zero streamline) is exported polewards throughout the lower half of the atmosphere, while the remainder, confined to much lower levels, is directed to the equatorial regions, mainly to feed the **ITCZ**. In Southern hemisphere low latitudes, a large part of the water vapor is exported to the north in a shallow layer near the earth's surface, some of it crossing the equator and falling as precipitation in the **ITCZ**. The other part is transported southwards, at much higher levels, up to 400 mb. Polewards of 40° latitude, in both hemispheres, the streamlines terminate at the surface, indicating the existence of an excess of precipitation over evaporation.

The interactions between the atmospheric branch and the terrestrial branch (see Appendix) are essential to achieve and maintain the state of quasi-equilibrium of the climatic system. Occasionally, however, imbalances may persist for an extended period of time, leading to drought or to flood conditions.

*
* *

Our results confirm the importance of the aerological branch in the dynamics of the hydrological cycle considered **in toto.**

Of course, solar radiation provides the bulk of the energy that sets the water circulation in motion, as a whole. Part of this solar energy is redistributed within the climatic system by the aerological branch of the hydrological cycle, when the phase transitions of the water substance occur. If we add the role of water vapor and clouds in the reflection, absorption and emission of both solar and terrestrial radiation, we may conclude that the hydrological cycle is one of the crucial factors in the dynamics of climate.

5. MATHEMATICAL SIMULATION OF THE HYDROLOGICAL CYCLE

Observational studies, such as those reported in this paper, go hand in hand with numerical model experiments of the general circulation of the atmosphere (GCM's), incorporating major elements of the hydrological cycle. The GCM's are three-dimensional time dependent global models, based on the dynamical equations of the atmosphere. They simulate explicitly, with as much fidelity as possible, the various processes occurring in the atmosphere and oceans. The GCM's with a hydrological cycle were developed by Manabe and Hollway (1975).

The GCM includes continents with mountains and soil, that has a uniform "field capacity" of 15 cm water. In other words, the maximum amount of water the soil can retain, without generating runoff, is 15 cm. In the case of the oceans they assumed an infinite field capacity but no redistribution by ocean currents, i.e. a swamp-type ocean model. The model computes basically the fields of temperature, wind and water vapor for the atmosphere, and derives the rate of precipitation (rain or snow) minus evaporation from the convergence of moisture transport in the atmosphere. The evaporation is inferred through a simple scheme which assumes that evaporation is a function of soil moisture and potential evaporation. The rates of change of soil moisture and snow depth are then computed keeping track of the water and heat budgets of the ground. Runoff occurs when the accumulated soil moisture exceeds the prescribed field capacity of 15 cm as if the "bucket" were full. The excess water is assumed to flow directly into the surrounding seas without further infiltration. The model also depicts the equatorial rain zone and the major subtropical deserts. The

total precipitation and other hydrological parameters compare well with
those observed in the real atmosphere as shown in Figs. 11 and 11a.

6. EPILOGUE

We see that the hydrological cycle is maintained by the general
circulation of the atmosphere. We cannot understand its mechanisms unless
we understand the mechanisms of the general circulation. Only the
consideration of the earth, oceans and atmosphere as a single interacting
system will lead us to a basic understanding of the global water cycle,
because

"all the rivers run into the sea; yet the sea is not full; into
the place from which the rivers come, to there and from there
they return again"
(Ecclesiastes 1:7).

APPENDIX

1. Precipitation, evaporation and runoff
1.1. Precipitation

Precipitation is one of the principal climatic elements. It is highly variable in space and time. Nevertheless its average values are fairly stable and can be represented well in map form.

The most significant feature of the distribution is the high rainfall that occurs in equatorial latitudes along the strong convection in the **ITCZ** which corresponds to the thermal equator. It is interesting to note the very high values of precipitation over the equatorial regions in South America, Africa, and Indonesia and in the equatorial Pacific Ocean where precipitation may exceed 3 m yr^{-1}. During the annual cycle the **ITCZ** migrates north and south in phase with the solar insolation which explains the shift of the observed maxima.

In subtropical regions under the influence of the large semi-permanent anticyclones where subsidence predominates the precipitation is low, less than 0.2 m yr^{-1}, and large parts of the subtropical continents are covered by deserts, where the precipitation is even lower, e.g., in Africa and Australia.

Over the midlatitude belts where the polar fronts with the associated disturbances predominate, the precipitation shows a secondary maximum. Here precipitation will be abundant in all seasons, except on their equatorial side where dryness would prevail during the summer season when the anticyclones have moved poleward, such as in the Mediterranean region.

Over the polar regions the moisture content of the atmosphere is very low, and the amounts of precipitation are less than 0.2 m yr^{-1} during all seasons.

The seasonal shifts of the **ITCZ** are found to be more pronounced over land than over the oceans. The latitudinal range of the shifts in the **ITCZ** is associated with the existence of the marginal climates bordering the arid and desert-like regions, such as occurs in the Sahel belt in Africa.

2. Evaporation

The evaporation rate depends on many factors. The most important ones are the temperature, vertical diffusion coefficients, the vertical gradient of humidity, wind speed and the availability of water. Evaporation is often measured with a shallow circular pan. However, such values are very much influenced by local conditions and local exposure. Thus these measurements, useful as they may be for local purposes, such as assessing evaporation from water reservoirs, small lakes, irrigated areas, etc., are of little use in computing the water budget for larger regions of the earth.

3. Surface runoff

After considering the precipitation and evaporation fields separately it is useful to compare the behavior of these two quantities, since they are closely related elements of climate and hydrology. Thus, we present in Table 1 the mean zonal averages of precipitation and evaporation for $10°$ latitude belts as well as the hemispheric and global averages according to Baumgartner and Reichel (1975). Similarly, annual mean values of the same quantities for individual continents and oceans are presented in Table 2 according to the same authors. The tables include also the P-E differences, the so-called discharge or runoff, and the values of the evaporation ratio **E/P** and the runoff ratio **(P-E)/P**. These last two quantities are of interest since they are sometimes used as climatic indices or in hydrology studies. Table 2 further includes the estimated river discharge R_o from the continents as measured by the peripheral runoff that reaches the oceans.

The **P-E** values in Table 1 show an excess of precipitation over evaporation at mid and high latitudes as well as in the equatorial zone between 10°S and 10°N, whereas a deficit of precipitation is found in the subtropics of each hemisphere between about 10° and 40° latitude. In the long-term mean, the excess or deficit in each belt has to be compensated by a net meridional divergence or convergence of water in the particular belt. The runoff ratio **(P-E)/P** gives an idea of the fraction of the precipitation that is involved in the runoff. The values of the evaporation ratio **E/P** in Table 1 show clearly the high aridity of the subtropics with ratios larger than one.

One must be aware that there is not always a close agreement between the values of **P** and **E** published by different authors, as demonstrated by the comparisons in Tables 1 and 2 with Sellers (1965) values based on similar observations.

Over the globe as a whole, evaporation must balance precipitation in the long-term mean. The precipitation in the two hemispheres is almost the same whereas a large difference exists for the evaporation. The higher value of evaporation in the Southern Hemisphere results because this hemisphere is largely covered by oceans. The Northern Hemisphere shows a positive water balance (P-E=73 mm yr^{-1}) whereas in the Southern Hemisphere a net negative value of -73 mm yr^{-1} occurs. Thus we are led to the conclusion that a flow of water in the liquid form must take place across the equator from the Northern into the Southern Hemisphere. As we have seen an equal amount of water in the vapor form has to be exported in the opposite direction to maintain the balance of the water substance.

As shown from Table 2, the quantities **P, E** and **P-E** are not the same for the different oceans and continents. This is not only due to physiographic differences between them but also due to the differences in aerial extent. For example, South America shows the highest **P, E** and **P-E** values in agreement with what we have seen before; on the other hand, Australia, Africa, and Antarctica show very small runoff **(P-E)** values. Overall, the precipitation and evaporation tend to be smaller over the continents than over the oceans, except for the extremely high values of **P** and **E,** over South America and the extremely low values of **P** and **E** over the Arctic Ocean. The mean value of **P-E** over all continents is estimated to be 266 mm yr^{-1}. This surplus of condensed water must be transported by rivers and glaciers from the continents into the oceans where a deficit of -110 mm yr^{-1} is found. When the surplus over land and the deficit over the oceans are multiplied by the appropriate areal factors they must balance. The table further shows that **P-E** is positive for the Arctic and Pacific Oceans and strongly negative for the Atlantic and Indian Oceans, (the "dry" oceans), leading to the net deficit for all oceans combined. The implications of these estimates of **P-E** taken together with the values of the observed river discharge are that a net transfer of water must occur from the Pacific and Arctic Oceans into the Atlantic and Indian Oceans. For example, water transport by the rivers from the surrounding continents into the Atlantic is estimated to be on the order of $R=197$ mm yr^{-1} so that the equivalent of $E-P-R=175$ mm yr^{-1} must come from the Pacific and Arctic Oceans. Further for the Indian Ocean which only receives 72 mm yr^{-1} from continental runoff an inflow of 179 mm yr^{-1} must take place from the Pacific Ocean.

The net excess of precipitation over evaporation over the continents must be maintained by a net influx of water in the vapor form from the large ocean sources.

REFERENCES

(1) **BAUMGARTNER,** and **E. REICHEL,** **1975:** The World Water Balance. Elsevier, Amsterdam, 179 pp.

(2) **CHENG,** **D.C.,** **1986:** On the Global water vapor flux and maintenance during FEGG. Month. Weather Review, 114.

(3) **MANABE,** S. and **HOLLOWAY,** J.L. Jr., **1975:** The Seasonal Variation of the Hydrologic Cycle as Simulated by a Global Model of the Atmosphere. Jour. of Geophys. Research, 80, 617-1649.

(4) **MOHAMNDY,** A. **HOLLINGSWORTH** and **S.K. DASH,** **1985:** The Asian Summer Monsoon Circulation Statistics. 1979-1984, Publication of the ECMWF.

(5) **MOLLER,** F., **1951:** Quarterly Charts of Rainfall for the Whole Earth. Petermanns Geographische Mitteilungen 95, 1-7 (in German).

(6) **OORT,** A.H., **1983:** Global atmospheric circulation statistics. 1958 1973. NOAA Profess. Paper, nº 14, U.S. Grov.'s Printing Office Washingto, 180 pp.

(7) **PEIXOTO,** J.P., **1973:** Atmospheric Vapour Flux Computations for Hydrological Purposes. WMO Publ. nº 357, Geneva, 83 pp.

(8) **PEIXOTO,** J.P. and **A.H. Ooort,** **1983:** The Atmospheric Branch of the Hydrological Cycle and Climate in Variations of the Global Water Budget. Reidel, London, 5-65.

(9) **PEIXOTO,** J.P. and **A.H. OORT,** **1984:** Physics of Climate. Rev. Modern Phys., 56, 365-429.

(10) **ROSEN,** R.D. et al., **1985.** Monthly Weather Review, 113, 201.

(11) **SELLERS,** W.D., **1965:** Physical Climatology. The Univ. of Chicago Press, 271 pp.

(12) **STARR,** V.P. AND **J.P. PEIXOTO,** **1958:** On the Global Balance of Water vapor and the Hydrology of Deserts. Tellus, 10, 189-194.

ACKNOWLEDGEMENTS

We wish to thank to Prof. A. Oort from GFDL (Princeton University) and to Drs. Carlos Marques, A. Tomé and Paulo Pinto from the Geophysical Institute for their collaboration.

Table 1. Estimated mean annual values of the precipitation **P**, evaporation **E**, runoff **(P—E)**, evaporation ratio **E/P** (an aridity index), and runoff ratio **(P—E)/P** for 10° latitude belts, the hemispheres and the globe from Baumgartner and Reichel (1975). For comparison, Seller's (1965) estimates for **P** and **E**, and Peixoto and Oort's (1983) independent estimates of **P—E** as computed from divergence are shown in parentheses.

Latitude	surface area	P	E	P—E	E/P	(P—E)/P
80°-90°N	3.9	46 (120)	36 (42)	10 (93)	0.78	0.22
70°-80°N	11.6	200 (185)	126 (145)	74 (124)	0.63	0.37
60°-70°N	18.9	507 (415)	276 (333)	231 (224)	0.54	0.46
50°-60°N	25.6	843 (789)	447 (469)	396 (250)	0.53	0.47
40°-50°N	31.5	874 (907)	640 (641)	234 (156)	0.73	0.27
30°-40°N	36.4	761 (872)	971 (1002)	-210 (23)	1.28	-0.28
20°-30°N	40.2	675 (790)	1110 (1246)	-435 (-435)	1.64	-0.64
10°-20°N	42.8	1117 (1151)	1284 (1389)	-167 (-322)	1.15	-0.15
0°-10°N	44.1	1885 (1934)	1250 (1235)	635 (478)	0.66	0.34
0°-10°N	44.1	1435 (1445)	1371 (1304)	64 (144)	0.96	0.04
10°-20°N	42.8	1109 (1132)	1507 (1541)	-398 (-342)	1.36	-0.36
20°-30°N	40.2	777 (857)	1305 (1416)	-528 (-312)	1.68	-0.68
30°-40°N	36.4	875 (932)	1181 (1256)	-306 (-128)	1.35	-0.35
40°-50°N	31.5	1128 (1226)	862 (895)	266 (150)	0.76	0.24
50°-60°N	25.6	1003 (1046)	553 (520)	450 (278)	0.55	0.45
60°-70°N	18.9	549 (418)	229 (174)	320 (245)	0.42	0.58
70°-80°N	11.6	230 (82)	54 (45)	176 (98)	0.23	0.77
80°-90°N	3.9	73 (30)	12 (0)	61 (32)	0.16	0.84
0°-90°N	255.0	990 (1009)	897 (944)	73 (39)	0.92	0.07
0°-90°S	255.0	975 (1000)	1048 (1064)	-73 (-39)	1.07	0.07
globe	510.0	973 (1004)	973 (1004)	—	1.00	—
units	10^6 Km^2	mm yr^{-1}	mm yr^{-1}	mm yr^{-1}	—	—

Table 2 — Estimated mean annual values of the precipitation P, evaporation E, runoff P-E, river runoff R_o from continents into the oceans, evaporation ratio E/P, and runoff ratio (P-E)/P for the various continents and oceans from Baumgartner and Reichel (1975). For comparison, estimates of P,E and P-E from Sellers (1965) have been added in parentheses.

Region	surface area (10^6 km^2)	P (mm yr^{-1})	E (mm yr^{-1})	P-E (mm yr^{-1})	R_o (mm yr^{-1})	E/P	(P-E)/P
Europe	10.0	657 (600)	375 (360)	282 (240)	-	0.57	0.43
Asia	44.1	696 (610)	420 (390)	276 (220)	-	0.60	0.40
Africa	29.8	696 (670)	582 (510)	114 (160)	-	0.84	0.16
Australia	7.6	447 (470)	420 (410)	27 (60)	-	0.94	0.06
North America	24.1	645 (670)	403 (400)	242 (270)	-	0.62	0.38
South America	17.9	1564 (1350)	946 (860)	618 (490)	-	0.60	0.40
Antarctica	14.1	169 (30)	28 (0)	141 (30)	-	0.17	0.83
All Land Areas	148.9	746 (720)	480 (410)	266 (310)		0.64	0.36
Arctic Ocean	8.5	97 (240)	53 (120)	44 (120)	307	0.55	0.45
Atlantic Ocean	98.0	761 (780)	1133 (1040)	-372 (-260)	197	1.49	-0.49
Indian Ocean	77.7	1043 (1010)	1294 (1380)	-251 (-370)	72	1.24	-0.24
Pacific Ocean	176.9	1292 (1210)	1202 (1140)	90 (70)	69	0.93	0.07
All Oceans	361.1	1066 (1120)	1176 (1250)	-110 (-130)	110	1.10	-0.10
Globe	510.0	973 (1004)	973 (1004)	0 (0)	-	1.10	0

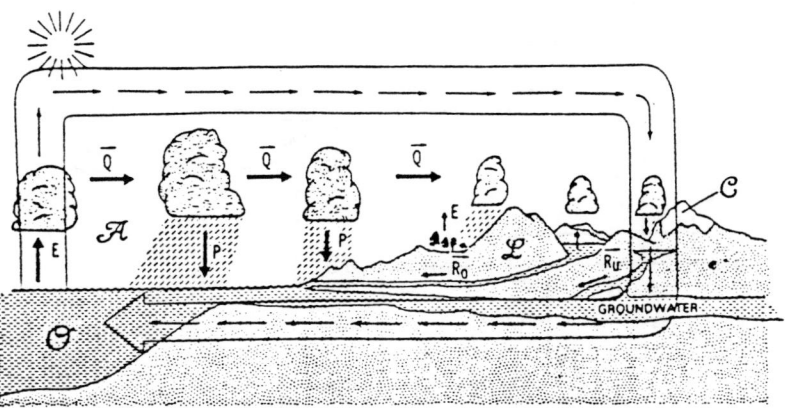

Fig.1 Schematic diagram of the atmospheric and terrestrial branches of the hydrological cycle showing the importance of evaporation E, advection of water vapor in atmosphere $\vec{Q}$, precipitation P, river runoff $\vec{R}_0$ and underground runoff $\vec{R}_u$.

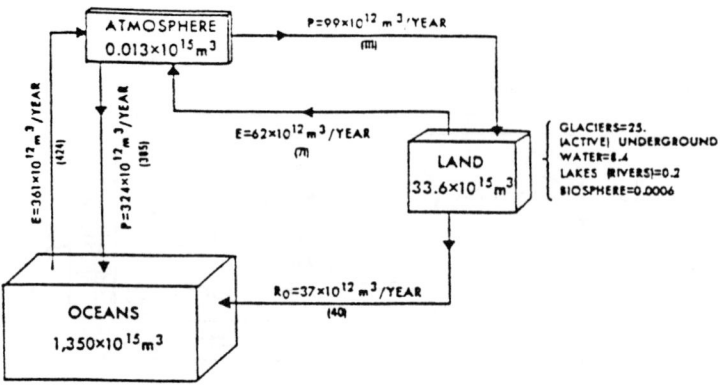

Fig.2 The amounts of water stored in the oceans, land and atmosphere. and the amounts exchanged annualy between the different reservoirs through evaporation, precipitation and runoff (estimates are from Peixoto and Kettani, 1973, and, in parentheses, from Baumgartner and Reichel, 1975).

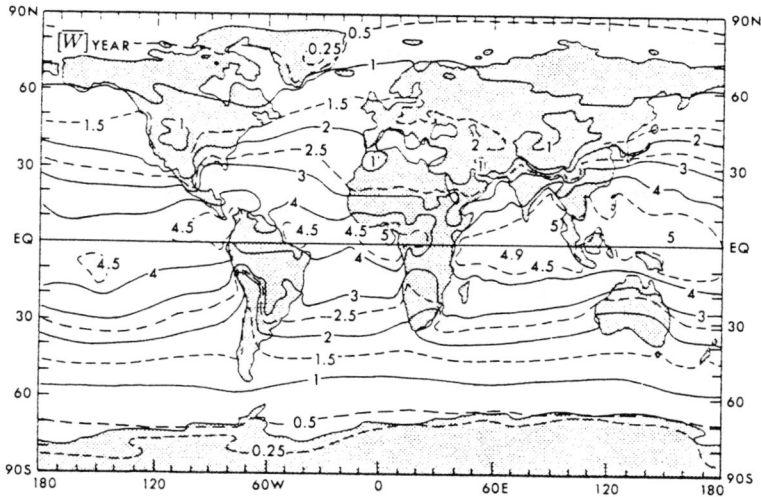

Fig.3 Global distribution of precipitable water in the atmosphere for annual mean conditions in units of 10 kgm^{-2} (Peixoto and Oort, 1983). Note that the global mean value is about 25 kgm^{-2}.

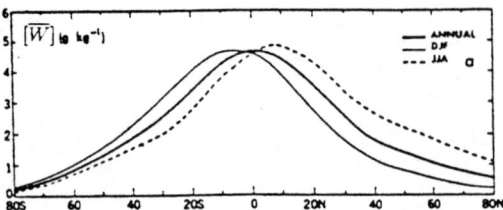

Fig.4 Meridional profiles of: a, the zonally-averaged precipitation water content of the atmosphere $[\overline{W}]$ (10 kgm^{-2}).

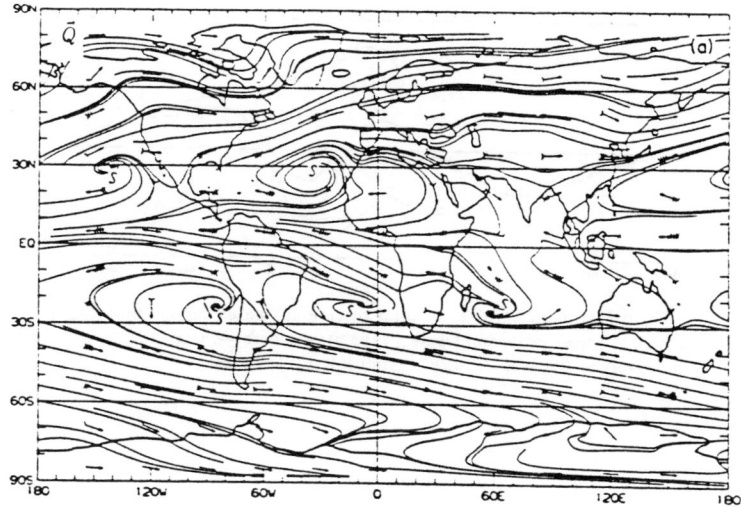

Fig.5 Global distributions of the total aerial runoff, Q, and some corresponding streamlines for annual mean conditions. Each barb on the shaft of an arrow indicates a value of 2 m s^{-1} g kg^{-1} (from Peixoto and Oort, 1983).

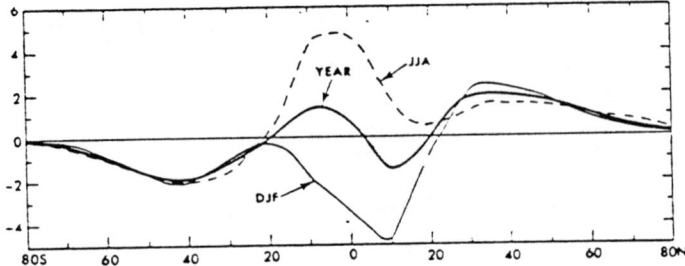

Fig.6 Meridional profiles of total meridional water vapour flux $[\overline{Q}_\phi]$ (10kg m^{-1} s^{-1}) across latitudinal walls in the atmosphere:—, year;—, DJF;- - -, JJA. To convert to units of 10^8 kg s^{-1}, multiply by 4 × $cos\phi$.

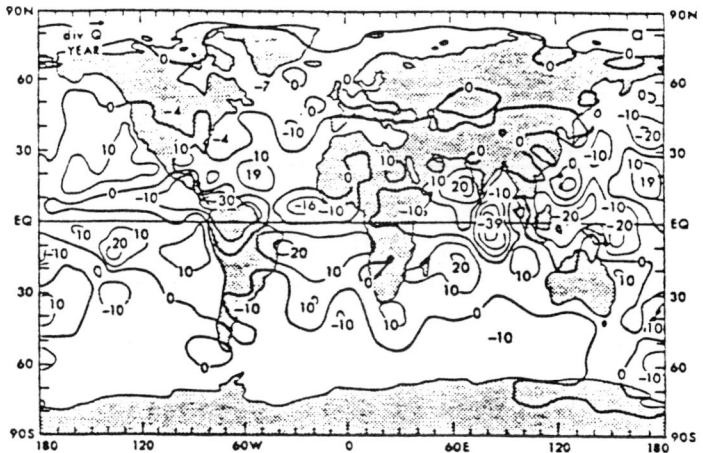

Fig.7 Global distribution of the horizontal divergence of the vertically integrated transport of water vapor, div $\overline{W}$, for annual mean conditions in units of m yr^{-1}. Positive values indicate areas of divergence or source regions ($\overline{E} - \overline{P} > 0$) for atmospheric water vapor, and negative value areas of convergence or sink regions ($\overline{E} - \overline{P} < 0$).

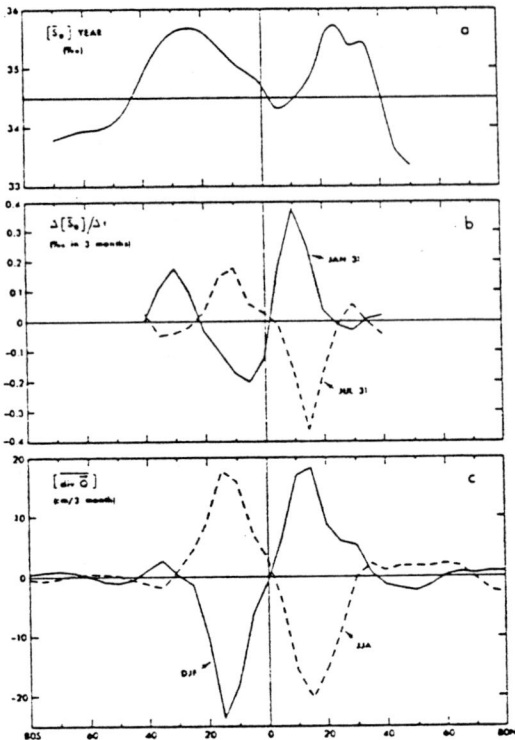

Fig.8 Profile of: a, the zonal-mean salinity (%) at the ocean surface ; b,
rate of change of zonal-mean salinity (% per 3-month period); c, zonal-mean
divergence of total atmospheric water vapour flux $[div\,\bar{Q}]$ (cm per 3-mounth
period) for ocean and land, the annual mean profile of $[div\,\bar{Q}]$ having been
removed from the seasonal profiles:—, DJF;---, JJA.

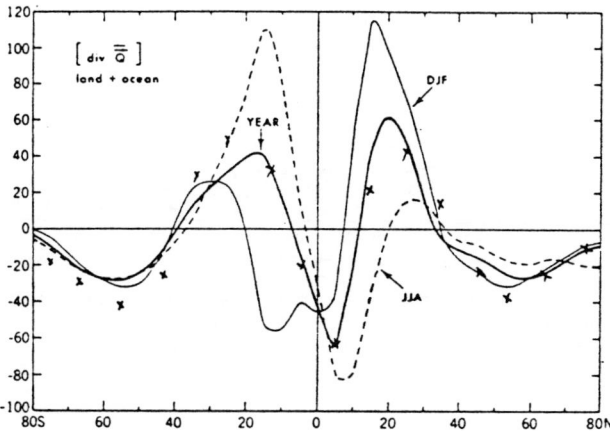

Fig.9 Meridional profiles of the zonal mean divergence of the total water vapor transport, $[div\,\bar{Q}] = [E - P]$, in 0.01 m yr^{-1} for annual, DJF and JJA mean conditions. Some estimates of the E-P by Baumgartner and Reichel (1975) are added for comparison.

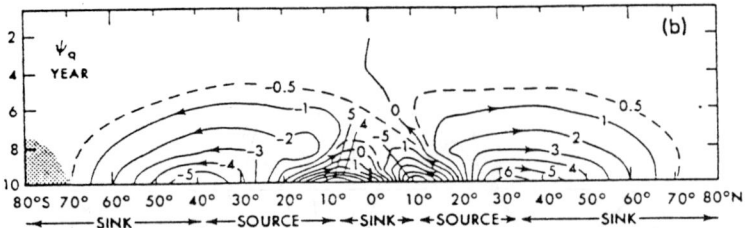

Fig.10 Zonal mean cross sections of flow of water substance, ψ_q (vapor plus liquid plus sólid), in the atmospheric branch of the hydrological cycle in units of $10^8 kgs^{-1}$ for annual mean condition, from Peixoto and Oort (1983).

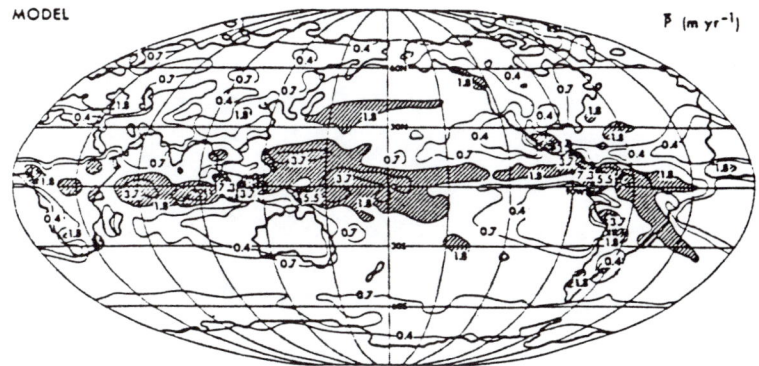

Fig.11 Global distribution of the annual mean rate of precipitation (m yr^{-1}) simulated by a general circulation model after Manabe and Holloway (1975). (Compare with observed precipitation rates in Fig 11a). Stippled areas indicate $\overline{P} < 0.4m\ yr^{-1}$ and shaded areas $\overline{P} < 1.8m\ yr^{-1}$.

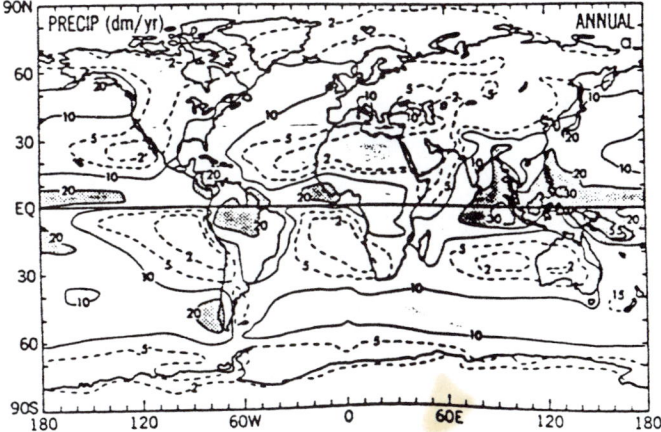

Fig.11a Global distributions of the precipitation rate for annual mean conditions in dm yr^{-1}.

BIOCLIMATIC BELTS OF WEST EUROPE
(Relations between Bioclimate and Plant Ecosystems)

S. RIVAS-MARTINEZ
Department of Plant Biology II, Complutense University, E-28040 Madrid

Summary

Bioclimatology is an ecological science which seeks to clarify existing relationships between living organisms (Biology) and climate (Physics). Among the bioclimatic models and indices which allow demonstration of this correlation, the thermicity index $[It = (T + m + M)10]$ and bioclimatic belts are of particular usefulness and predictive value. In our view, the bioclimatic belts are primarily each one of the thermoclimatic spaces delimited by bioindicators, recognizable in the terrestrial environment in altitudinal or latitudinal zonation. Such thermoclimatic spaces are delimited on determined values of the thermicity index or thermotypes (e.g., thermo-, meso-, supra-, oro-, etc.). In each of these, with respect to precipitation, diverse rainfall types or ombric variants (e.g., arid, semiarid, dry, subhumid, etc.) must be recognized. In Western Europe, fifteen bioclimatic belts (thermotypes) are accepted, each with their own characteristic rainfall types, whose thermic values expressed as It units are given as follows. In the Eurosiberian region: thermocoline (370 to 310), coline (310 to 180), montane (180 to 50), subalpine (50 to -50) and alpine (-50 to -170); in the Mediterranean region: thermomediterranean (450 to 350), mesomediterranean (350 to 210), supramediterranean (210 to 70), oromediterranean (70 to -10) and cryoromediterranean (-10 to -100); finally, in the Macaronesian region (Canary Islands): infracanarian (580 to 460), thermocanarian (460 to 320), mesocanarian (320 to 200), supracanarian (200 to 80) and orocanarian (80 to -10).

Chapters
1. Introduction, 2. Bioclimatic belts (thermotypes), 3. Bioclimatic horizons (thermotypes), 4. Ombric types (ombric variants), 5. Altitudinal zonations, 6. Climatic diagrams (ombrothermoclimatic diagrams), 7. Bioindicators, 8. Biogeography of West Europe. Bibliography.

1. INTRODUCTION

Bioclimatology is an ecological science which tries to clarify existing relationships between living organisms (Biology) and climate (Physics). It differs from Climatology because its descriptive models and units have been delimited according to species and biocoenoses (bioindicators) wich, given their stability, have been to the present time principally botanical.

Among climatic factors, rainfall and temperature have been considered as the most directly responsible for the distribution of ecosystems on Earth. For that reason, several numeric and graphic indices have been proposed during this century in order to establish such relationship between the climate and living organisms with emphasis on vegetation.

2. BIOCLIMATIC BELTS (THERMOTYPES)

By belts, primarily we mean each thermoclimatic type or space which is produced in an altitudinal or latitudinal cliserie (zonation). In practice, these bioclimatic units are conceived and delimited according to phytocoenosis which offer obvious correlations with specific thermoclimatic intervals or cesurae. The phenomenon of altitudinal or latitudinal thermic zonation has universal jurisdiction, and in every biogeographical region or group of similar regions there are special bioclimatic belts with their own particular thermic values, which may be calculated by using thermicity indices (thermotypes). Every type of belt has obviously different ombric types or variants according to its particular amount precipitation and seasonal distribution of rainfall (ombrotypes, see chapter 4 and table 1).

It should be remembered that the thermicity index (It) is the value or figure produced by adding up in tenths of degrees centigrade: T (mean annual temperature), m (mean minimum temperatures of the coldest month) and M (mean maximum temperatures of the coldest month); this is expressed as It = (T+m+M)10. The advantage of this universal application index centres on the fact that the value of the minimum temperatures of the coldest month (m) and its correction, due to the time it lasts, with average maximum temperatures also for the coldest month (M), works as a limiting factor in the law of minimum.

So far we have established correlations between climate and living organisms, in particular with vegetation in Europe, North Africa, the Himalayas and Tropical America (Peru, Ecuador, Bolivia). Together, we know with a certain amount of detail about the bioclimatic belts which correspond to the Mediterranean regions or groups of regions (6 belts), Eurosiberian region (5 belts), Macaronesian region (5 belts), Andine region (6 belts), Amazon region (3 belts), Hindustan region (3 belts) and Himalayan region (4 belts).

As far as Western Europe is concerned, five of the existing six bioclimatic belts have been delimited in the Mediterranean region: (infra-), thermo-, meso-, supra-, oro- and cryoromediterranean; five if we consider as a belt the thermocoline horizon in the Eurosiberian region: thermocoline, coline, montane, subalpine and alpine; and five in the Macaronesian area (Canary Islands): infra-, thermo-, meso-, supra-, and orocanarian belts. Each of these belts, which may be divided up into horizons, possesses its own particular species and types of vegetation, according to its particulars ombric variants.

In this bioclimatic approach --the eigth produced since 1981-- we calculate that It values are only subject to a margin of error of no more than 10 (one degree centigrade) in relation to the cesurae in the ecosystems and discriminant types of vegetation. The limits of the average temperatures and its reciprocal relations are given in the tables 1 and 2.

Eurosiberian region

Alpine .... T -1 to 3, M -4 to 0, m -12 to -8, It -170 to -50
Subalpine .... T 3 to 6, M 0 to 3, m -8 to -4, It -50 to 50
Montane .... T 6 to 10, M 3 to 8, m -4 to 0, It 50 to 180
Coline .... T 10 to 14, M 8 to 12, m 0 to 5, It 180 to 310
Thermocoline .... T 14 to 16, M 10 to 14, m 5 to 7, It 310 to 370

Mediterranean region

Cryoromediterranean .... T 2 to 4, M -3 to 1, m -9 to -6, It -150 to -10
Oromediterranean .... T 4 to 8, M 1 to 3, m -6 to -4, It -10 to 70
Supramediterranean .... T 8 to 13, M 3 to 9, m -4 to -1, It 70 to 210
Mesomediterranean .... T 13 to 16, M 9 to 14, m -1 to 5, It 210 to 350
Thermomediterranean .... T 16 to 18, M 14 to 18, m 5 to 9, It 350 to 450
Inframediterranean .... T 18 to 20, M 18 to 20, m 9 to 10, It 450 to 500
(North Africa only)

Macaronesian region (Canarian subregion)

Orocanarian .... T 3 to 6, M 2 to 4, m -5 to -2, It -100 to -10
Supracanarian .... T 6 to 10, M 4 to 8, m-2 to 2, It 80 to 200
Mesocanarian ....T 10 to 14, M 8 to 13, m 2 to 5, It 200 to 320
Thermocanarian .... T 14 to 18, M 13 to 18, m 5 to 10, It 320 to 460
Infracanarian .... T 18 to 21, M 18 to 21, m 10 to 16, It 460 to 580

BIOCLIMATIC BELT	T	M	m	It
Alpine	(-2)-1 to 3 (4)	(-6)-4 to 0 (2)	(-14)-12 to -8(-6)	-170 to -50
Subalpine	(2) 3 to 6 (7)	(-2) 0 to 3 (5)	(-10) -8 to -4(-2)	-50 to 50
Montane	(5) 6 to 10(11)	(1) 3 to 8(10)	(-6) -4 to 0 (2)	50 to 180
Coline	(9)10 to 14(15)	(6) 8 to 12(14)	(-2) 0 to 5 (7)	180 to 310
Thermocoline	(13)14 to 16(17)	(10)12 to 14(16)	(3) 5 to 7 (9)	310 to 370

	T	M	m	It
Cryoromedit.	(1) 2 to 4 (5)	(-5)-3 to 1 (3)	(-11) -9 to -6(-4)	-100 to -10
Oromedit.	(3) 4 to 8 (9)	(-1) 1 to 3 (5)	(-8) -6 to -4(-2)	-10 to 70
Supramedit.	(7) 8 to 13(14)	(1) 3 to 9(11)	(-6) -4 to -1 (1)	70 to 210
Mesomedit.	(12)13 to 16(17)	(7) 9 to 14(17)	(-3) -1 to 5 (7)	210 to 350
Thermomedit.	(15)16 to 18(19)	(12)14 to 18(20)	(3) 5 to 9(10)	350 to 450
Inframedit.	(17)18 to 20(21)	(16)18 to 20(22)	(7) 9 to 10(12)	450 to 500

	T	M	m	It
Orocanarian	(2) 3 to 6 (7)	(0) 2 to 4 (6)	(-7) -5 to -2 (0)	-10 to 80
Supracanarian	(5) 6 to 10(11)	(2) 4 to 8(10)	(-4) -2 to 2 (4)	80 to 200
Mesocanarian	(9)10 to 14(15)	(6) 8 to 13(15)	(0) 2 to 5 (7)	200 to 320
Thermocanarian	(13)14 to 18(19)	(11)13 to 18(20)	(3) 5 to 10(12)	320 to 460
Infracanarian	(17)18 to 21(22)	(16)18 to 21(23)	(8) 10 to 16(18)	460 to 580

Table 1. Bioclimatic belts: Thermotypes (thermicity index unities and limit values).

3. BIOCLIMATIC HORIZONS (THERMOTYPES)

Within bioclimatic belts, it is possible to identify horizons or sub-belts that usually reveal changes in the distribution of bioindicators as species or series of vegetation, sub-series and plant communities. Intentionally, these horizons also coincide with the distribution limits of many natural or cultivated species. The It interval (thermicity index) is the most important figure for establishing the approximate bioclimatic limits of horizons or sub-belts. Names between brackets are synonyms that we have habitually used for some Eurosiberian horizons.

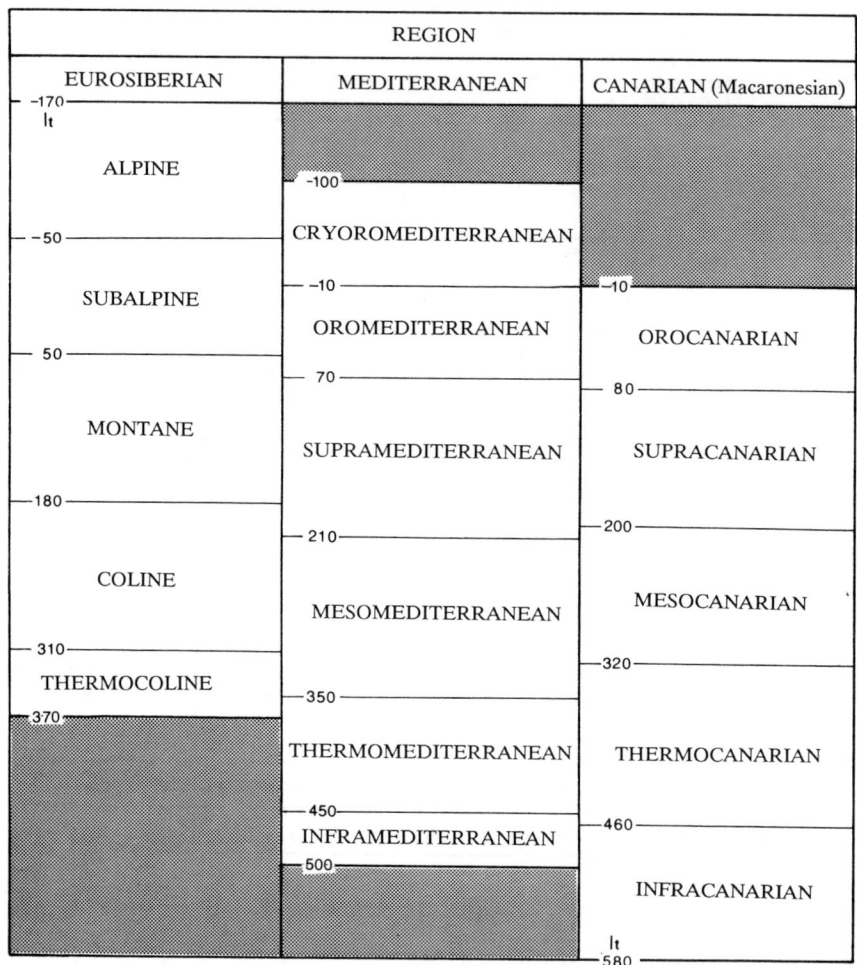

Table 2. Bioclimatic belts (Thermotypes)

Eurosiberian region: Upperalpine (subnival): -111 to -170; Loweralpine: -51 to -110; Uppersubalpine: -1 to -50; Lowersubalpine: 49 to 0; Uppermontane (altimontane): 114 to 50; Lowermontane (middlemontane): 179 to 115; Uppercoline (submontane): 244 to 180; Lowercoline (eucoline): 309 to 245; Thermocoline: 370 to 310.

Mediterranean region: Uppercryoromediterranean: -56 to -100; Lowercryoromediterranean: -11 to -55; Upperoromediterranean: 29 to -10; Loweroromediterranean: 69 to 30; Uppersupramediterranean: 119 to 70; Middlesupramediterranean: 163 to 120; Lowersupramediterranean: 209 to 164; Uppermesomediterranean: 256 to 210; Middlemesomediterranean: 303 to 257; Lowermesomediterranean: 349 to 304; Upperthermomediterranean: 401 to 350; Lowerthermomediterranean: 449 to 400; Inframediterranean: 500 to 450.

Canarian subregion (Macaronesian region): Upperorocanarian: 34 to -10; Lowerorocanarian: 79 to 35; Uppersupracanarian: 139 to 80; Lowersupracanarian: 199 to 140; Uppermesocanarian: 259 to 200; Lowermesocanarian: 319 to 260; Upperthermocanarian: 384 to 320; Lowerthermocanarian: 459 to 385; Upperinfracanarian: 519 to 460; Lowerinfracanarian: 580 to 520.

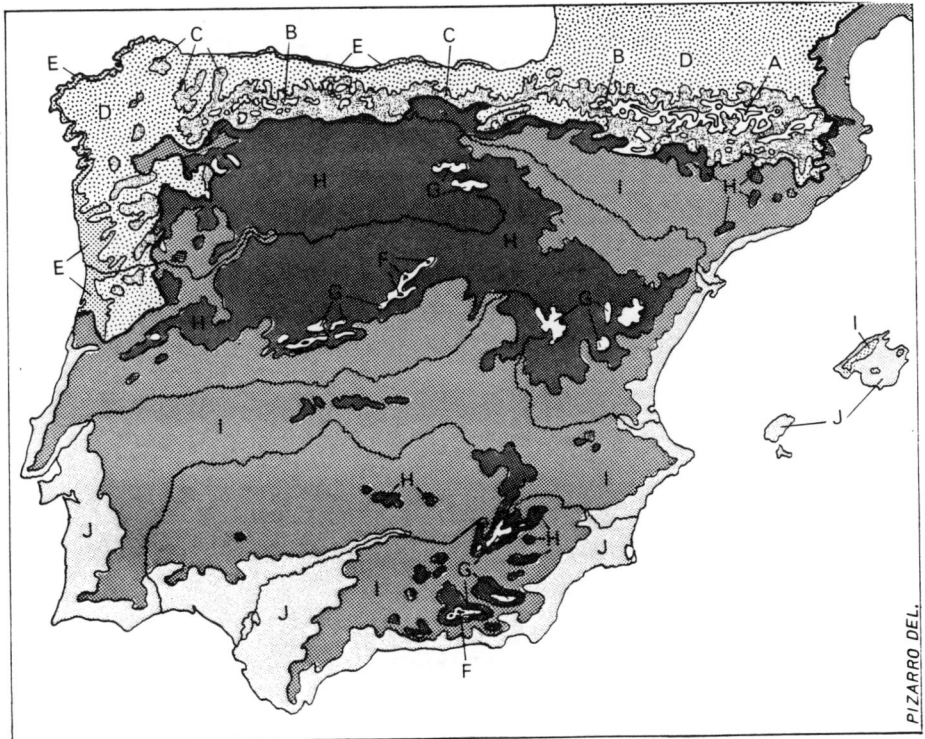

MAP 1.- Bioclimatic belts (thermotypes) of Iberian Peninsula: A. Alpine, B. Subalpine, C. Montane, D. Coline, E. Thermocoline, F. Cryoromediterranean, G. Oromediterranean, H. Supramediterranean, I. Mesomediterranean, J. Thermomediterranean.

Belt	1 Arid	2 Semiarid	3 Dry	4 Subhumid	5 Humid	6 Perhumid	7 Ultra-perhumid
A Alpine	-	-	-	-	A5	A6	A7
B Subalpine	-	-	-	-	B5	B6	B7
C Montane	-	-	-	C4	C5	C6	C7
D Coline	-	-	-	D4	D5	D6	D7
E Thermocoline	-	-	-	E5	E5	E6	-
F Cryoromediterranean	-	-	[F3]	[F4]	F5	F6	F7
G Oromediterranean	-	[G2]	[G3]	G4	G5	G6	G7
H Supramediterranean	[H1]	H2	H3	H4	H5	H6	H7
I Mesomediterranean	[I1]	I2	I3	I4	I5	I6	-
J Thermomediterranean	J1	J2	J3	J4	J5	-	-
K Inframediterranean	[K1]	[K2]	[K3]	-	-	-	-
L Orocanarian	-	-	L3	-	-	-	-
M Supracanarian	-	-	M3	M4	-	-	-
N Mesocanarian	-	N2	N3	N4	N5	-	-
O Thermocanarian	-	O2	O3	-	-	-	-
P Infracanarian	P1	P2	P3	-	-	-	-

Table 3. Bioclimatic belts (Thermo-ombric types) existing in West Europe
(Between brackets, types that only exist in North Africa).

Ombric types: 1. Arid, 2. Semiarid, 3. Dry, 4. Subhumid, 5. Humid, 6. Perhumid, 7. Ultraperhumid.

Alps	Pyrenees	P.Europa	Guadarrama	S.Nevada	Teide	ALTITUDE
						3700 m
(shaded)				IT	−10 / 2	3600
	IT			−100	30 (Orocan.)	3400
IT	−158			−70	58	3200
−170	−132			−44	86	3000
−132	−106 (Alpine)			−18	114 (Supracanarian)	2800
−104 (Alpine)	−80	IT		8	142	2600
−76	−54	−84 (Alpine)	IT (Alpine)	34	170	2400
−48	−28	−54	−44 (Cryo.)	60	200	2200
−24	−2	−24 (Subalpine)	−10 (Oromed.)	86 (Oromed.)	230	2000
2 (Subalpine)	24	6	21	118	260 (Mesocanarian)	1800
28	50	36	49	150	290	1600
54	76	66	77	182 (Supramediterranean)	320	1400
82	108	99 (Montane)	105 (Supramediterranean)	214	350 (Thermocanarian)	1200
110	140 (Montane)	132	139	246	380	1000
138 (Montane)	172	165	173	278 (Mesomediterranean)	412	800
166	204	198	207	316	440	600
194 (Coline)	236 (Coline)	231 (Coline)	241 / 258 (Meso.)	354	486 (Infracanarian)	400
222	268 (Mesom.)	264		392 (Thermomed.)	532	200
		297		430	578	0 m
		330 (Ther.)				

Table 4. Altitudinal zonation of bioclimatic belts in Western Europe. Variations of decreasing It values every 100 meters of altitude: <u>Alps</u> (continental, Wallis): 400 m to 1400 m, It = 14/100 m, 1400 m to 3200 m, It = 13/100 m, <u>Pyrenees</u> (continental, central): south, 200 m to 1400 m, It = 16/100 m, north 100 m to 1400 m, It = 15/100 m. <u>Picos de Europa</u> (North Spain, oceanic): north, 0-1600 m, It = 16,5/100 m, south 700 m to 1600 m, It = 14/100 m, 1600 m to 2648 m, It = 15/100 m. <u>Sierra de Guadarrama</u> (Central Spain, Continental): south 500 m to 1400 m, It = 17/100 m, 1400 m to 2430 m, It = 14/100 m. <u>Sierra Nevada</u> (South Spain, oceanic and continental) northoccidental, 0 to 800 m, It = 19/100 m, 800 m to 2000 m, It = 16/100 m, south, 0 to 2000 m, It = 18/100 m, 2000 m to 3481 m, It = 13/100 m. <u>Teide</u> (Canary Islands, oceanic): north, 0 to 600 m, It = 17/100 m, south, 0 to 600 m, It = 23/100m, 600 to 2400 m, It = 15/100 m, 2400 m to 3717 m It = 14/100 m.

4. OMBRIC TYPES (OMBRIC VARIANTS)

The possible ombric types and their annual average values in the Eurosiberian, Mediterranean and Macaronesian regions belonging to Western Europe, North Africa and Canary Islands are the following (in every ombrictype can be recognized if necessary two subtypes (upper & lower) dividing by half its rainfall values:

Eurosiberian region

Subhumid	500 to 900 mm
Humid	900 to 1400 mm
Perhumid	1400 to 2100 mm
Ultraperhumid	> 2100 mm

Mediterranean region

Arid	100 to 200 mm
Semiarid	200 to 350 mm
Dry	350 to 600 mm
Subhumid	600 to 1000 mm
Humid	1000 to 1600 mm
Perhumid	1600 to 2300 mm
Ultraperhumid	> 2300 mm

Canarian subregion (Macaronesian region)

Arid	100 to 200 mm
Semiarid	200 to 350 mm
Dry	350 to 550 mm
Subhumid	550 to 800 mm
Humid	> 800 mm

In summary, we can specify that only 52 among 112 possibles bioclimatic belts (thermo-ombric types) existing in West Europe, Canary Islands included (see table 3).

5. ALTITUDINAL ZONATIONS

The distribution of flora and vegetation in changing belts or stages related to decreasing temperatures due to changing altitude is an universal phenomenon recognized and described in the last two centuries.The higher and abrupter the mountains are the greater changes in vegetal ecosystems. Diversity of flora and climate also play an essential role. These belts of vegetation that change with altitude in their structure, biotic composition and mesological features have been the basis not only for the definition of bioindicators but also for understanding and delimitation of bioclimatic belts, that is to say, of each one of the types or groups of changing environments in any altitudinal or latitudinal zonation.

We will now try to outline in a brief and comparative way the impact over the potential vegetation of the climatic changes associated with altitude in mountains of western Europe. Because the abstract must necessarily brief, comparative and systematic, we will use models that represent the maturity stage of climatophilous vegetation series in these territories.

The altitude versus distribution of bioclimatic belts of Alps, Pyrenees, Picos de Europa, Sierra de Guadarrama, Sierra Nevada and Teide (Canary Islands) is represented comparatively in table 4, and its natural potential vegetation, head of the series or *sigmetum* in the illustrations A-G.

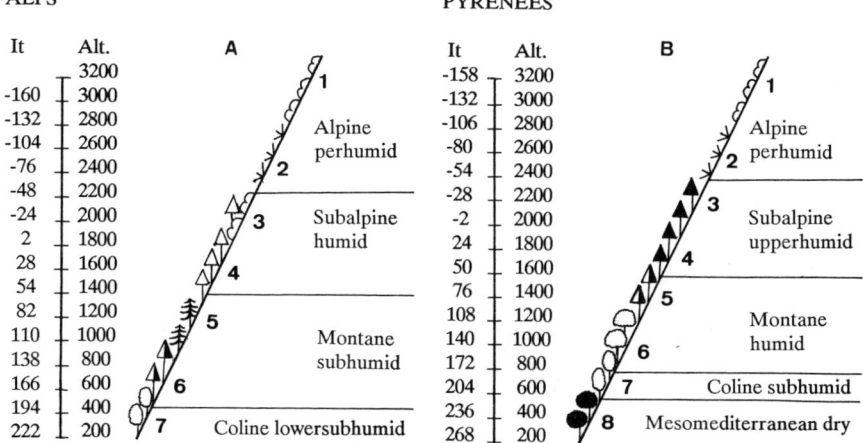

ALPS

PYRENEES

A. Altitudinal zonation of vegetal ecosystems (natural potential vegetation) of the <u>Alps</u>: from river Rhone, Valais (350 m) to Monte Rosa (4650 m, lim. subnival 3100 m). Acidophilous-continental zonation.

1. *Artemisio genipi-Saxifragetum muscoidis* (subnival, open stony communities); 2. *Caricetum curvulae* (loweralpine cryophilous grasslands); 3. *Junipero nanae-Arctostaphyletum uva-ursi pinetosum cembrae* (uppersubalpine sunny matorral and pine forest); 4. *Vaccinio myrtilli-Pinetum cembrae* (subalpine pine forest), 5. *Luzulo luzulinae-Piceetum abietis (=Piceetum subalpinum)* (lowersubalpine and altimontane spruce forest); 6. *Ononido rotundifoliae-Pinetum sylvestris* (mesomontane, xerophilous pine forest); 7. *Campanulo boloniensis-Quercetum pubescentis* (submontane, thermophilous oak forest).

B. Altitudinal zonation of vegetal ecosystems (natural potential vegetation) of the <u>Pyrenees</u>: From Ebro Valley (250 m) to Monte Perdido (3371 m). Basophilous zonation.
1. *Saxifrago iratianae-Androsacetum ciliatae* (subnival, open stony communities); 2. *Carici rosae-Elynetum myosuroidis* (loweralpine basophilous grasslands); 3. *Rhododendro ferruginei-Pinetum uncinatae* (subalpine perhumid shady pine forest); 4. *Arctostaphilo uva-ursi-Pinetum uncinatae* (subalpine sunny pine forest); 5. *Echinosparto horridi-Pinetum sylvestris* (altimontane sunny pine forest); 6. *Buxo-Fagetum sylvestris* (mesomontane beech forest); 7. *Buxo-Quercetum pubescentis* (coline-montane deciduous oak forest); 8. *Quercetum rotundifoliae* (mesomediterranean evergreen oak forest).

PICOS DE EUROPA

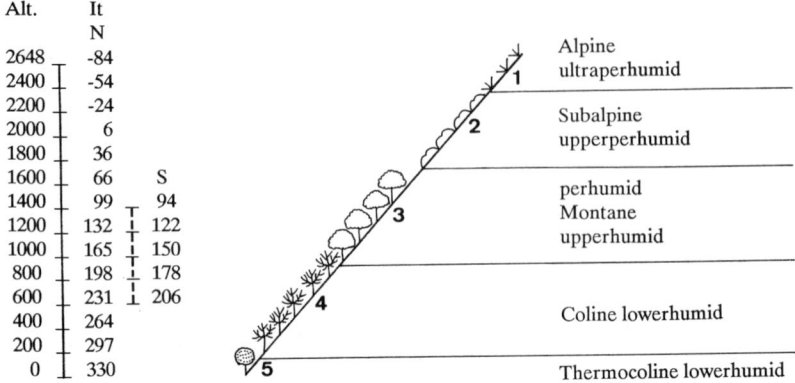

Alt.	It
	N
2648	-84
2400	-54
2200	-24
2000	6
1800	36
1600	66	S
1400	99	94
1200	132	122
1000	165	150
800	198	178
600	231	206
400	264
200	297
0	330

Alpine ultraperhumid

Subalpine upperperhumid

perhumid Montane upperhumid

Coline lowerhumid

Thermocoline lowerhumid

C. Altitudinal zonation of vegetal ecosystems (natural potential vegetation) of the Picos de Europa: from Cantabrian sea to Torre de Cerredo (2648 m) (North of Spain). Basophilous, oceanic type).
1. *Oxytropi pyrenaicae-Elynetum myosuroidis* (alpine grassland); 2. *Daphno cantabricae-Arctostaphiletum uva-ursi* (subalpine scrub); 3. *Carici sylvaticae-Fagetum sylvaticae* (montane beech forest); 4. *Polysticho setiferi-Fraxinetum excelsioris* (coline mixed oak forest); 5 *Lauro nobilis-Quercetum ilicis* (thermocoline evergreen oak forest). The vegetation in the southern exposition to the Duero river from 1000 to 650 m is supramediterranean.

SIERRA DE GUADARRAMA

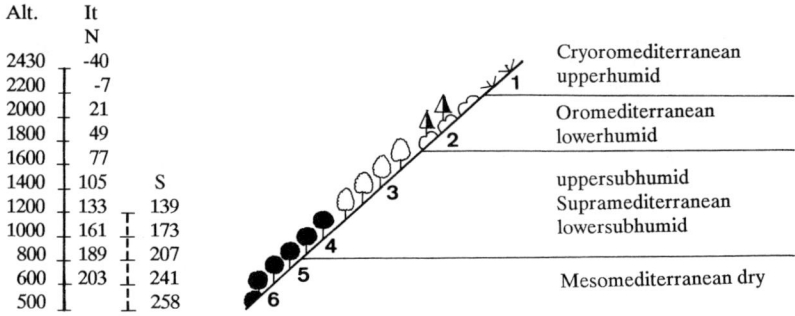

Alt.	It
	N
2430	-40
2200	-7
2000	21
1800	49
1600	77
1400	105	S
1200	133	139
1000	161	173
800	189	207
600	203	241
500		258

Cryoromediterranean upperhumid

Oromediterranean lowerhumid

uppersubhumid Supramediterranean lowersubhumid

Mesomediterranean dry

D. Altitudinal zonation of vegetal ecosystems (natural potential vegetation) of the Sierra de Guadarrama: from Tajo valley (New Castilla) to the peak of Peñalara (2430 m), south exposition (Central Spain). Silicicolous, continental xerophytic type.
1. *Hieracio myriadeni-Festucetum aragonensis* (cryoxerophilous grasslands); 2. *Senecioni carpetani-Cytisetum oromediterranei* (oromediterranean scrub and pine forest); 3. *Luzulo forsteri-Quercetum pyrenaicae* (supramediterranean decidous oak forest); 4 & 5. *Junipero oxycedri-Quercetum rotundifoliae* (supra and mesomediterranean silicicolous evergreen oak forest); 6. *Quercetum rotundifoliae* (mesomediterranean basophilous evergreen oak forest). The supramediterranean overtakes an altitude of 650 m in North exposition to Duero river.

234

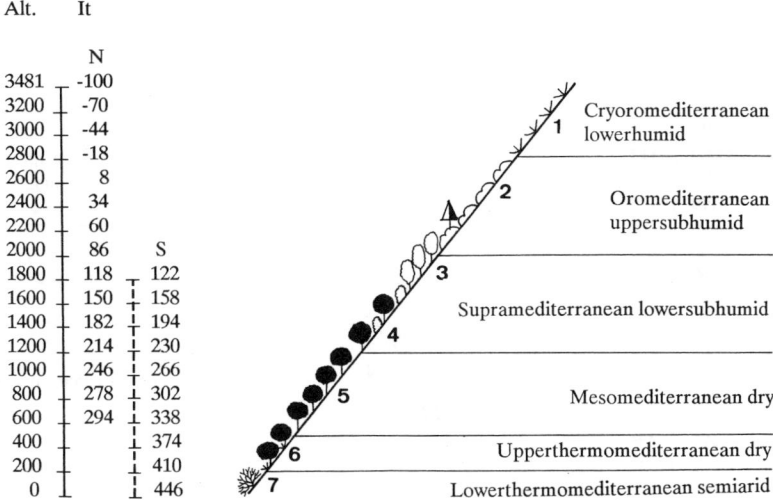

E. Altitudinal zonation of vegetal ecosystems (natural potential vegetation) of the Sierra Nevada: from Mediterranean Sea (Cabo Sacratif) to the Mulhacen peak (3481 m) (South of Spain). Basiphilous, silicicolous, semioceanic and continental type.

1. *Erigeronto frigidi-Festucetum clementei* (cryoromediterranean, cryoxerophilous grassland); 2. *Genisto versicoloris-Juniperetum nanae* (oromediterranean silicicolous scrub); 3. *Festuco elegantis-Quercetum pyrenaicae* (supramediterranean silicicolous decidous oak forest); 4. *Berberido hispanicae-Quercetum rotundifoliae* (supramediterranean basophilous evergreen oak forest); 5. *Paeonio coriaceae-Quercetum rotundifoliae* (mesomediterranean basophilous evergreen oak forest); 6. *Smilaci asperae-Quercetum rotundifoliae* (thermomediterranean basophilous evergreen oak forest); 7. *Rhamno angustifoliae-Maytenetum europeae* (lowerthermomediterranean semiarid scrub).

TEIDE (CANARY ISLANDS)

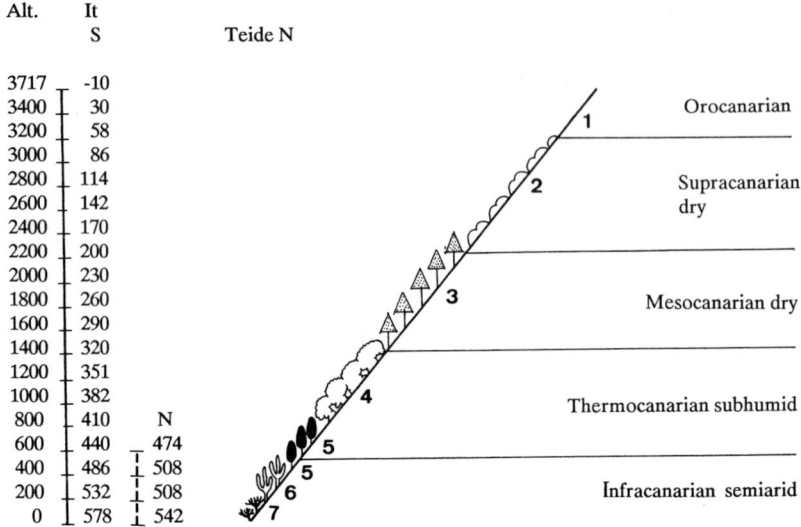

Alt.	It				
	S		Teide N		

3717 -10
3400 30 ················ Orocanarian
3200 58 ················ 1
3000 86
2800 114 ··············· Supracanarian
2600 142 ··············· dry
2400 170 ··············· 2
2200 200
2000 230
1800 260 ··············· Mesocanarian dry
1600 290 ··············· 3
1400 320
1200 351
1000 382 ··············· Thermocanarian subhumid
800 410 ··············· 4
600 440 474 ··············· 5
400 486 508
200 532 508 ··············· Infracanarian semiarid
0 578 542 ··············· 6, 7

F. Altitudinal zonation of vegetal ecosystems (natural potential vegetation) of <u>Teide</u> from the north littoral to the peak of Teide 3717 m. Canary Islands, Tenerife, volcanic oceanic type.

1. Orocanarian semiabiotic volcanic lavas; 2. *Spartocytisetum supranubii*, dry supracanarian scrub "retamar del Teide"; 3. *Cisto symphytifolii-Pinetum canariensis*, dry mesocanarian pine forest "pinar"; 4. *Perseo indicae-Lauretum azoricae*, foggy subhumid thermocanarian *Laurus* forest "laurisilva"; 5. *Mayteno canariensis-Juniperetum phoeniceae*, dry thermo-infracanarian *Juniperus* forest "sabinar"; 6. *Euphorbio lindleyii-Euphorbietum canariensis*, semiarid infracanarian succulent *Euphorbia* semidesert "cardonal"; 7. *Astydamio latifoliae-Euphorbietum balsamiferae*, arid-semiarid infracanarian aerohaline littoral succulent *Euphorbia* community "tabaibal litoral"; 8. *Ceropegio fuscae-Euphorbietum balsamiferae*, arid infracanarian succulent *Euphorbia* semidesert, in the south of Tenerife between 0-300 m.

6. CLIMATIC DIAGRAMS (OMBRO-THERMOCLIMATIC DIAGRAMS)

The climatic diagrams are very useful as graphic expressions of the climate of any territory. In order to explain the bioclimatic necesities, the ombrothermoclimatic diagrams initially suggested by Gaussen, later popularized by Walter & Leith and slightely modified and amplified by ourselves, can be successfully employed

In the Gaussen ombrothermoclimatic diagram, average monthly 2T and P values are plotted versus time (in months) in such a way that P=2T for each point in the ordinate scale. In dry period, the rainfall (P) line will be placed under the one corresponding to temperature (2T). In such a plot, the higher the area between both lines, the higher the aridity of the climate.

Thermicity negative index (Itn): ten times the result of adding together the mean monthly absolute minimum temperatures that are below zero in degrees centigrade.

Continentality index (Ic): This is obtained by adding Ma plus 0.6A divided into one hundred minus ma. The Ic values that we believe could define the different types of oceanity-continentality are the following: Ultraperoceanic < 10; Peroceanic 10 to 20; Oceanic 20 to 33; Semi-oceanic 33 to 43; Semi-continental 43 to 52; Continental 52 to 65; Percontinental 65 to 80; Ultrapercontinental > 80.

Summer mediterraneity index (Imv): This is obtained dividing summer months evapotranspiration into summer rainfall.

KEY TO THE CLIMATIC DIAGRAMS

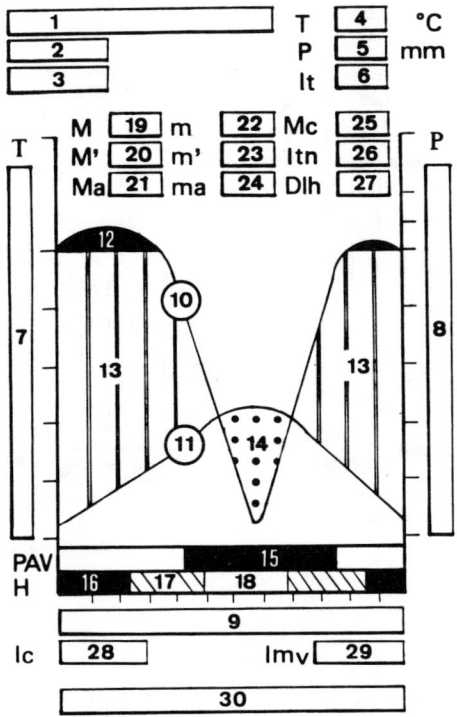

1. Meteorological station.
2. Altitude above mean sealevel.
3. Duration of observations in years.
4. T: Mean annual temperature in °C.
5. P: Mean annual rainfall in mm.
6. It: Thermicity index = (T+M+m)10.
7. Temperature scale in °C.
8. Rainfall scale in mm.
9. Month scale.
10.Mean monthly rainfall line.
11.Mean monthly temperature line.
12.Mean monthly rain > 100 mm (black scale reduced to 1/10).
13.Humid period.
14.Dry period.
15.PAV: Period of vegetative activity (tm > 7.5 °C).
16.Sure frost period (mean monthly absolute minimum temperatures < 0°C).
17.Probably frost period (mean monthly absolute minimum temperatures < 2°C).
18.Frost-free period.

19.M: Mean maximum temperatures of the coldest month.
20.M': Mean absolute maximum temperatures of the warmest month.
21.Ma: Mean annual absolute maximum temperatures.
22.m: Mean minimum temperatures of the coldest month.
23.m': Mean absolute minimum temperatures of the coldest month.
24.ma: Mean annual absolute minimum temperatures.
25.Mc: Mean maximum temperatures of the warmest month.
26.Itn: Negative thermicity index.
27.Dlh: Number of frost-free days.
28.Ic: Continentality index.
   (Ic = Ma - ma + (0.6 A/100)
29.Imv: Summer mediterraneity index =
   ETP(evapotranspiration)/Pv(summer rainfall)
30.Bioclimatic diagnosis.

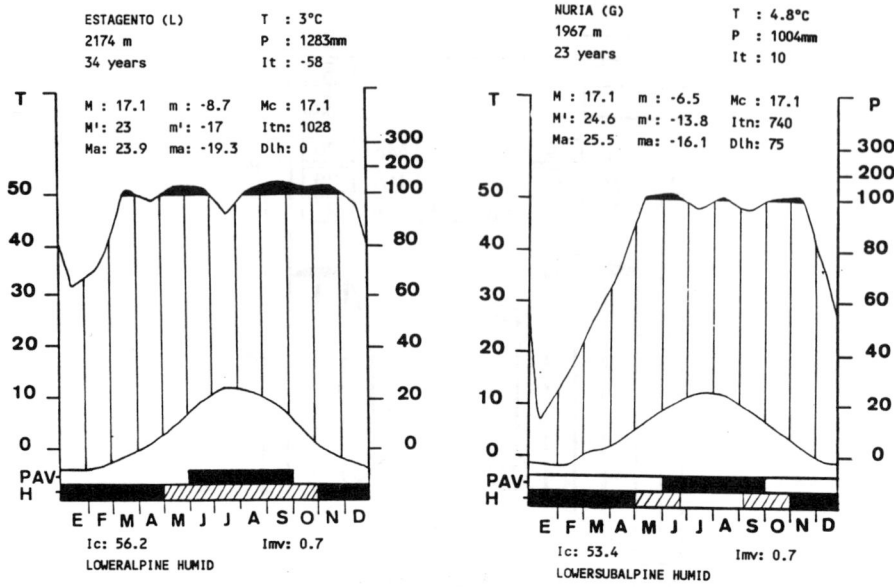

ESTAGENTO (L)   T : 3°C
2174 m   P : 1283mm
34 years   It : -58

T   M : 17.1   m : -8.7   Mc : 17.1   P
     M': 23   m': -17   Itn: 1028
     Ma: 23.9   ma: -19.3   Dlh: 0

Ic: 56.2   Imv: 0.7
LOWERALPINE HUMID

NURIA (G)   T : 4.8°C
1967 m   P : 1004mm
23 years   It : 10

T   M : 17.1   m : -6.5   Mc : 17.1   P
     M': 24.6   m': -13.8   Itn: 740
     Ma: 25.5   ma: -16.1   Dlh: 75

Ic: 53.4   Imv: 0.7
LOWERSUBALPINE HUMID

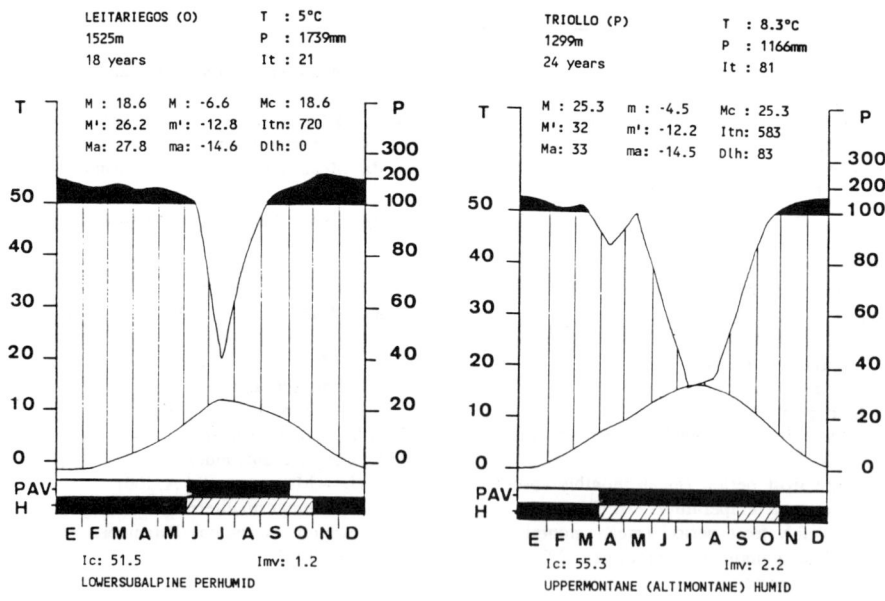

LEITARIEGOS (O) T : 5°C
1525m   P : 1739mm
18 years   It : 21

T   M : 18.6   M : -6.6   Mc : 18.6   P
     M': 26.2   m': -12.8   Itn: 720
     Ma: 27.8   ma: -14.6   Dlh: 0

Ic: 51.5   Imv: 1.2
LOWERSUBALPINE PERHUMID

TRIOLLO (P)   T : 8.3°C
1299m   P : 1166mm
24 years   It : 81

T   M : 25.3   m : -4.5   Mc : 25.3   P
     M': 32   m': -12.2   Itn: 583
     Ma: 33   ma: -14.5   Dlh: 83

Ic: 55.3   Imv: 2.2
UPPERMONTANE (ALTIMONTANE) HUMID

238

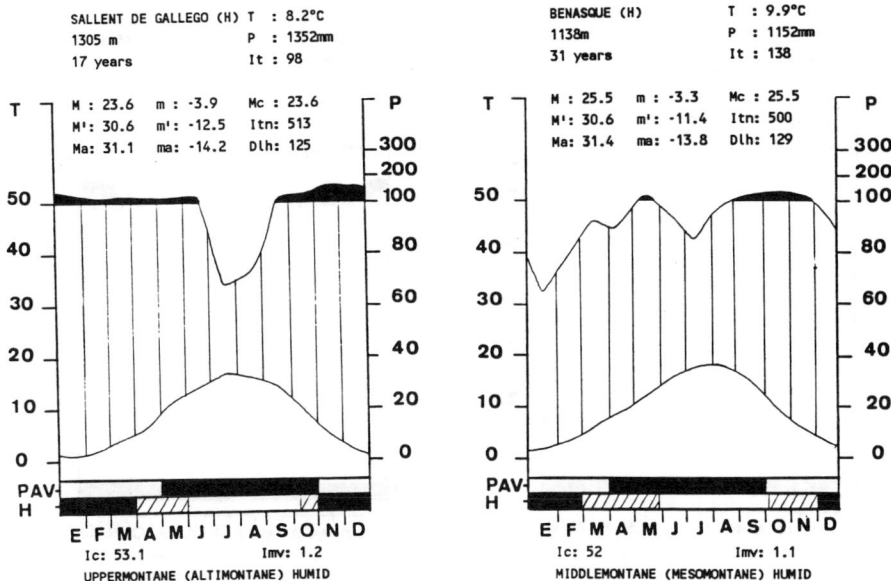

SALLENT DE GALLEGO (H) T : 8.2°C
1305 m   P : 1352mm
17 years   It : 98

T   M : 23.6   m : -3.9   Mc : 23.6   P
     M': 30.6   m': -12.5   Itn: 513
     Ma: 31.1   ma: -14.2   Dlh: 125

Ic: 53.1   Imv: 1.2
UPPERMONTANE (ALTIMONTANE) HUMID

BENASQUE (H)   T : 9.9°C
1138m   P : 1152mm
31 years   It : 138

T   M : 25.5   m : -3.3   Mc : 25.5   P
     M': 30.6   m': -11.4   Itn: 500
     Ma: 31.4   ma: -13.8   Dlh: 129

Ic: 52   Imv: 1.1
MIDDLEMONTANE (MESOMONTANE) HUMID

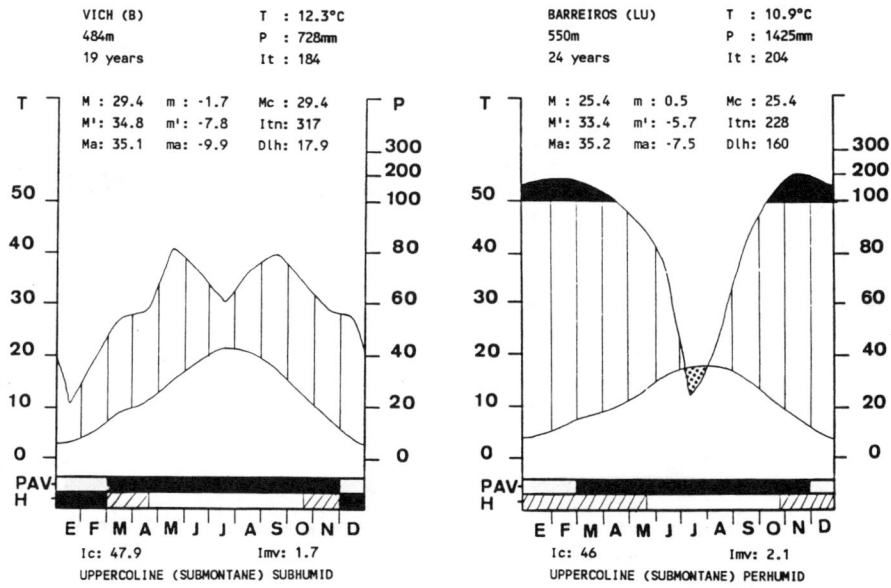

VICH (B)   T : 12.3°C
484m   P : 728mm
19 years   It : 184

T   M : 29.4   m : -1.7   Mc : 29.4   P
     M': 34.8   m': -7.8   Itn: 317
     Ma: 35.1   ma: -9.9   Dlh: 17.9

Ic: 47.9   Imv: 1.7
UPPERCOLINE (SUBMONTANE) SUBHUMID

BARREIROS (LU)   T : 10.9°C
550m   P : 1425mm
24 years   It : 204

T   M : 25.4   m : 0.5   Mc : 25.4   P
     M': 33.4   m': -5.7   Itn: 228
     Ma: 35.2   ma: -7.5   Dlh: 160

Ic: 46   Imv: 2.1
UPPERCOLINE (SUBMONTANE) PERHUMID

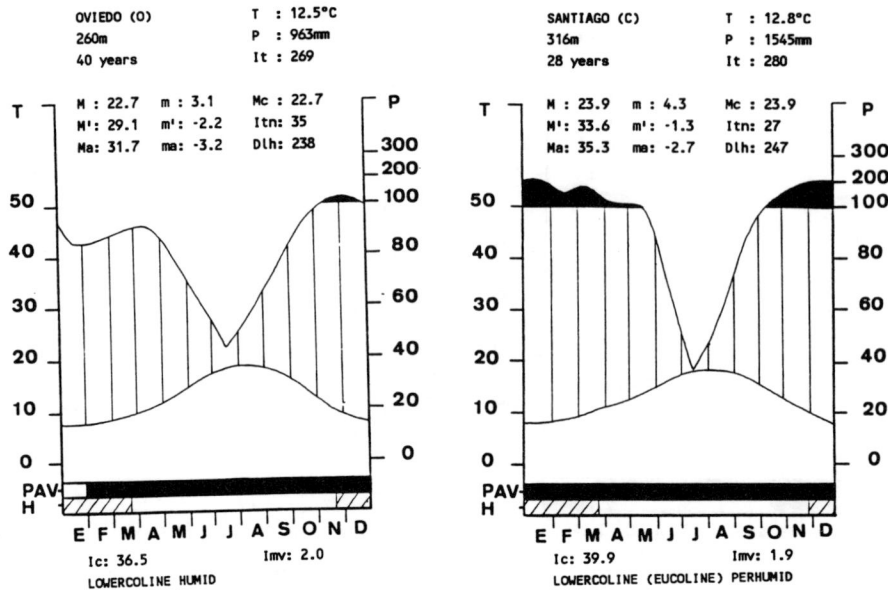

OVIEDO (O)   T : 12.5°C
260m   P : 963mm
40 years   It : 269

M : 22.7   m : 3.1   Mc : 22.7
M': 29.1   m': -2.2   Itn : 35
Ma: 31.7   ma: -3.2   Dlh: 238

Ic: 36.5   Imv: 2.0
LOWERCOLINE HUMID

SANTIAGO (C)   T : 12.8°C
316m   P : 1545mm
28 years   It : 280

M : 23.9   m : 4.3   Mc : 23.9
M': 33.6   m': -1.3   Itn: 27
Ma: 35.3   ma: -2.7   Dlh: 247

Ic: 39.9   Imv: 1.9
LOWERCOLINE (EUCOLINE) PERHUMID

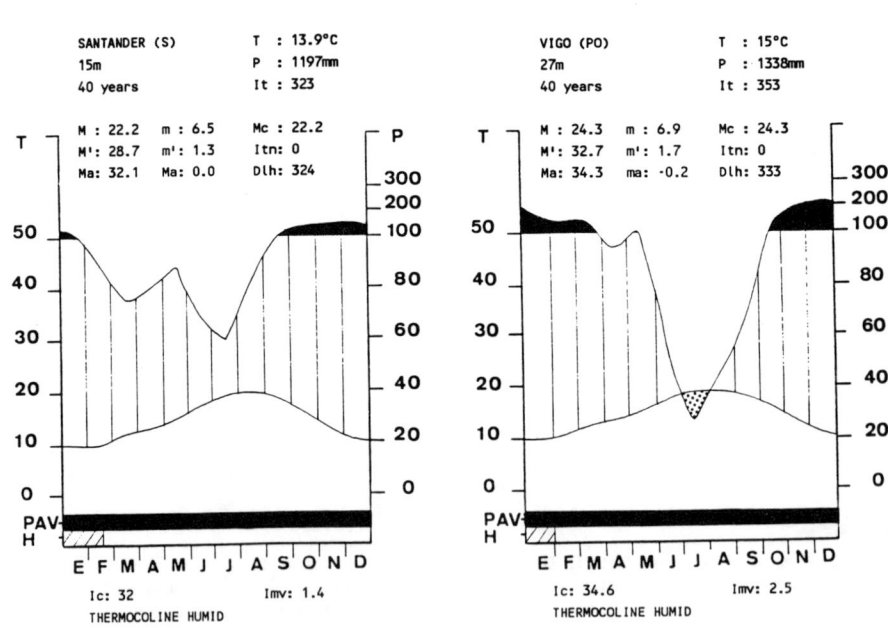

SANTANDER (S)   T : 13.9°C
15m   P : 1197mm
40 years   It : 323

M : 22.2   m : 6.5   Mc : 22.2
M': 28.7   m': 1.3   Itn: 0
Ma: 32.1   Ma: 0.0   Dlh: 324

Ic: 32   Imv: 1.4
THERMOCOLINE HUMID

VIGO (PO)   T : 15°C
27m   P : 1338mm
40 years   It : 353

M : 24.3   m : 6.9   Mc : 24.3
M': 32.7   m': 1.7   Itn: 0
Ma: 34.3   ma: -0.2   Dlh: 333

Ic: 34.6   Imv: 2.5
THERMOCOLINE HUMID

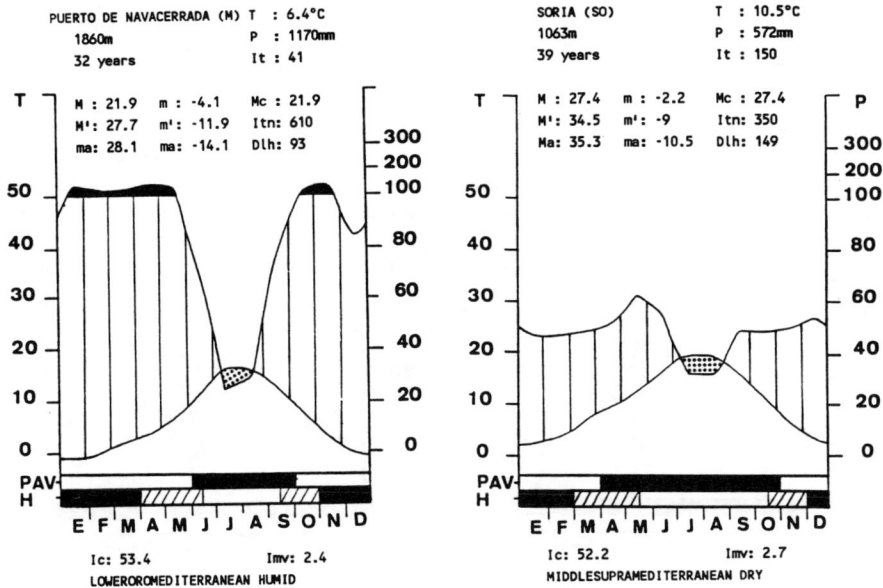

PUERTO DE NAVACERRADA (M) T : 6.4°C
1860m   P : 1170mm
32 years   It : 41

M : 21.9   m : -4.1   Mc : 21.9
M': 27.7   m': -11.9   Itn : 610
ma: 28.1   ma: -14.1   Dlh: 93

Ic: 53.4   Imv: 2.4
LOWEROROMEDITERRANEAN HUMID

SORIA (SO)   T : 10.5°C
1063m   P : 572mm
39 years   It : 150

M : 27.4   m : -2.2   Mc : 27.4
M': 34.5   m': -9   Itn : 350
Ma: 35.3   ma: -10.5   Dlh: 149

Ic: 52.2   Imv: 2.7
MIDDLESUPRAMEDITERRANEAN DRY

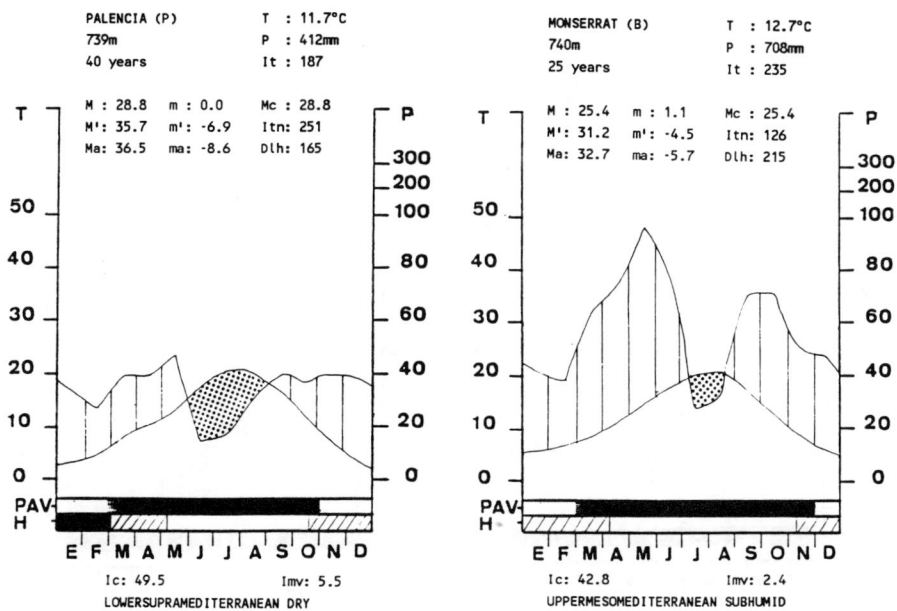

PALENCIA (P)   T : 11.7°C
739m   P : 412mm
40 years   It : 187

M : 28.8   m : 0.0   Mc : 28.8
M': 35.7   m': -6.9   Itn : 251
Ma: 36.5   ma: -8.6   Dlh: 165

Ic: 49.5   Imv: 5.5
LOWERSUPRAMEDITERRANEAN DRY

MONSERRAT (B)   T : 12.7°C
740m   P : 708mm
25 years   It : 235

M : 25.4   m : 1.1   Mc : 25.4
M': 31.2   m': -4.5   Itn: 126
Ma: 32.7   ma: -5.7   Dlh: 215

Ic: 42.8   Imv: 2.4
UPPERMESOMEDITERRANEAN SUBHUMID

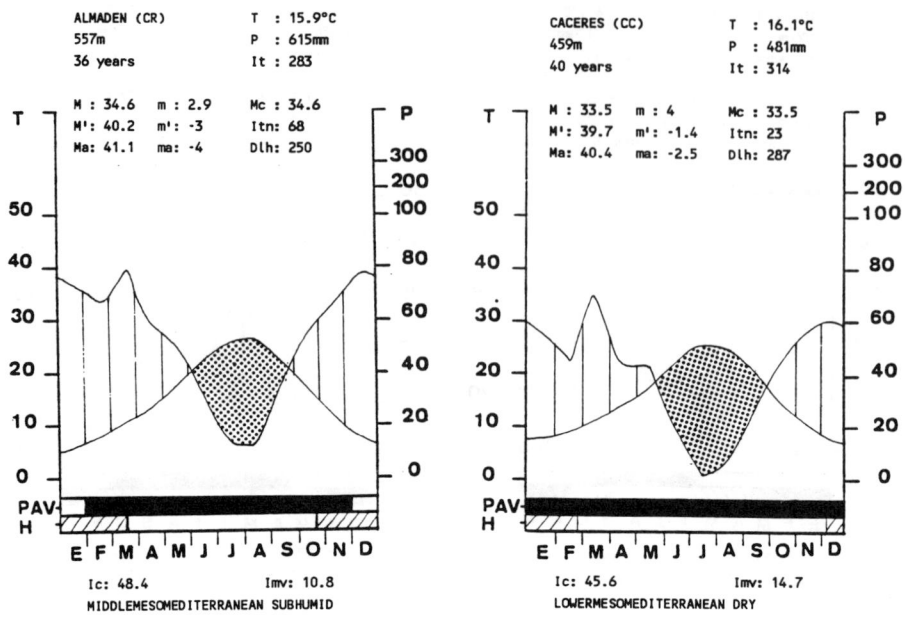

ALMADEN (CR)   T : 15.9°C
557m   P : 615mm
36 years   It : 283

M : 34.6   m : 2.9   Mc : 34.6
M': 40.2   m': -3   Itn: 68
Ma: 41.1   ma: -4   Dlh: 250

Ic: 48.4   Imv: 10.8
MIDDLEMESOMEDITERRANEAN SUBHUMID

CACERES (CC)   T : 16.1°C
459m   P : 481mm
40 years   It : 314

M : 33.5   m : 4   Mc : 33.5
M': 39.7   m': -1.4   Itn: 23
Ma: 40.4   ma: -2.5   Dlh: 287

Ic: 45.6   Imv: 14.7
LOWERMESOMEDITERRANEAN DRY

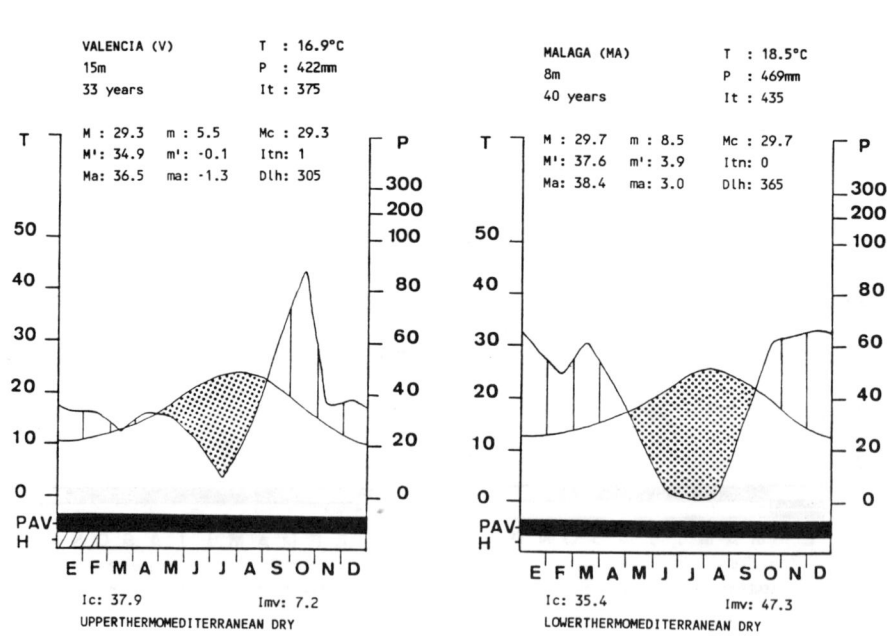

VALENCIA (V)   T : 16.9°C
15m   P : 422mm
33 years   It : 375

M : 29.3   m : 5.5   Mc : 29.3
M': 34.9   m': -0.1   Itn: 1
Ma: 36.5   ma: -1.3   Dlh: 305

Ic: 37.9   Imv: 7.2
UPPERTHERMOMEDITERRANEAN DRY

MALAGA (MA)   T : 18.5°C
8m   P : 469mm
40 years   It : 435

M : 29.7   m : 8.5   Mc : 29.7
M': 37.6   m': 3.9   Itn: 0
Ma: 38.4   ma: 3.0   Dlh: 365

Ic: 35.4   Imv: 47.3
LOWERTHERMOMEDITERRANEAN DRY

242

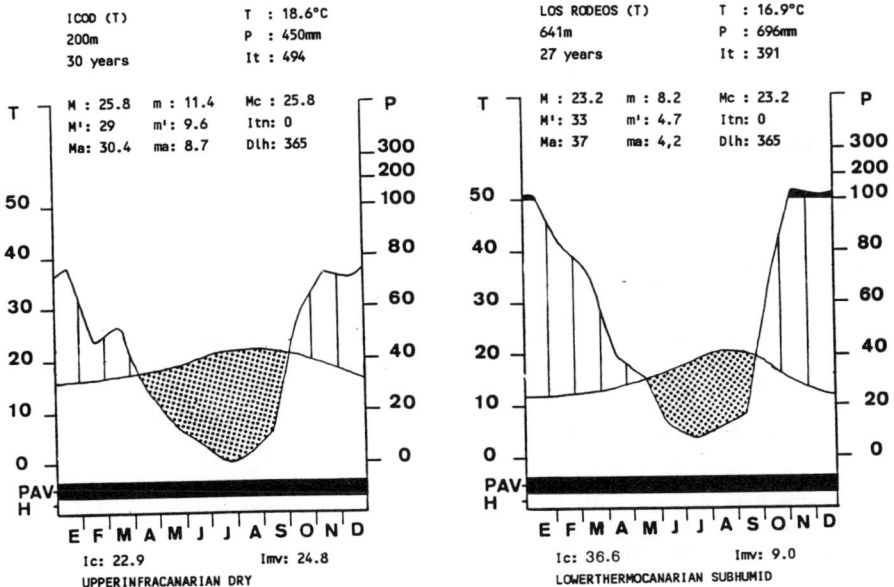

ICOD (T)   T : 18.6°C
200m   P : 450mm
30 years   It : 494

M : 25.8   m : 11.4   Mc : 25.8
M': 29   m': 9.6   Itn: 0
Ma: 30.4   ma: 8.7   Dlh: 365

Ic: 22.9   Imv: 24.8
UPPERINFRACANARIAN DRY

LOS RODEOS (T)   T : 16.9°C
641m   P : 696mm
27 years   It : 391

M : 23.2   m : 8.2   Mc : 23.2
M': 33   m': 4.7   Itn: 0
Ma: 37   ma: 4,2   Dlh: 365

Ic: 36.6   Imv: 9.0
LOWERTHERMOCANARIAN SUBHUMID

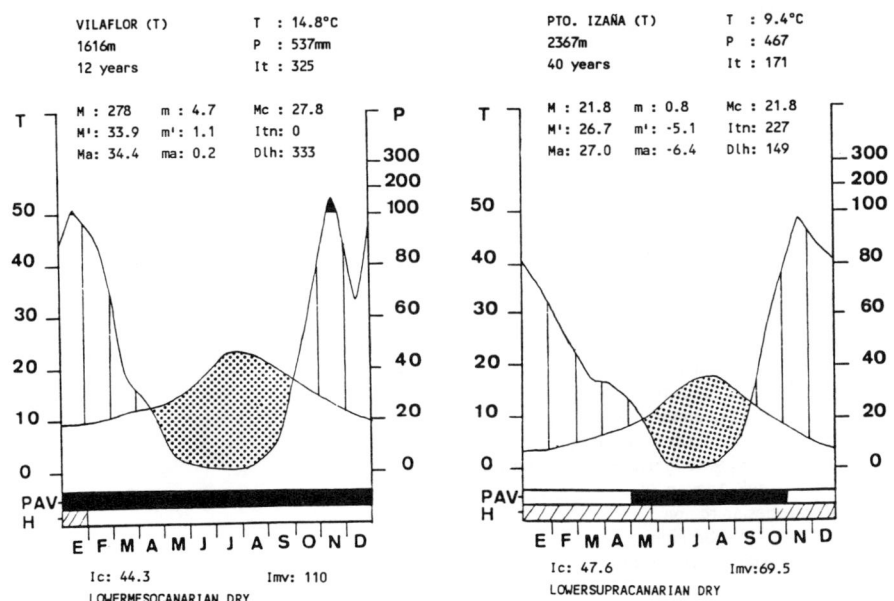

VILAFLOR (T)   T : 14.8°C
1616m   P : 537mm
12 years   It : 325

M : 278   m : 4.7   Mc : 27.8
M': 33.9   m': 1.1   Itn: 0
Ma: 34.4   ma: 0.2   Dlh: 333

Ic: 44.3   Imv: 110
LOWERMESOCANARIAN DRY

PTO. IZAÑA (T)   T : 9.4°C
2367m   P : 467
40 years   It : 171

M : 21.8   m : 0.8   Mc : 21.8
M': 26.7   m': -5.1   Itn: 227
Ma: 27.0   ma: -6.4   Dlh: 149

Ic: 47.6   Imv:69.5
LOWERSUPRACANARIAN DRY

243

7. BIOINDICATORS

Bioindicator refers to both taxa and syntaxa which can be used for the purpose of identifying characteristics of the environment or site. According to the mesological or territorial factors that they reveal, bioindicators can be classified into: 1. Climatics (thermoclimatics, ombroclimatics), 2. Biogeographics, 3. Edaphics, 4. Phytocoenotics, 5. Aquatics (freshwaters, marines), 6. Paleoclimatics, etc.

8. BIOGEOGRAPHY OF WESTERN EUROPE

Biogeography is a branch of Geography which deals whith the distribution of living organisms on Earth. Within this science relating Physics and Biology, Chorology has greatly developed in the last years. By Chorology, we mean, not only the speciality that studies the distribution and localization of species and communities but the subject which bearing in mind, through the knowledge of relics, the past and the present areas of taxa and syntaxa, as well the information from other sciences (such as Physics, Geography, Geology, Geobotany, Ecology, Bioclimatology, Edaphology, Zoology, etc.. Chorology aims at establishing a typology or biogeographic systematic in the emerged territories of our planet. In this way Chorology and Biogeography show a tendency to become synonyms. Finally, Biogeography such as we consider and treat it, can be also considered as a part of terrestrial ecology which analizes and clasifies the territorial biocoenoses and biotopes.

According to Braun-Blanquet, Takhtajan, Meusel, etc., the great ranges or hierarchies commonly accepted in Biogeography are the following: kingdom, region, province and sector. All of these unities must be geographic territories of continuous surface which may include orographic accidents and lithological diversity. In a recent approach to the European biogeographic typology, I have recognized 3 regions, 7 sub-regions and 43 provinces, 37 among which can be grouped in 12 superprovinces (map 1).

The separation between Eurosiberian and Mediterranean regions is mostly based on floristic, phytocoenological (particular series of vegetation) and bioclimatic (ombroclimatic indexes) criteria. The range of unitary region (Macaronesian) has been maintained for Atlantic archipelagos: Azores, Madeira and Canary Islands, especially because of its particular flora and vegetation, and in spite of their undeniable relationship to Atlantic-Middle European subregion (Azoric subregion), as ell as Western Mediterranean subregion (Canarian subregion).

A. Eurosiberian region
  Aa. Artic subregion
  Ab. Boreo-Continental subregion
  Ac. Atlantic-Middle-European subregion
    Ac1. Alpine-Pyrenean superprovince
    Ac2. Middle-European superprovince
    Ac3. Atlantic superprovince
    Ac4. Carpathian superprovince
    Ac5. Pontic-Pannonic superprovince
    Ac6. Illyric superprovince
B. Mediterranean region
  Ba. Western Mediterranean subregion
    Ba1. Mediterranean-Iberian-Levantine superprovince
    Ba2. Mediterranean-Iberian-Atlantic superprovince
    Ba3. Italic-Tyrrenian superprovince
  Bb. Estern Mediterranean subregion
    Ba4. Adriatic-Mediterranean superprovince
    Ba5. Greek-Aegean Superprovince
C. Macaronesian region
  Ca. Canarian subregion
  Ca1 Canarian superprovince

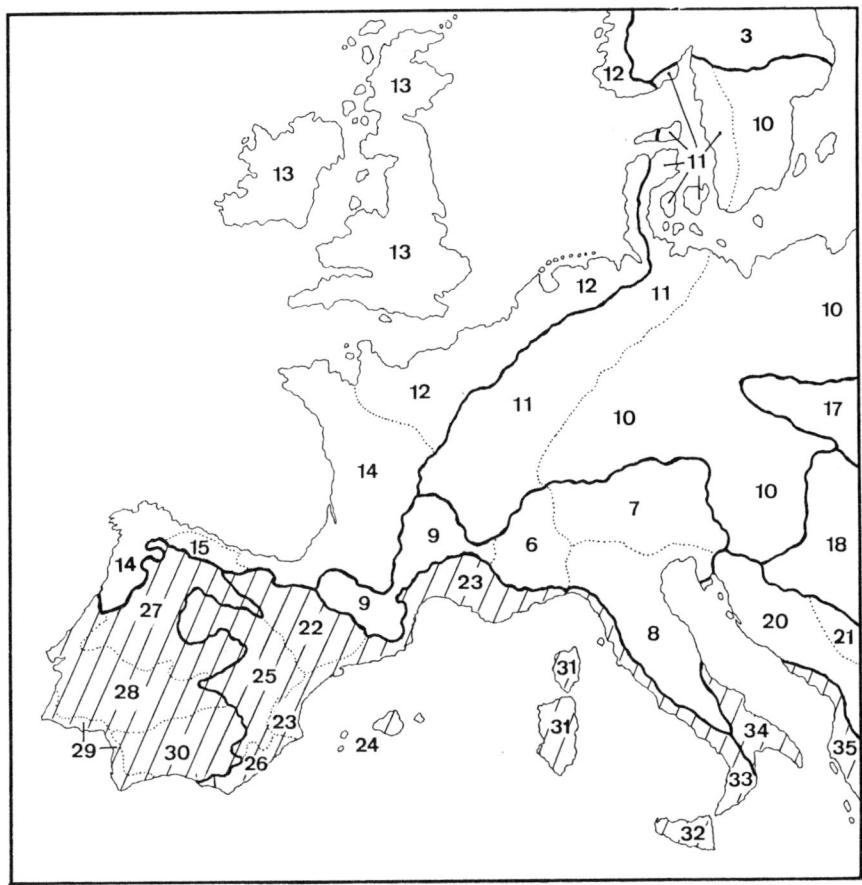

MAP 2.- Biogeographics provinces of Middle, Western and Southern Europe. A. Eurosiberian region. Aa. Artic subregion; Ab. Boreo-Continental subregion: 3. Boreal-European province; Ac. Atlantic-Middle-European subregion, Ac1. Alpine-Pyrenean superprovince: 6. Western-Alpine province, 7. Middle-Eastern-Alpine province, 8. Apennine-Padane province, 9. Pyrenean province (included Cevenian); Ac2. Middle-European superprovince: 10. Middle-European province, 11. Sub-Atlantic province; Ac3. Atlantic superprovince: 12. North-Atlantic province, 13. Britannic province, 14. Cantabrian (Cantabrian-Atlantic) province, 15. Oro-Cantabrian province; Ac4. Carpathian superprovince: 17. Tatric province; Ac5. Pontic-Pannonic superprovince: 18. Pannonic province; Ac6. Illyric superprovince: 20. Illyric-Bosnian province, 21. Servian-Macedonian province. B. Mediter-ranean region. Ba. Western Mediterranean subregion, Ba1. Mediterranean-Iberian-Levantine superprovince: 22. Aragonese province, 23. Valencian-Catalonian-Provençal province, 4. Balearic province, 25. Castilian-Maestracense-Manchegan province, 26. Murcian-Almeriense province, Ba2. Mediterranean-Iberian-Atlantic superprovince: 27. Carpetanic-Iberian-Leonese province, 28. Lusiatanian-Estremenian province, 29. Gaditanian-Onubian-Algarvian province, 30. Betic province; Ba3. Italic-Tyrrenian superprovince: 31. Corsican-Sardinian province, 32 Ligurian-Roman-Calabrian province, 33. Sicilian province; Bb. Eastern Mediterranean subregion, Ba4. Adriatic-Mediterranean superprovince: 34. Puglic province; Ba5. Greek-Aegean Superprovince: 35. Etolic-Epirotic province.

BIBLIOGRAPHY

Allué, A. J.L.: Subregiones fitoclimáticas de España, Ifie. Ed. Ministerio de Agricultura, Madrid, 1966.

Bagnouls, F., & Gaussen, H.: Les climats biologiques et leurs classifications. Ann. de Géogr., 66: 193 -220, 1957.

Bolós, O.: Corología de la flora dels Paisos Catalans. Institut d'Etudis Catalans, Secc. Ciencias. Orca 1: 48 pp. Barcelona, 1985.

Braun-Blanquet, J.: La végétation alpine des Pyrénées orientales. Barcelona, 1948.

Cabrera, A.L., & Willink, A.: Biogeografía de América Latina. Organización de los Estados Americanos, Ser. Biología. Monogr., 13: 1-120. Washington, 1973.

Daget, Ph.: Le bioclimatit méditerranéen: caracteres généraux, modes de caracterisation. Vegetatio, 34 (1): 1-20, 1977.

Díaz, T.E., & Penas, A.: Bases para el mapa fitogeográfico de la provincia de León. Diputación Provincial de León, 101 p., 1984.

Elías, C.F., & Ruiz, B.L.: Agroclimatología de España. Cuadernos Inia, núm. 7. Ministerio de Agricultura. Madrid, 1977.

Emberger, L.: Un projet d'une classification des climats du point de vue phytogéographique. Bull. Soc. Hist. Nat. Toulouse, 77: 97-124, 1942.

Emberger, L.; Gaussen, H. & De Phillipis, W.: Carte bioclimatique de la région méditerranéenne. Unesco. París, 1963.

Gaussen, H.: Théorie et classification des climats et des microclimats du point de vue phytogéographique. VII Congr. Inst. Bot., 125-130, 1954.

Géhu, J.M. & Rivas-Martínez, S. Notions fondamentales de phytosociologie, in Syntaxonomie (Red. H. Dierschke): 5-33. Ed. J. Cramer. Vaduz, 1981.

Géhu, J.M.; Géhu-Frank, J., & Bournique, C.: Sur les étages bioclimatiques de la région Eurosiberien française. Documents phytosociologiques NS 8: 29-43, 1984.

Meusel, H.; Jager, E., & Weinert, E.: Vergleichande chorologie der zentraleuropaischen flora. Gustav Fischer, Verlag. Jena, 1965.

Montero de Burgos, J.L., & González Rebollar, J.L.: Diagramas bioclimáticos. Icona. Ministerio de Agricultura. Madrid, 1974.

Quézel, P.; Barbero, M.; Bonin, G., & Loisel, R.: Essais de correlations phytosociologiques et bioclimatiques entre quelques structures actuelles et passées de la végétation méditerranéenne Naturalia Monspeliensia (Colloque de la Fondation L. Emberger, 9-10 abril 1980): 101-126, 1980.

Rivas Goday, S., & Rivas Martínez, S.: Vegetación potencial de la provincia de Granada. Trab. Dep. Bot. Fisiol. Vegetal, 4: 2-85, 1971.

Rivas-Martínez, S.: Phytosociological and corological aspects of the Mediterranean region. Documents phytosociologiques, 15-18: 137-145, 1976.

Rivas-Martínez, S.: Les étages bioclimatiques de la végétation de la peninsule Ibérique. Anales Jard. Bot. Madrid, 37 (2): 251-268, 1981.

Rivas-Martínez, S., & Tovar, O.: Vegetatio Andina, I. Datos sobre las comunidades vegetales altoandinas de los Andes centrales del Perú. Lazaroa, 4: 167-187, 1983.

Rivas-Martínez, S.: Biogeografía y Vegetación. Real Acad. Ciencias Exactas, Físicas y Naturales: 103 pp. Madrid, 1985.

Rivas-Martínez, S.: Memoria del Mapa de series de vegetación de España. Icona. Serie Técnica: 208 pp. Madrid.

Sánchez-Egea, J.: El clima. Los dominios climácicos y los pisos de vegetación de las provincias de Madrid, Avila y Segovia: ensayo de un modelo fitoclimático. Anales Inst. Bot. Cavanilles, 32 (2): 1039-1078, 1975.

Takhtajan, A.: The floristic regions of the world. Ed. Nauka. Leningrad, 1978.

Tuhkanen, S.: Climatic parameters an indices in plant geography, Acta Phytogeogr. Suec., 67: 105 pp., 1980.

Tüxen, R.: Die heutige potentielle naturliche vegetation als gegenstans der vegetationskartierung. Angewandte pflanzensoziologie, 13: 5-42. Hannover, 1956.

Walter, H., & Lieth, H.: Klimadiagram weltatlas. Ed. G. Fischer. Jena, 1960.

CLIMATE GLOBAL CHANGE AND LAND DEGRADATION

A. C. IMESON

Landscape and Environmental Research Group, University of
Amsterdam

Summary

The first part of the paper introduces land degradation
within the framework of eco-pedological landscape systems,
outlines the processes included under land degradation and
considers its historical context. The main part of the paper
deals with monitoring and predicting land degradation. The
current state of land degradation is firstly examined and
the relationship of climate to a number of selected climate-
sensitive key processes is examined. This section then goes
on to consider monitoring and prediction with respect to two
examples from Europe. Lastly land degradation is briefly
considered as a factor of global change considering how land
degradation influences ground surface processes.

1. INTRODUCTION AND BACKGROUND INFORMATION

1.1 Definition and scope of paper

This paper introduces the problem of monitoring and forecasting the
effects of global change on land degradation. Land degradation, however,
in turn influences global change; the soil is an important source and
sink for many radiative gasses, it has important regulating effects within the hydrological cycle and it influences the albedo. Details on the
role of the soil in the various global cycles can be found in Bouwman
(1989); the impact of land degradation on these cycles is not included in
this introduction.

Land degradation is used as a general term to describe the effects
of accelerated processes of soil erosion, soil degradation and desertification on the landscape. It is expressed in higher rates of soil loss,
lower soil fertility and productivity and sometimes in a loss of biological diversity. A useful framework for considering land degradation is
provided by Naveh (1987a 1987b) who considers the landscape as a complex
system to which concepts of biocybernetic regulation can be applied.
Against this background, in which self-organisation and self-regulation
take place, a distinction is made by Naveh into those natural landscapes
in which soils and plants have co-evolved under very little human influence and semi-natural landscapes that have had their structures and
functions changed by man and which may be dependent on repeated human
perturbations. (Naveh 1988). This implies that, although in many cases
land degradation is a direct or indirect result of ecological stress, in
other cases ecosystems are maintained by a certain amount of disturbance,
for example by fire. When land degradation and climate are approached in
this way, it is clear that the types of response that global change might
produce will be non-linear and evolutionary in character.

Although very many interacting processes and phenomena could be included under the subject of land degradation, emphasis is given here to those that are related to the soil. Because the soil, vegetation and landscape have co-evolved over long periods of time, land degradation concerns concepts of time in landscape evolution and depends on perceived values of the past and present degradation status. With respect to areas currently undergoing land degradation, there is a general association with "cultural" landscapes, even though human-activity does not necessarily result in degradation. Many areas affected by land degradation in the past, have been so seriously lost or altered that there is little possibility of further degradation in the coming centuries. This is the case in many parts of upland and Mediterranean Europe. Areas not undergoing land degradation are mostly characterized by "non-erodible" soils. Because of vegetational and other effects such soils have evolved properties that can resist mechanical and physico-chemical processes of erosion under the prevailing extreme climatic conditions.

1.2 Component processes of land degradation

For convenience a distinction can be made between processes operating at at the scale of the soil profile, field or slope and those that can be considered in the context of the drainage basin. In the second category are included transport and deposition processes in river channels and flood plains. At the scale of the field or slope, on cultivated or overgrazed land, **physical soil degradation,** whereby a soil develops unfavorable properties (for example loss of aggregation, dispersive behavior and low permeability), can result from lower organic matter contents, the impoverishment of the soil fauna and compaction by heavy machinery. Lower water availability, surface sealing and erosion can result. In regions with a seasonal rainfall distribution, poor physical properties may result from the dispersion of clay due to an overabundance of monovalent cations in the soil solution at certain times of the year.

There are many forms of **chemical soil degradation,** for example caused by nutrient depletion or pollution, which can effect both cultivated and (semi)natural ecosystems. The **biological degradation** of the soil often occurs in association with physical and chemical degradation.

The loss of the soil by **erosion** can take on many forms and is frequently a response to soil degradation. In particular soil degradation influences 1) the partitioning of rainfall between infiltration and overland flow, 2) the soil shear strength and surface roughness and 3) crop growth and the duration of soil exposure to rainfall. At the scale of the slope and zero-order drainage basin, rill, gully and tunnel erosion can be major causes of degradation, especially on tilled soils and where soils are potentially dispersive.

Particularly in mediterranean and subtropical regions with seasonal climates, salt accumulation is a major problem, as for example in Alberta and Western Australia. This can also occur in higher order basins and may result from increased groundwater salinity. In higher order riparian zone river channels, erosion can occur because of increased rates of runoff (for example, due to soil acidification in forests lowering transpiration rates) and alterations in channel sediment inputs that disturb prevailing equilibria.

Badland areas occur in many regions where sediments and regolith materials are either very highly erodible and/or occur or produce cond-

Figure 1 A scheme illustrating the way in which climatic change could be translated through the soil system to affect hillslope hydrology ans erosion in the mediterranean region (Eijbergen and Imeson 1989).

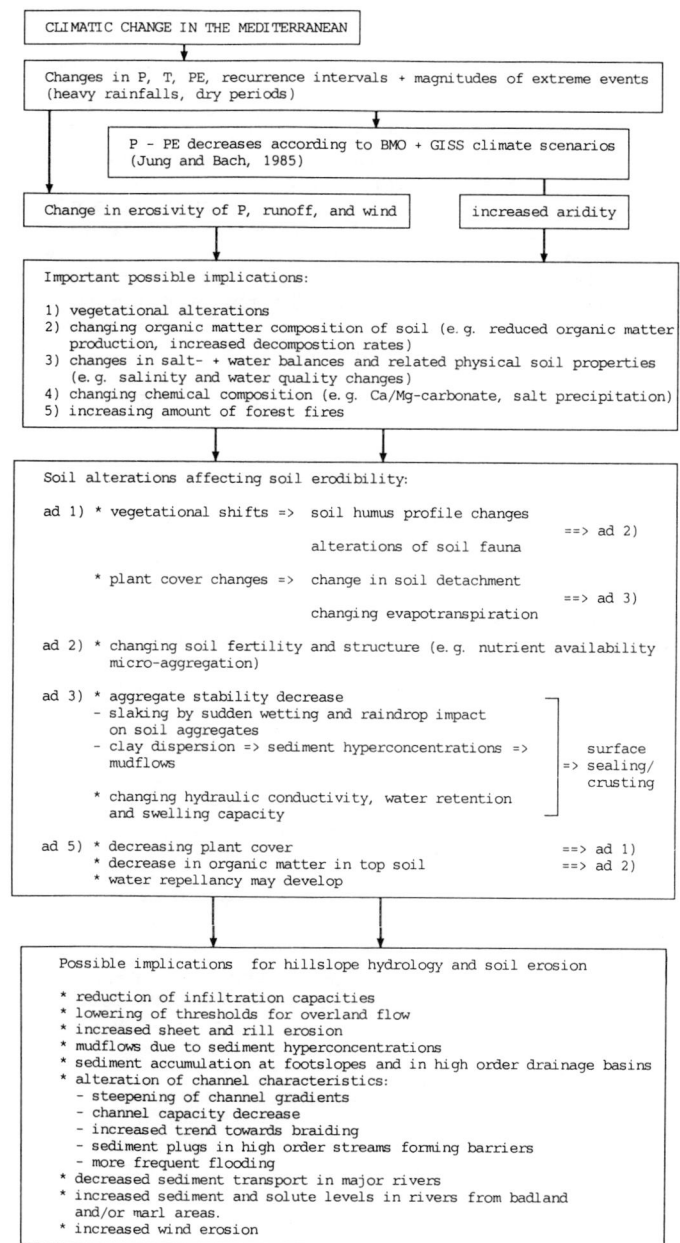

itions unfavorable for crop growth (Bryan and Yair 1982). Although the location of badland areas is determined by geological conditions, whether or not badlands occur, and the type of erosion that takes place is in many ways related to climate. The implications of climate change for badlands in SE-Spain has recently been examined by Harvey and Calvo-Cases (1989).

A major form of land degradation is by mass-wasting processes (landslides mudflows etc). The importance of extreme precipitation in triggering landslides is well documented (for example, Gallart and Clotet-Perarnau 1988). Mass-wasting is most dominant in, but not confined to, non-cultivated upland areas. Recently, abandoned terraces in southern Europe, have been affected by this form of degradation.

In Figure 1, a scheme is given of how climate-sensitive soil properties, in the mediterranean region, could change under altered climatic conditions and how these changes in turn are translated into increases in erosion and degradation. The table is at a high level of generalisation and ignores the great range of conditions in the mediterranean region that warrant a fuller consideration (see Eijbergen and Imeson (1989) Imeson and de Groot (1989) and Imeson and Emmer (1988).

1.3 Some important climate-sensitive parameters and processes

From the above, two important problems can be highlighted. The first concerns the great variability of land degradation resulting from the effects of geology, climate and land-use history. The response of present day eco-geomorphological systems to global change will reflect the memory of past changes retained to a greater or lesser extent in the system. Most slopes and soils are probably not in equilibrium with the prevailing climate (Kirkby 1989a). The second problem concerns the relative rates of soil formation (or soil structure development) in relation to the rate of soil loss. This can be considered with respect to the balances in the input and output flows in the soil system. Measured rates of soil loss, leaching and weathering, needed to establish such balances have been reported for a number of soils or drainage basins.

Of particular importance for land, degradation is the water-storage capacity of the soil in relation to precipitation. Its importance has been demonstrated in mathematical simulation of the effects of climate on erosion in Spain by Kirkby (1989a). This ratio influences percolation, leaching and weathering (in the case of permeable substrata) and overland flow and soil loss from soils or rocks having horizons or layers of low permeability. Except on valley bottoms and footslopes, where eroded soil accumulates as colluvium, the loss of storage capacity by erosion is invariably many times greater on cultivated land than the rate of storage capacity gain by weathering. Under semi (natural) conditions the rate of weathering and erosion are both very slow. Kirkby (1989b) has modelled the effect of increasing regolith depth on erosion and hillslope hydrology, from which it appears that in the long-term over many thousands of years pedological and hydrological processes result in equilibrium hillslope development having fairly uniform regolith profiles.

The ratio of storage capacity to precipitation is also important for land degradation through its influence on the soil salt balance. When winter precipitation is too low to flush highly water soluble salts from the soil profile or slope the ability of the soil to resist erosion (soil erodibility) can be greatly reduced. Adverse soil properties caused by water-soluble salts are both a major cause and result of soil degradation.

Where soil degradation does not occur, the soil can resist erosional forces because of the effects of organic matter on soil structure. The structure of a soil is the term used to describe the agglomeration and arrangement of primary soil particles into larger units. An important parameter is the ratio between a) soil aggregate formation and stabilisation by organic matter, and b) soil structure destabilisation. In this context the nature, and not necessarily the quantity, of organic matter that is important. In mediterranean and semi-arid environments, subsurface biomass production in the form of fine roots and fungal hyphae can be considered as particularly important, in terms of its effect on the large soil structural units, while fungal and bacterial activity are important for stabilising smaller particles. Climate has an important effect on these balances and thus has a great potential effect on soil aggregation. Examples of some of these effects on various aspects of land degradation are considered in section 2.

It has been mentioned that water soluble salts, which are highly responsive to small changes in climate, influence soil erodibility. Generalizing, it can be stated that two sets of conditions exist , separated by a threshold, in which clay particles are either dispersed or flocculated (for a full discussion see Bolt and Bruggeweert 1976). Flocculated soils have much better physical characteristics. If during a rainfall event the threshold between flocculation and dispersion is crossed, as the soil solution is diluted and rainfall energy applied to the ground surface, clay dispersion can occur and lower the rate of infiltration to a fraction of what it would have been had the soil been flocculated.

Runoff transporting dispersed clay can become hyperconcentrated, carrying 40-70 % by weight sediment (Costa 1988) and having a viscosity (100-400 dynes/cm2) at least ten times higher than that of non-hyperconcentrated flow (Beverage and Culbertson 1964, Imeson and Verstraten 1981). It can lead to very high transport rates of bed-material. Many major difference in the sediment transport characteristics of rivers draining the dryer parts of the mediterranean with those in more humid areas is related to this way to sediment-water interactions.

1. 4 The historical context of land degradation

Processes of erosion, weathering and soil formation occur in all landscapes and these have varied greatly in the past both in terms of their relative importance and intensity. Often the distinction is made between periods in the past that were dominated by either erosion (stability) or by soil formation (stability; k-cycles of Butler 1959) and which may have lasted for hundreds of years. The switch from one phase to the other has been usually attributed to climate or man by most geomorphologists, but in fact intrinsic geomorphological or ecological thresholds could be involved that do not require external changes (Schumm 1973) Changes from periods of erosion to periods of soil formation indicate that the landscape has an ability to recover from periods of degradation. Recovery does not imply that lost soil profiles are recreated but that new soil conditions, probably qualitatively quite different, develop under the altered situation with properties that resist erosion. Within a few years, various organic matter effects can greatly modify the soil.

Kochel (1987) has studied the effect of major floods on river channels and the time needed for recovery. For river channels, recovery does not imply increasing resistance to erosion. On the Pecos River, Kochel (1987) describes the recovery time as being greater than 500 years. As a channel recovers to low flow conditions, catastrophic adjustments in morphology become more likely in response to high flows. The influence of climatic change on the Mississippi river, where there are more than one hundred years of data, has been extensively studied by Knox (for example 1987). One of his conclusions is that peak flood magnitudes are highly associated with amount of winter rainfall and early summer precipitation.

Also on slopes, in the longer term, resistance to erosion does not always increase with time. Some soils evolve properties that can ultimately lead to their cyclic erosion. For example soil formation in colluvium in topographic depressions in California has been shown by Dietrich et al (1986) to steadily increase the risk of landslides and gully erosion. As clay weathers and accumulates in B horizons the shear strength is lowered and the critical rainfall and soil moisture conditions required for saturation and overland flow generation are lowered. Soil development and weathering in mine spoil heaps can decrease the hydraulic conductivity and lead to catastrophic erosion after only a few tens of years. The cyclic accumulation of highly soluble salts in slope foot sediments and alluvium is probably also important. Small amounts of salt greatly influence the resistance to erosion and are sensitive to small changes in the water balance. DePloey (1989) has recently drawn attention to the inevitability of erosion that is somehow related to the total erosion potential in a landscape. If erosion does not occur by one process, say overland flow, it will be caused by another, for example by weathering and mass-movements.

2. CURRENT DEGRADATION AND THE IMPACT OF GLOBAL CHANGE

2.1 Introduction

This section firstly considers the distribution of areas currently suffering from degradation processes. These may not necessarily coincide with the areas of past or future degradation. Next, a number of key climate-sensitive processes are considered in case studies, from two regions in Europe, where some of the greatest problems of land degradtion are found today.

2.2 Present land degradation

That land degradation and desertification are affecting an increasingly large area of the world is evident from sequential satellite images and from the multitude of reports from governmental and non-governmental organizations. Although the location of the affected areas is known, there is little quantitative information that would allow a global survey to be made. Although the causes of degradation are many, a major factor is the physical degradation of the soil.

One way of obtaining a global assessment of land degradation is to look at sediment yield data. From a relatively recent global survey by Jansson (1988) an impression can be obtained of current rates of sediment transport in relation to climate and land use. Central, southern and southeast Asia, China, southern Africa and the Andes emerge as areas of high sediment loss. Also evident is the contrast between the high sediment loads of Mediterranean and southern European rivers compared to those in northern Europe. Relationships between sediment yield data and erosion are only very general and only a partial impression can be obtained concerning land degradation. Most eroded sediment is transported only short distances and never reaches rivers. Of the sediment that is transported from fields perhaps no more than 40-60 percent reaches most rivers where it is joined by sediment eroded from river channels. Sediment transport must be seen in the context of river channel sediment budgets over long periods of time and considered in the context of complex response (Schumm 1978). To relate current land-surface processes to sediment transport in large river channels, very detailed studies are needed of sediment budgets and very few of these have been made. With respect to climate, one difficulty is that sediment yield data greatly underestimate the erosion taking place in arid and semi-arid basins. Nevertheless, in spite of these limitations, information such as that reported and analysed by Jansson (1988) is very useful. Relationships between sediment yield and effective precipitation, first described by Langbein and Schumm (1958) indicate peak erosion rates at an effective precipitation of 300-400 mm. Yair en Enzel (1987) have shown that surface processes make this relationship much more variable in semi-arid and arid regions and discuss this in detail.

Another way of obtaining a global estimation of the state of land degradation is to systematically collect data from field locations that can be input into a soil and terrain digital data base. UNEP and ISRIC have initiated the GLASOD (Global Assessment of Soil Degradation) project to do this. Field data are being collected at a scale of 1:10 and 1 to 1 million in pilot areas and the first results should appear soon. The study is trying to overcome the difficulty of collecting data across national boundaries (UNEP 1988) and link in to existing soil maps and classification schemes. Data from studies such as GLASOD could provide an important base-level with which future assessments could be compared. However, the assessments being made at the scale of GLASOD are probably not sensitive enough to enable the rather small effect of climate and global change to be followed. There are in fact very few quantitative data available on rates of erosion or degradation with which assessments made in the field can be evaluated.

Although there is a lack of quantitative information on current erosion rates, soil erosion and conservation experts are well aware of the factors influencing erosion and of its distribution. With respect to Europe de Ploey (1988) has recently compiled a map showing the occurrence of erosion in Europe. This map is based on expert knowledge of soil conservation scientists consulted in the various countries. It emerges that erosion is concentrated in areas most subject to soil degradation, namely in the silty (loess) belt of northwest Europe and in the Mediterranean region. Erosion by mass movements is a problem in certain mountain areas. In the next section, special attention will be given to examples taken from the two areas in Europe where erosion and soil degradation is most serious.

2. 3. Predicting the effects of global change on land degradation: loess soils in The Netherlands

Th silty (loess) soils of northern Europe are sensitive to erosion when organic matter and clay contents are low. Erosion and soil degradation occur particularly under maize, sugar beet and potatoes. Loess soils may suffer under intensive agriculture due to subsoil compaction and organic matter and soil faunal depletion. De Ploey et al (1990), Imeson (1988) and Kwaad (1989) have looked at the relationship with climate. Of interest is the relationship of soil strength to soil moisture. A peak stability is reached at moderate moisture contents so that erosion might be expected to occur when the soil is either very dry or very wet. In fact almost all erosion in the loess belt in The Netherlands occurs in the summer months so that dry periods are critical. Most erosion takes place during one or two rainfall events each year (Kwaad 1989) and the greatest risk is when rain falls on dry soil between April and August.

The limitations of GCM's are well known but it is nevertheless interesting to consider the significance, for erosion in the nearby Netherlands loess belt, of changes predicted by Bultot et. al. (1988) under a 2*CO2 scenario, on the water balance of three small drainage basins in Belgium. The changes in precipitation predicted for the critical period of erosion show an increase of 10 mm in March and April but a decrease of a few mm each month between May and August. The calculated temperature increases range from 3.4 °C in March to 2.3 °C in August and September. It is assumed that the frequency of rain days remains unchanged under the 2*CO2 scenario, although this is of great significance for erosion. Of particular interest for soil erodibility is the implications of these changes for soil moisture. At Dyle, where the PE today is similar to that at Maastricht Airport in Limburg. the hypothesized change in PE is greatest in June (7.4 mm) and ranges between 5 and 6 mm during the other summer months. During the autumn the predicted PE values will be only 1 or 2 mm higher than today. Changes in the effective evapotranspiration are indicated as being about 5.5 mm in March but only 2.1 and 1.5 mm in July and August respectively. Simulated soil moisture contents in the aeration zone show a decrease of only 0.6 mm in April and 0.8 mm in March but between 9 and 10 mm in July and August. The direct negative impact on soil erosion, of such quantitatively small changes in soil moisture, is likely to be small. Between June and August, when the decreases are at a maximum these are only about 5 per cent lower than those calculated for the present time. In fact when monthly rainfall records from South Limburg are compared for wet or dry years, these show changes an order of magnitude greater than those considered above. When evaluating the significance of the predicted changes for erosion, the assumption that the frequency of rain days will remain unchanged and the lack of information on rainfall intensity pose restrictions. It is the time between showers and the intensity of rainfall events that are of major interest and about which little can be said.

The climatic changes will, however, mean that the sowing of row crops such as maize, potatoes and sugar beet can be advanced and that these will provide a protective cover several weeks earlier than is the case today. During 1989 the mild winter (about 3.5 °C warmer than average) and earlier spring were possibly a major factor in limiting the amount of erosion in that year, although the lack of heavy storms was another. This, together with the more active soil fauna may more than compensate for any change in precipitation.

Rainfall simulation experiments can also be used to examine the likely impact of different conditions on soil degradation. An application of this approach is described for South Limburg by Imeson (1988) who emphasized the importance of climate-sensitive soil properties. The general conclusion is that the impact of climate on land degradation is likely to be beneficial in the loess belt of The Netherlands, provided that the frequency of high magnitude rainfall events and the number and intensity of droughts does not increase.

2.4 Case studies from the mediterranean area

The potential impact of climatic change on erosion in the Mediterranean region has recently been reviewed by Imeson et. al. and by Imeson and Emmer (1988), where more information can be found. In contrast to the loess belt of northern Europe, the impact of global change could be great. Firstly, although there appears to be little information about the changes in precipitation that are expected in the future under 2*CO2 scenario for the Mediterranean (see, for example, Wigley 1988), it is predicted by global climatic models that temperature will increase by about 3°C and potential evapotranspiration by about 400 mm. This is considerably greater than for the example from Belgium described above and coul alone, result in a potentially large increase in aridity, especially in the spring and autumn. Secondly drought and irregular but sometimes high amounts of winter precipitation make the soils very vulnerable to degradation and erosion. Th effect on soil organic matter (production, amount, composition and form and mineralization. have been already mentione. The simulation model of Kirkby (1989a) indicates that organic matter effects will depend on the degree to which precipitation can be stored in the soil and on the general amount of rain. Indirect impacts on erosion that can be expected, include, for example, those resulting from more frequent forest fires and the longer recovery time.

2.4.1. Highly calcareous soils in Alicante and Valencia

In southeast Spain extensive areas of highly calcareous soils occur on marls in regions receiving between 350 and 500 mm of rainfall per year. Where these soils are under cultivation, soil erodibility is very high (Imeson and Verstraten 1985). About 60 per cent of the soil consists of inert silt sized calcium carbonate particles and because clay contents are low, organic matter in the form of fungal hyphae, fine roots and polysaccharides are required for the formation of stable aggregates and to enhance the ater-holding capacity of the soil. Today on tilled soils organic matter contents are often so low that there is little possibility of further decrease should the climate change.

At slope foot positions material eroded in the past has accumulated as colluvium. Here, gentle slopes and stones reduce the erosion hazard and soils are slightly less erodible than on the marl slopes. At a few locations water soluble salts have accumulated in the colluvial deposits and these are subject to gully erosion. An increase in the area of soils affected by water soluble salt concentrations could lead to an extension of gully prone area. On the higher rocky slopes of the region former erosion has lead to the removal of erodible material so that further erosion is unlikely. The main areas therefore likely to be directly affected by climatic change are the sediments deposited during previous erosional phases at slope foot positions. As in The Netherlands infiltration envelopes for different soils, that show the relationship between rainfall intensity and time to ponding, can be used to model the effect of altered rainfall intensities and durations on runoff. In Figure 2 infiltration envelopes for marls and alluvial soils are shown. It can be seen that for soils with crusts very short rainfall durations (< 3 minutes) at intensites as low as 3-5 cm/h. are enough to pond these soils.

Soils in the Selva Region of Catalonia

Soil erosion in the Selva region of the Province of Gerona is locally severe, particularly in areas of granitic rock, on granodiorite and on the Pliocene-Pleistocene deposits derived from them. Soil erosion is especially prominent where soils having a texture-contrast or duplex profile are found. These soils suffer erosion when the Mediterranean forests have been replaced be eucalyptus, when the soils are tilled on slopes and sometimes after forest fires.

Erosion on cultivated soils is a result of the poor structure of Ap horizons which develop in the albic horizons of the original soil, and because of the ease with which saturated overland flow develops on the impermeable B horizons in the winter even at low rainfall intensities. Subsoil compaction also occurs. Erosion following fire is often the result of the construction of access roads, or e of Eucalyptus terrace construction. Whilst low intensity rainfall events produce much erosion, so also do the infrequent high intensity summer thunderstorms. In the forest and burnt forest sites infiltration rates are very low due to water repellency.

Several types of data are being collected to examine the potential impact of climatic change on erosion. These include information on the relationship between weather and fire frequency, on the relationship between site aridity and the recovery of burnt forests, and on the infiltration characteristics of soils along climatic gradients. Several relationships between fire frequency and climate have been described by various workers in Gerona. It is known that fire frequency increases with drought and is influenced by humidity.

Infiltration envelopes have been established to look at relationships with climate but the use of these is made difficult because of the hydrophobic soil properties and because of storage-controlled runoff development in the winter. So far measurements along climatic gradients (between 1200 and 500 m have not enabled relationships to be demonstrated between soil organic matter and infiltration characteristics (see Imeson 1990, in press).

Soils in the Judean Desert

In the Judean Desert an extremely sharp climatic gradient occurs between Jerusalem and The Dead Sea. Due to tectonic uplift and relatively similar pastoral land-use, it is possible to find comparable sites on similar lithological formations. Along a transect on comparable sites at which precipitation ranged from about 430 to 160 mm. and the mean annual temperature from 18 to 23 °C, Lavee et al compared the response of soils to simulated rainfall and studied soil erodibility. They found very clear relationships with climate that can be of great use in paramatising predictive models. The trend is for soil erodibility to increase with aridity and for infiltration rates to decrease. A summary of the infiltration results is given in Table 1 for sites at located at elevations of 640m (Site A), 320 m (site B) and 20 m (site C) having respectively mean annual precipitations of 430, 330 and 160 mm.

Table 1. Effect of similar infiltration experiments using the Morin rainfall simulator at three sites along a climatological gradient in the Judean Desert.

	Site A	Site B	Site C
Parameters			
plot gradient	10	10	9
soil depth (cm)	60	25	14
rainfall itnensity (mm/h)	34.8	38	36
rainfall duration (min)	225	80	61
results (*)			
time to ponding(min)	14	6.5	1.2
time to overland flow (min)	25	8	3.3
wetting depth immediately after rainfall (cm)	60	4-11	7-8
notes		small cracks disappeared after 7 min; big cracks after 19 min.	cracks disappeared after 1.2 min.

Figure 2. Results of rainfall simulation experiments on highly calcareous soils. Very small amounts of rainfall produce ponding and runoff on the soils which are sensitive to slaking and crusting (surface sealing).

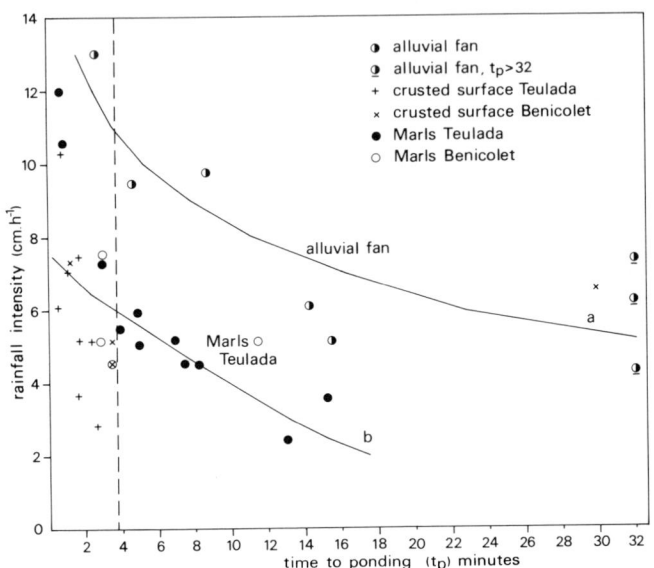

3. 0. PREDICTION AND MONITORING

3. 1. Predicting the effects of global change

The effect of climatic change on component processes of land degradation can be predicted with models that rely on relationships that describe how key processes are influenced by climate parameters. These depend on field or laboratory measurements such as those shown in figure 3. A good example of this approach is given by Kirkby(1989a). In this paper Kirkby applies his slope evolution model (SLOPEC Kirkby 1989b) to look at the effect of a 2*CO 2 scenario with a 3 degree rise in temperature on sediment production for hypothetical sites at Barcelona, Malaga and Almeria. He staggered his climatic change in four pulses at intervals of 25 years so that the responses could be examined. Kirkby's simulations raise too many interesting points to be included in this summary. Amongst his most important conclusions, it is noted that at all sites sediment yields from slopes will increase by 40-400 per cent but that except in the dryer sites where storage capacity is low, there will be a gradual adjustment to new equilibria with lower sediment yields over several hundreds of years. One of Kirkby's examples for a soil profile with a low storage capacity is shown in Figure 4. This shows how the loss of soil from the slope by rainwash responds to the staggered simulated climatic change. The model is interesting in that it predicts changes in erosion rates of a magnitude that can be measured. The model provides a useful framework for collecting data in the context of monitoring programmes. A major problem is the lack of field data on many climate-sensitive component processes.

To obtain such data there are several approaches. These include the use of spatial or temporal analogues and controlled laboratory and field experiments. At the moment there are no validated models of soil erosion that have the sensitivity to be used to predict the effects of global change on erosion.

3. 2 Monitoring

There are ways that land degradation is and could be monitored but because of noise, making the association with climate is not simple. At the LICC meeting in Lunteren (1989), a general conclusion was that priority should be given to establishing harmonized measurement stations across climatic gradients concentrating on ecotones. Measurements should be made of key indicative parameter values at regular intervals to establish trends. Soil aggregation and subsoil biomass characteristics, for example, might give an early response to climate change. The data collection could take place in the context of existing programmes, such as EPOCH (EEC) or the Mediterranean Action Plan, Priority Action Programme on rainfall induced erosion (UNEP, 1989). This second programme stresses the need to collect information according to harmonized programmes so that data and information can be transferred.

Remote sensing, and sequential photos of various types can be used to follow land degradation. As calibration improves, the potential of remote sensing to monitor degradation will become increasingly valuable. An example of the application of remote sensing to soil erosion monitoring is that by Bocco and Valenzuela (1988). The main task is to collect data to validate the models that are applied in the interpretations.

Figure 3. Relationship between mean annual rainfall and organic C content of virgin soil and adjacent cultivated soil (20 years under wheat). Southern Queensland Australia (from Dalal and Mayer, 1986).

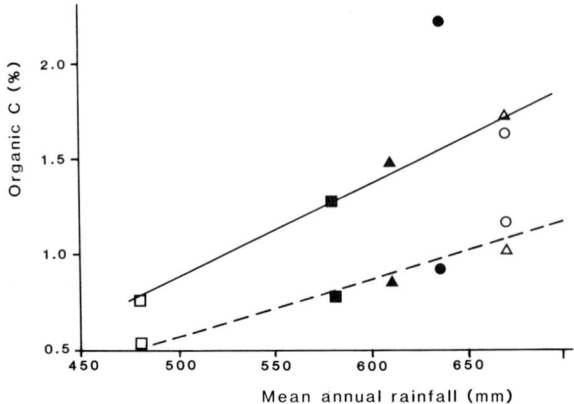

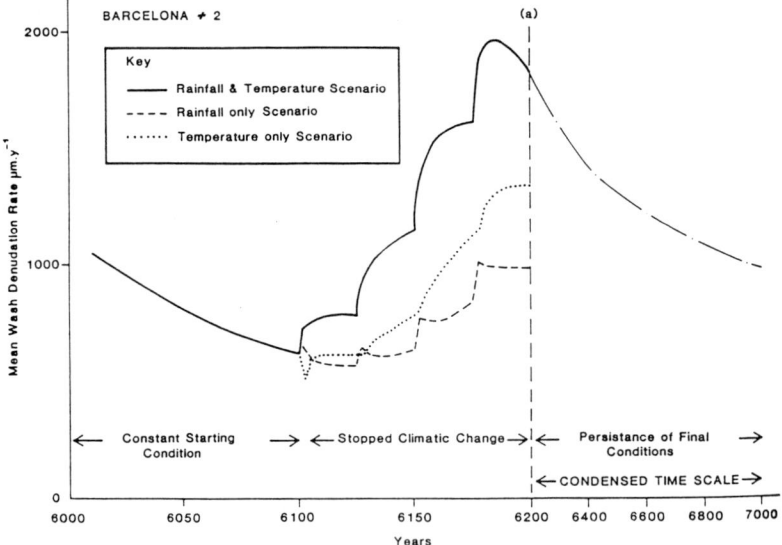

The simulated response of wash erosion to a staggered climatic change of a 3° C increase in temperature and/or a 100 mm increase in rainfall. (From Kirkby 1989a)

References

Beverage, J. P and J. K. Culbertson (1964) Hyperconcentrations of suspended sediment. J. Hydraulics Division ASCE 90 HY6 117-128

Bocco, G. and C. R. Valenzuela 1988 Integration of GIS and image processing in soil erosion studies using ILWIS. ITC Journal 1988-4, 309-319.

Bolt, G. H. and M. G. M. Bruggenwert 1976 Soil chemistry A: basic elements Elsevier 281p

Bouwman, A. F. 1989 The role of soils and land use in the greenhouse effect. Background Paper "International Conference on soils and the Greenhouse effect", Wageningen The Netherlands, 14-18 th August 1989, International Soil Reference and Information Centre. Wageningen 148p.

Bryan R. B. and A. Yair (eds.) 1982 Badland Geomorphology and Piping. Geobooks, Norwich.

Butler, B. E. 1959 Periodic phenomena in landscapes as a basis for soil studies. C. S. I. R. O. Australia, Soil Publication No 14 20p.

Costa J. E. Rheologic, geomorphic and sedimentologic differentiation of water floods, hyperconcentrated flows and debris flows. pp 113-122 in Baker, V. R, R. C. Kochel and P. C. Patton Flood Geomoprhology, Wiley.

de Ploey J. (1990) Modelling the erosional susceptibility of catchments in terms of energy, Catena in Press.

dePloey, J. Imeson, A. C. and Oldeman. (In Press) Soil erosion, soil degradation and climatic change. in IASA meeting proceedings on Land Use changes in Europe, Warsaw Sept. 1988

Dietrich, W. E, C. J. Wilson and S. L. Reneau. 1986 Hollows, colluvium and lanslides in soil-mantled landscapes, pp 361-388 in (A. D. Abrahams (ed.) Hillslope Processes, Allen and Unwin, London.

Eijbergen, F. and A. C. Imeson, Geomorphological processes and climatic change. Catena 16, 307-319.

EPOCH (1989) European programme om climatology and natural hazards 1989-1992, Directorate General for Science Research and Development Brussels

Gallart, F and N. Clotet-Perarnau 1988 Some aspects of the geomorphic processes triggered by an extreme rainfall event: The November 1982 Flood in the Eastern Pyrenees. in A. M. Harvey and M. Sala (ed.) Geomorphic processes in environments with strong seasonal contrasts Catena, Supplement 13, 79-95

Harvey A. M. and A. Calvo-Cases, The distribution of badlands in southeast Spain: implications of climatic change. 13 pp. in LICC Discussion Report, Mediterranean Region. Landscape Ecological Impact of Climatic

Change, Conference Lunteren Dec 3-7 1989

Imeson, A. C. and I. Emmer, 1988 Implications of climatic change for land degradation in the Mediterranean. UNEP, Mediterranean Action Plan

Imeson, A. C. and de R. S. Groot R. 1989 Landscape-ecological impact of climatic change on the Mediterranean region. Introduction to Discussion report pp 1-19. LICC Workshop Lunteren December 3-7th 1989.

Imeson, A. C. and J. M. Verstraten 1981 Suspended solids concentrations and river water chemistry. Earth Surface Processes and Landforms 6, 251-263 .

Imeson, A. C. and J. M. Verstraten, 1985, The erodibility of highly calcareous soil material from Southern Spain, Catena, 12, 291-306

Jansson, M. 1988. A global survey of sediment yield. Geogr. Ann 70A 81-98

Knox, J. C. 1988. Climatic influence on upper Mississippi valley floods. pp. 279-300 in Baker, V. R , R. C. Kochel and P. C. Patton Flood Geomoprhology, Wiley.

Kochel, R. C. 1988 Geomorphic impact of large floods: review and new perspectives on magnitude and frequency pp 169-187 in. Baker, V. R , R. C. Kochel and P. C. Patton Flood Geomoprhology, Wiley.

Kwaad, F. J. P. M. Precipitation parameters and soil erosion on arable land on loess soils in South Limburg (The Netherlands) Paper presented in LICC DIscussion Report on Fluvial Systems (ed R. H. G. Jongman and M. M. Boer.

Langbein and Schumm (1958) Yield of sediment in relation to mean annual precipitation. Transactions. Amer. Geophysics Union 39, 1076-1084

Lavee, H. Imeson, A. C. Pariente, S. and Benyamini. The response of soils to simulated rainfall along a climatological gradient in an arid and semi-arid region. manuscript submitted to Catena

LICC (1989) Landscape Ecological Impact of Climatic Change, Conference Lunteren, The Netherlands December 3-7th 1989. Discussion Reports

Schumm, S. A. 1973. Geomorphic thresholds and complex response of drainage systems pp 293-310 in Morisawa (ed.) Fluvial Geomorphology, Binghampton, New York.

Naveh, Z. 1987a Biocybernetic and thermodynamic perspectives of landscape functions and land use patters. Landscape Ecology, 1, 75-83.

Naveh, Z 1987b Biocybernetic perspectives of landscape ecology and management. in Landscape Ecology and Management, Proc. First Sympossium of the Canadian Society for Landscape Ecology and Management, Guelph May 1987.

Naveh, Z. 1988 Multifactorial Reconstruction of Semiarid Mediterranean
    Landscape for multipurpose land uses. pp 234-256 in Allen, E. B. (ed.)
    The reconstruction of disturbed arid lands- and ecological approach.
    AAAS Selected Symposium 109, Westview Press.

UNEP 1989 Report on the implementation of the pilot project on soil eros-
    ion mapping and measurement in the Mediterranean coastal zones. PAP-8
    /EM. 8/1, PAP/RAC Malaga 1989

Yair, A. and Y. Enzel 1987. The relationship between annual rainfall in
    arid and semi-arid areas. The case of the Northern Negev. Catena
    Supplement 10, 121-136.

Wigley, T. M. L. 1988 Future climate of the Mediterranean basin with par-
    ticular emphasis on changes in precipitation. UNEP (OCA)/WG. 2/6 26p.

Commission of the European Communities
EPOCH
European School of Climatology and Natural Hazards
Course on
"Climate and Global Change"
Arles 4-12 April 1990

HUMAN RESPONSE TO GLOBAL CHANGE

R. Frassetto
Director of Research - National Research Council ISDGM - Venice -
Italy

Summary

The growth of population and of its welfare with the economy of waste
of modern society have been principal causes of the climatological and
environmental alterations of our planet.
The rate of increase in production and emission of heat trapping gases
has assumed an inertia which risks to create future effects similar to
natural catastrophies of geological or historical times.
Alertness of the problem has been raised. The role of science is now to
reduce uncertainties in understanding and predicting. The role of
policy makers is to properly interpret the preliminary scientific
information and deal with preparedness, prevention, defenses and
with the analysis of socio-economic impacts of the Global Change.
There is no more justification for the policy of "wait and see".
Education of the new generations as well as of the present managers, of
policy makers and administrators is the major task of the academic
milieu and of the large International Organizations of the United
nations.

Introduction

Alertness of the global climate and environment change triggered by the effects of the economy of waste of industrial modern society has been raised at all levels from the Governments to the newspapers-reading population.

The response of the heads of developed as well as of developing Countries is continuosly kept alive by the intensive scientific-tecnical information and operational activity of several United Nations Organizations and WCRP (World Climate Research Program) of WMO.

The present large international research effort will produce, through the IPCC (Intergovernmental Panel of Climate Change) a "State of the Art" report for the world climate meeting in Geneva, Switzerland on October-November 1990.

On the basis of this document world-wide agreements and protocols will be improved for action in two major issues:
Limitation and Adaptation.

Limitation

By limitation it is meant the reduction of the causes of the major changes of climate of our planet which are the atmosphere and ocean warming , the intensification of storms-precipitations-draughts, the Sea Level Rise, the increase in ultraviolet radiation, all of which create large and costly impacts on the life and socio-economic activity of modern society which must be evaluated before it is too late.

At the origin of this planet's climate alteration is the growth of population, of its demands and welfare and the economy of waste of modern society, as well as the uncontrolled and irresponsible use of those renewable and non-renewable resources of the earth which are interconnected with the natural environment and the climatic equilibrium.

Increased industrial production and pollution have generated the alterations of the physical and bio-geochemical spheres as the first lectures of this school have illustrated.

The necessary human response is to create gradually, but with no delays, all the premises for a global strategy and national policies of limitation of the causes.

There are priority actions such as:

Elimination of CFC's use by the year 2000. This is already a goal recommended in Geneva in March 1990 which dealth with the Ozone depletion and CFC's effects on the troposphere as heat trapping gases.

Limitation of CO2, metane and other combustion gasses will require longer time as it involves large economical interests in production and energy and generates conflicts between industrially advanced nations and developing countries. These problems are being analised in international and national scientific, tecnical and political milieus and will be the subject of long term educational programs at all levels, from first grade schools to policy makers advanced training.

Since a decade, all possible impacts, strategies, options have been the subject of extensive approach studies and will continue to be elaborated and improved in the next decade of space tecnology.

It appears evident that some sound decisions can be taken since now, such as: elimination of the use of CFCs, reduction of pollution and waste by increasing the efficiency of industrial production, planning healthier urban life and better hydrological managements and studying new economical methods to recycle wastes and manage reforestation. Perhaps an extensive and intelligent "wood" economy might bring large benefits to the natural recycling of the present concentration of antropogenic CO2 in the troposphere while natural phytoplancton growth might be intensifying the carbon cycle in the oceans as a biological pump since the onset of the CO2 increase. All of these clever possible human or natural processes should be the subject of scientific research and analysis. On a long term they could produce also economical benefits to be exploited. A compensation for the risks of hazards?

Of the many options of policies to be analysed and weighted in terms of cost and benefit are:

- Reduction of emissions:

Recovering waste heat and intensive saving and conservation   of energy could reduce carbon emissions from 5 to 1 Gigaton/year in few decades if all available options of conservation were taken. This policy is costly and difficult to accept entirely.

- Increasing energy efficiency is a realistic policy option.

Substitution of fossil fuel with other energy sources (solar radiation, wind-wave power, nuclear fusion) are costly promising, but far from practical today on large scale. A safer nuclear fission is probably possible and realistic.

- Delay of major climatic changes can be obtained by the extensive use of natural gas instead of oil or coal or synthetic fuels (methanol).

- Filtering greenhouse gasses is costly today but advanced technologies may be studied to face this problem.

A US study to remove 90% of CO_2 from effluents of power station shows that it would double the cost of stations, and would increase the cost of electricity by 1,5 to 2 times; would increase the cost of station cleaning process, would take many decades for the change and would have effects later in the next century.

- Recovery of greenhouse gasses can be obtained through: massive planned reforestation and management, recycling solid - liquid - gas wastes, (including the biological methane in cattle and suine farms) and with advanced tecniques of urban and industrial waste disposal and treatments.

Adaptation

Following the assumption that a global change is likely to occur in the near future, adaptation changes means creating the strategies to face the variety of future possible impacts of global and regional climatic alterations on specific areas for the period of time (50-100 years?) before effective results from the global "Limitation" policies can be obtained.

It involves national and municipal rational preparedness to increases in frequency and intensity of climatic hazards such as floods from land or ocean, storms, draughts, coastal erosion, degradation of historical city structures, and damages to navigation, recreation, business, commerce, industry , agricolture, forest.

Supported research on the socio economic impacts of environmental changes and information gathering should become an intensified integral part of the policy development while preventive measures are being engaged.

Politicians and lawmakers are up to face the responsability for introducing the necessary modifications also in the behaviour of society to counteract the undesirable effects and to prevent costly catastrophies and loss of lives.

The time needed to reach results of human adaptation policies may be of the same order of magnitude of the time taken by nature to respond to the anthropogenic alteration of the environment.

For this reason problem analysis and actions must take place in time with a cautious and flexible interpretation of the range of predictions to allow for fine tuning of policies as more accurate informations become available on the climatic variabilities and trends.

As the expected global greenhouse warming should have very different rates in function of latitude, as at the equator, the increase of temperature may be relatively small while in polar regions of several degrees Celsius (high scenarios reach over 10°C) the offsetted climatic equilibrium will create winners and loosers.

In the tropical and subtropical regions some areas may benefit from small temperature rise and increase of precipitation due to increased evaporation over the tropical oceans.
In temperate regions most areas could suffer draughts and disconfort for temperature rise of 1.5 to 3°C and increase of humidity.

In subpolar and polar regions large temperature rise (>8°C/century?) and increase in precipitation may create beneficial effects in agricolture and forestry but undesirable acceleration of ice melting and Sea level rise for thermal expansion of water masses and consequent floods-storms increase in frequency and strenght.

A great deal of research is needed and is underway or planned for this decade, to assure that by the year 2000 we will have a significant reduction of present uncertainties of predictive models. At that time perhaps sufficient knowledge will have been acquired to move the center of action away from the research laboratories and onto the desks of policy makers and lawyers.

But since now research is underway also to assess the range of options and practical policies. Analyses are underway to assess effectiveness cost, the risk involved, and side effects in lessening the dangers of local and regional climate change impacts.

Global warming will arrive slowly (in human life not in geological time scales) but once started, could accelerate.

If adaptive strategies are adopted in time based on scientific warnings, damages to human life, activity and resources, may be minimized. The task is not that easy for the reluctance of policy makers to deal with medium and long term plans. But private enterprises may invest spontaneously in preventive programs for long termeconomical returns or savings, By the time the global warming was to show its trend agriculture couldhave already adapted to most of a local temperature rise. Different crops varieties and different farming techniques could help absorb the impact.

It takes about a decade to develop and introduce a new crop variety for a sudden significant increase of temperature such as 2°C.

The risks of draughts and watershortage may at the present time force administrators to plan water saving, water storage, efficient water use in agriculture (Israel examples of creating farms from deserts).

Adaptation to excess of precipitation can be planned following historical examples such as the subtropical regions in rainy seasons. Relative Sea Level Rise impacts on coastal regions and ocean cities in productive estuaries, deltas and low lands are being analysed in several advanced countries for defense or retreat the economical penalties of such hazards must be simultaneously analyzed to propose optional solutions in terms of cost and benefit. In conclusion: while preventive measures should be analized Scientific and economical research is the best present investment and should be accelerated and widened, along with training and education, to save this planet earth.

EUROPEAN SCHOOL OF CLIMATOLOGY AND NATURAL HAZARDS

Course on

"Climate and Global Change"

STUDENTS' PAPERS

DRAINAGE BASINS FEATURES AND HIDROLOGICAL BEHAVIOUR. RIVER MINATEDA BASIN

F. ALONSO-SARRIA
Physcal Geography Department
University of Murcia

Summary

In this paper, nine (shape, size and topological) basin variables have been analized in four little basins whith non-permanent run off. These geomorphological variables have been selected because of their high correlation with the Instantaneous Unit Hydrograph (IUH) parameters, calculated by theoretical formulae and estimated by direct perceptions of the land people who live in the countryside when the streams are in spate. The analized basins are within the River Segura basin (SE of Spain).
We can see that the variables can change from one little basin to another within a very short area; because of it, we can't make generalizations about the behaviour of the run off.
We conclude that the variations in geomorphological aspects betwen different basins, caused mainly by geological constraints, are one very important factor to control in a study of geoecological change derived from climatic change.

1. INTRODUCTION

If we want to modelize the environmental problems associated with a hypothetical climatic change, we must know, as well as possible, which are the land features that can be affected by it, or can affect other land features. Using geomorphological variables is the best approach to this problem.

In this way, we can study the influences of the variations of rainfall and potential evapotranspiration on run off; paying also attention to the morphological, litological, and biological features of the drainage units affected by these changes.

This kind of research must be done using a systemic approach, that is to say we have to consider the drainage basins like open systems. Then we can study the rainfall as a material input, the temperature as energetical input and run off, and evapotranspiration, as material and energetic output. The relations between input and outputs are given by some processes related with the internal features of the drainage basins, considered like "process-response systems" (CHORLEY and KENNEDY, 1971).

The main aim of this paper is to show that quite close drainage basins can have very differents values for some geomorphological variables, that can be found in a lot of drainage system's papers, that characterize the morphological features of the basin, very well corelated with the kind of run off that can be expected for any kind of rainfall. In this way, it was demonstrated that any one trying to modelize hidrological changes must pay a lot of attention to these geomorphological aspects.

2. LOCATION

The research has be done at four small basins. All them tributarys of the river Minateda (Albacete province), the northern tributary of the river Segura (principally developed in Murcia province). The fluvial network of the river Segura basin is considered one of the most torrentials in the Iberian peninsula, even one of the most torrentials in Europe.

3. METHODOLOGY

1:50.000-scale topographical maps have been used. Basin boundaries and geometry of the network were determined using the contour criterion. A study on the accuracy of the stream channel representation revealed thet the mapping based on this scale and contour criterion results in reliable data (EBISEMIJU, 1976).

Some shape, size and topology variables have been measured, or calculated. These geomorphologicalvariables have a high correlation index with the parameters that define the Instantaneous Unit Hidrograph (IUH) (calculated by empirical equations that relate IUH parameters with length of main channel and the unevnness within the basin. In a previous paper (ALONSO-SARRIA, 1990) the correlation indeces betwen 41 basin parameters and variables from a sample of 21 basins were calculated (all tributarys of the river Minateda). From that paper's results, the variables with a higher correlation index with IUH parameters were selected (table 1). The geomorphological variables selected were:

      (A) Basin's area (Km^2)
      (P) Basin's perimeter (Km)
      (L) Basin's length (Km)
      (Ic) Compacity index ($P/2\pi A$) (Km^{-1})
      (N_1) Number of first order reachs*
      (N_3) Number of third order reachs
      (Nt) Total number of reachs
      (Rb) Bifurcation ratio ($\Sigma (N_i/N_{i+1})$)/(basin order-1)
      (L_2) Length of second order reachs

The used IUH parameters are:

      (I) Lag time ($a(L\ Lg)^{0.2}$). Defined as the time from the gravity center of pluviograph and the IUH's. It's expresed in hours. L is the length of main channel, Lg the length from hidrological gravity center of the basin to the mouth, a is a constant that ranges to 1.1 to 1.4.

      (II) Peak (275 Cp/lag). It's the greatest height of the water at the outlet, from a 1 mm high rainfall event. Cp is an empirical coefficient.

* All references to the stream order of the basin, are refered to the Stralher's stream ordenation system (DOORNKAMP & KING, 1971)

(III) Concentration time ($0.067\ L^{1.55}\ H^{-0.385}$). It's the time that most of the run off uses to concentrate at the basin's outlet.

The drainage basins chosen to do the comparative study are (figure 1):

(A) "Rambla del Mullidar" basin
(B) "Rambla de Lomas de Buhos" basin
(C) "Rambla de la Plata" basin
(D) "Rambla del Arroyo de la Cañada de Albatana" basin

The values of the studied variables for these basins appear in table 2.

4. RESULTS

In table 1 the values of the correlation index for the selected (shape, size and topology) variables and the IUH parameters appear . The values were obtained in the previous work mentioned above.

	I	II	III	
A	M	M	M	M=Middle (r =0.50-0.62)
P	H	MH	VH	MH=Middle-High (r =0.62-0.75)
L	H	MH	VH	H=High (r =0.75-0.87)
Ic	MH	M	M	VH=Very High (r =0.87-1.00)
N_1	MH	M	H	
N_3	MH	M	H	
Nt	M	M	H	
Ib	MH	M	H	
L_2	H	MH	H	

Table 1: Correlation indeces betwen geomorphological variables and IUH parameters

Concentration time is the IUH parameter with a higher correlation index with geomorphological variables, lag time appears in second place with a lower correlation index.

Basin's length and compacity index show us the basin's length. The effect of this length is that if L and Ic increases, lag and concentration time increase, and peak decreases.

The total number of reachs, and especially the number of first and third order reachs, have a lot of importance, but all topological features of the basin can be expressed with only the bifurcation index. If Ib increases, lag and concentration time increases, but peak decreases.

The length of all reachs of second order is another good estimation of the network organization. If it increases, concentration and lag time increases and peak decreases

These four studied basins, are very close and have a common characteristic: They are fifth order basins using the Stralher's stream ordenation systems. This fact, that give a hidrological coherence to the four basins, is the main reason that they have been studied together.

[	A	B	C	D	]
[					]
[A (Km^2)	134.51	13.35	30.57	152.75	]
[P (Km)	85	14.75	27.50	72.50	]
[L (Km)	30.15	4.65	9.85	22.2	]
[Ic (Km^{-1})	2.07	1.14	1.40	1.65	]
[N_1	707	54	123	476	]
[N_2	35	7	7	28	]
[NT	923	81	163	641	]
[Rb	5.45	2.78	3.47	4.74	]
[L_2 (Km)	48.74	3.88	11.70	33.75	]
[I (hours)	4	1.71	2.06	2.17	]
[II (l/s/Km^2)	38.62	90.21	74.91	70.9	]
[III (hours)	20.98	1.75	4.62	17.89	]

Table 2: Values of the basin variables and the IUH parameters for the four studied basins

In table 2 we can see the very great differences that these basins present for the variables studied. This means that these basins have very differents hidrological behaviours.

The origin of all the differences that we have detected, are the lithological and tectonic characteristics of the area. In this area, two main structural units appear. The union and interaction of them, give very different characteristics to the drainage basins developed in that area.

We can also see in table 2, that the IUH parameters, calculated by equations, have a high correlation with the geomorphological variables studied. The IUH parameters have a good degree of signification, the descriptions made by people who live in the countryside, about high run off episodes, support the results obtained with equations

5. CONCLUSIONS

1- The hidrological behaviour of drainage basins depends not only of rainfall and evapotranspiration; geomorphological features, also have great importance.

2- A great correlation can be found betwen geomorphological variables and IUH parameters, i. e. between basin features and hidrological response.

3- The most important geomorphological variables, in terms of run off influence, show great variations within small areas.

The final objective of this kind of research must be the modelization of water behaviour on the land surface (run off) from the basin geomorphological features. One interesting project would be to link this kind of model with atmospheric models in order to make investigations about global change.

6. REFERENCES

-ALONSO-SARRIA,F (1990): *River Minateda basin: Morphometric features*; graduate thesis non published.

-CHORLEY, R.J. & KENNEDY, B.A. (1971): *Physical geography: a systems approach*; Prentice all, London.

-DOORNKAMP, J.C. & KING, C.A.M. (1971): *Numerical Analysis in Geomorphology*; Arnold, London.

-EBISEMIJU, F.S. (1976): "Morphometric work with Nigerian topographical maps" *Nigerian Geographical Journal*; n. 19, pp. 65-77.

CLIMATIC FLUCTUATION OF TEMPERATURE AND AIR CIRCULATION IN THE MEDITERRANEAN

A. BARTZOKAS and D.A. METAXAS
Lab. of Meteorology, Dept. of Physics, University of Ioannina, Greece

Summary

The global warming is tested in the Mediterranean area during the last 45 years using SST and 1000/500 mb thickness data, in order to avoid urban effects. The area is divided in two parts vis. the West and the East Mediterranean. Since warming is not apparent in the East Mediterranean but only during the last years, the time series of surface pressure and relative geostrophic vorticity were examined for possible explanations. It is found that there exists a continuous and considerable increase of the mean vorticity in the extreme East Mediterranean, meaning increased frequency and/or strength of N. winds there, causing a delay of the warming. The opposite phenomenon appeared in the extreme West Mediterranean where increased S. winds may contribute to the warming there.

1. DATA AND METHOD

In this paper the climatic fluctuation of temperature in the Mediterranean is studied using sea surface temperature (SST) data and 1000/500 mb thickness data, in order to avoid the urban effect which has been occasionally found to be too strong (see e.g.(1)). Surface pressure data have been used as well, in order to explain some of the results. All the data are monthly grid point values and they have been kindly provided by the UK Meteorological Office. The period for the thickness data is the total available, 1946-1988. The same period has been used for SST, due to many missing data during the war. However in the post-war period some values were missing too. We estimated them and corrected some others (outliers) using stepwise regression analysis, since the horizontal multiple correlation coefficients were found very high (mostly above 0.9). Pressure data were available since 1874 and all of them have been used. We computed seasonal and annual means and we smoothed them using 11 years moving averages. In order to see the most recent tendency we estimated the 10, 9, 8,..,1 year moving averages using the last values. Using pressure data we estimated the geostrophic relative vorticity $\{\zeta = (1/\rho f) \nabla^2 p\}$, taking into account that the distance between two grid points on a meridian ($5°$ difference) is not equal to the distance between two grid points on a parallel ($10°$ difference), which also depends on the latitude φ. So, instead of using the formula $\nabla^2 p = 4(\bar{p}-p_0)/d^2$, we used the

$\nabla^2 p = \{p_E+p_W+k^2(p_N+p_S)-2p_0(1+k)\}/\{kd\}^2$, where $k = 2\sin(90°-\varphi)$ and d is the meridional grid length ($5°$). The formula is linear so we are allowed to use mean monthly data.

The study has been done for two parts of the Mediterranean, the West and the East, because a local approach can be more appropriate for analysing climate changes (2). SST is estimated as the mean of SST values of all

the "squares" of each part, while thickness as the mean of the values at two
grid points in each area (shown in fig. 1).

2. RESULTS AND DISCUSSION

In figures 2 we can see the marching of SST and thickness. Since thickness values are related to the mean temperature of the layer 1000/500 mb,
$\{\Delta z = (R\bar{T}/g)\ln(1000/500)\}$, we evaluated the mean temperature of the layer
(taking into account the dependence of g on latitude and height) and we estimated that $1^{o}C$ corresponds to 20.3 gpm. So on each graph we can see two
temperature curves, for the SST and for the lower troposphere. The two
curves show a maximum around 1965 and two minima around 1955 and 1970-75.
Although the minima and the maxima of the two curves coincide and they are
in agreement with results of other researchers (3, 4) the tendencies are different. This is especially true for the West Mediterranean where the curves
appear to diverge, mainly during the winter (fig.2e). The 1000/500 mb thickness shows a considerable increase after 1970-75 in the West Mediterranean
while in the East this increase appears during the very recent years only.
In SST too we note that there is a difference between the West and the East
Mediterranean. During the last years, its values in the West are found above
the 1945-88 mean while in the East, although there is a tendency towards increasing, they remain below the mean in 3 out of 4 seasons.

In order to study the atmospheric circulation in the Mediterranean, we
examined the pressure fluctuation since 1874 at four grid points shown in
figure 1. Generally, the minima and the maxima of the four curves coincide
(figures 3) but the main characteristic of the marching is the continuous
falling of pressure at the extreme East grid point (EE) for the last 80
years, especially in summer (fig. 3c). This means that there is an increase
of pressure difference between the two grid points of the East Mediterranean which corresponds to an increase of the northerly geostrophic wind component between them. On the contrary, the pressure difference between the
two grid points of the West Mediterranean, after the war rather decreases,
which means an increase of the southerly geostrophic wind component between
them.

For a better view of these pressure change results, we estimated the
long term vorticity fluctuation at the above 4 points. The position of the
4 curves on the graphs (figures 4) depends on the season due mainly to thermal effect of the sea and land. In summer, for example (fig.4c), the curve
of the East Mediterranean is located above the others because of the west-
wards extension of the SW Asia low. In the marching an interesting maximum
appears around 1890, then an abrupt fall until about 1910, and an almost
stable situation until 1960. After 1960 a very strong increase appears in
the East Mediterranean, especially in summer. During the same period a
rather falling of vorticity in the Centraleast Mediterranean is seen, especially after 1966, when the circulation changes from cyclonic to anticyc-
lonic (fig. 4a). In the extreme West Mediterranean an increase of vorticity
is observed after the war, while a decrease is observed in the Centralwest
Mediterranean. This decrease is so strong that in the recent years the two
curves tend to cross each other for the first time (fig. 4a, 4d, 4e). Another
interesting feature of figure 4e is that, after 1975, the vorticity of the
East Mediterranean overtakes the one of the Centralwest.

This continuous increase of the vorticity in the extreme East Mediterranean, combined with the falling in the Centraleast, leads us to the same
conclusions as in the pressure differences, i.e. an increase of the frequency and intensity of northerly winds in the Aegean sea. On the contrary,
the vorticity change in the West and the Centralwest Mediterranean implies
an increase of the southerly winds there after the war.

3. CONCLUSIONS

The study of the long term fluctuation of SST, and 1000/500 mb thickness in the Mediterranean, has shown that the global warming does not appear everywhere in the Mediterranean. The warming has not been observed in the East Mediterranean (fig. 2) but only during the last years, so a strengthening of the northerly wind force and/or frequency must have occured. This strengthening is shown by the study of surface pressure and relative geostrophic vorticity time series. Thus we can assume that local atmospheric circulation changes can support or oppose the global warming in some places, as in the West and the East Mediterranean respectively.

REFERENCES

(1) KARL, T.R. and JONES, P.D. (1989). Urban Bias in Area-averaged Surface Air Temperature Trends. Bulletin of the American Meteorological Society. Vol. 70, No 3.
(2) WEBER, G.R. (1990). Tropospheric Temperatures Anomalies in the Northern Hemisphere 1977-86. International Journal of Climatology. Vol. 10, No 1.
(3) PAINTING, D.J. (1977). A study of some Aspects of the Climate of the Northern Hemisphere in Recent Years. HMSO. Meteorological Office scientific paper No 35.
(4) REPAPIS, C.C. and PHILANDRAS, C.M. (1988). A Note on the Air Temperature Trends of the last 100 Years as Evidenced in the Eastern Mediterranean Time Series. Theoretical and Applied Climatology. Vol. 39.

CAPTIONS

Figure 1. The areas and the grid points used.
Figure 2a. SST and 1000/500mb thickness fluctuation. 11years moving averages (year).
Figure 2b. SST and 1000/500mb thickness fluctuation. 11years moving averages (spring).
Figure 2c. SST and 1000/500mb thickness fluctuation. 11years moving averages (summer).
Figure 2d. SST and 1000/500mb thickness fluctuation. 11years moving averages (fall).
Figure 2e. SST and 1000/500mb thickness fluctuation. 11years moving averages (winter).
Figure 3a. Surface pressure fluctuation. 11 years moving averages (year).
Figure 3b. Surface pressure fluctuation. 11 years moving averages (spring).
Figure 3c. Surface pressure fluctuation. 11 years moving averages (summer).
Figure 3d. Surface pressure fluctuation. 11 years moving averages (fall).
Figure 3e. Surface pressure fluctuation. 11 years moving averages (winter).
Figure 4a. Relative Geostrophic Vorticity fluctuation. 11 years moving averages (year).
Figure 4b. Relative Geostrophic Vorticity fluctuation. 11 years moving averages (spring).
Figure 4c. Relative Geostrophic Vorticity fluctuation. 11 years moving averages (summer).
Figure 4d. Relative Geostrophic Vorticity fluctuation. 11 years moving averages (fall).
Figure 4e. Relative Geostrophic Vorticity fluctuation. 11 years moving averages (winter).

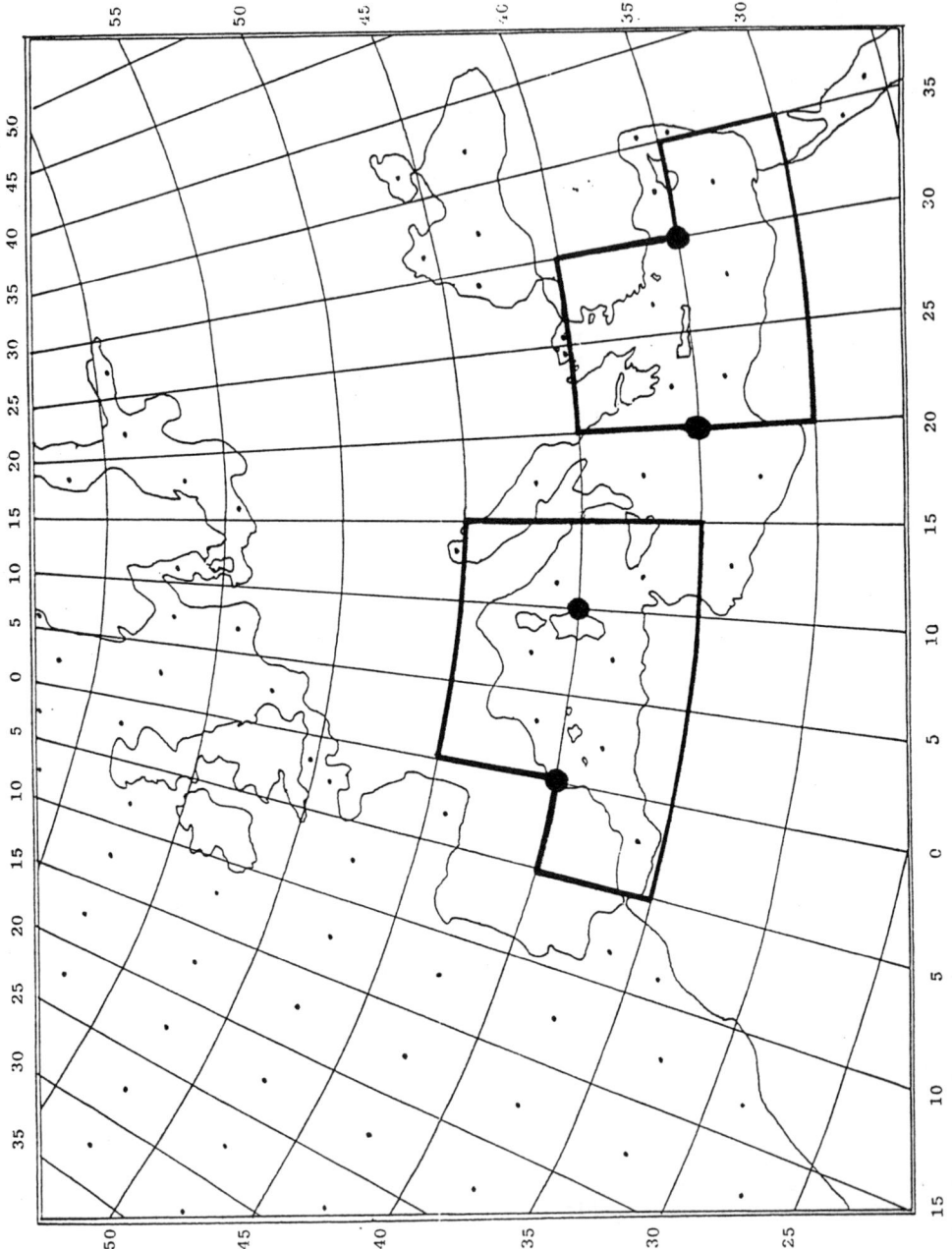

Figure 1

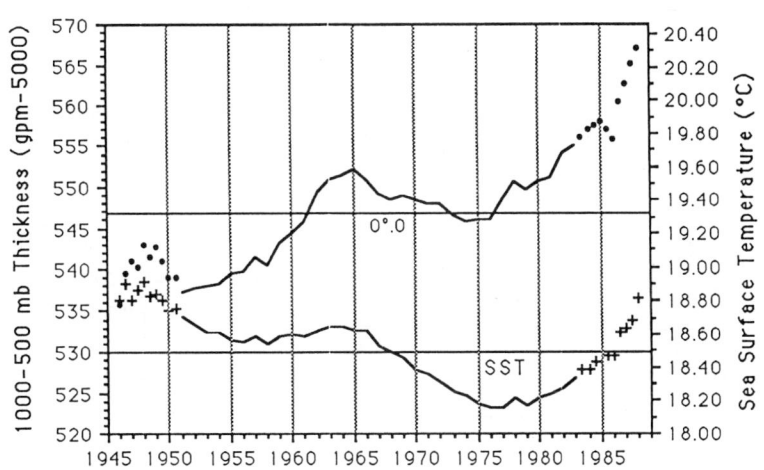

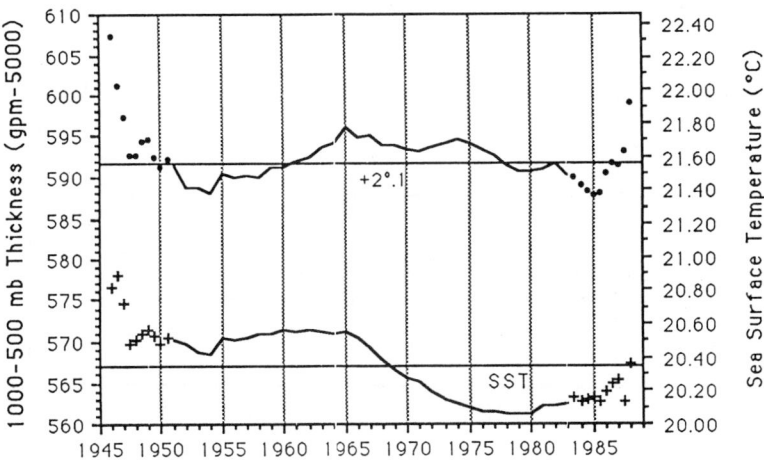

Figure 2a

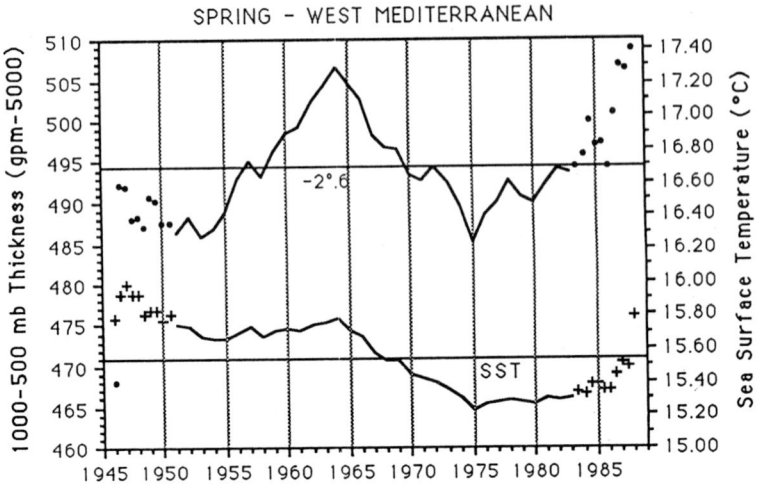

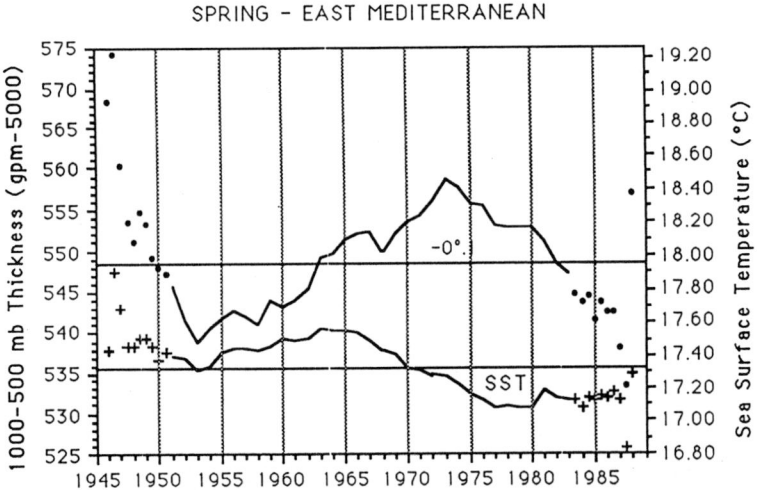

Figure 2b

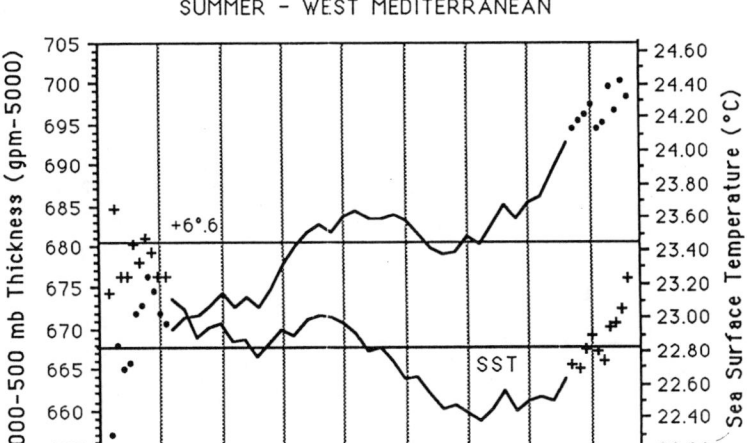

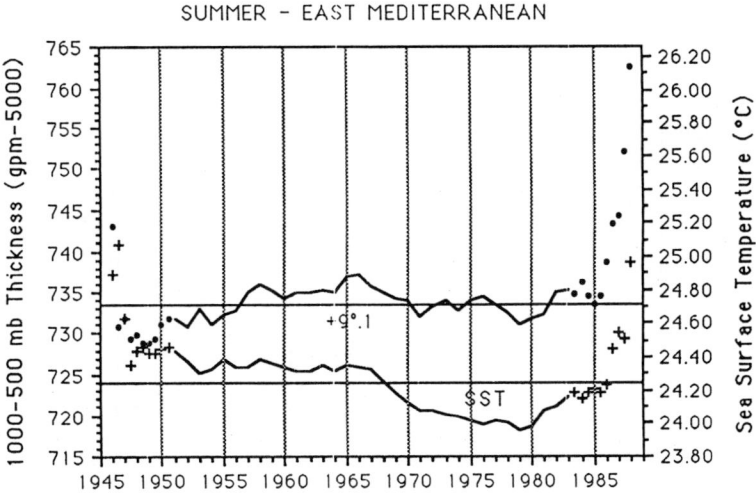

Figure 2c

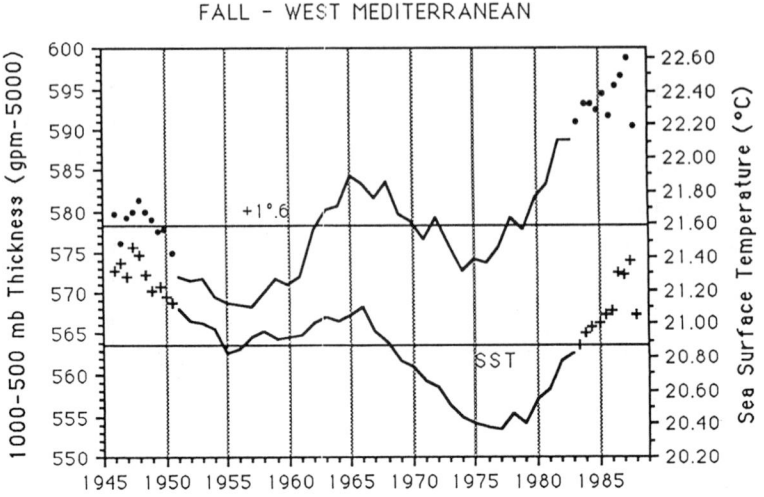

FALL - WEST MEDITERRANEAN

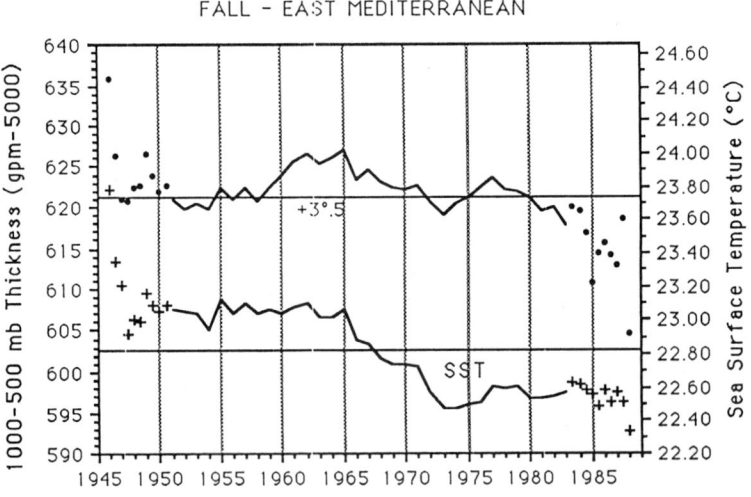

FALL - EAST MEDITERRANEAN

Figure 2d

WINTER - WEST MEDITERRANEAN

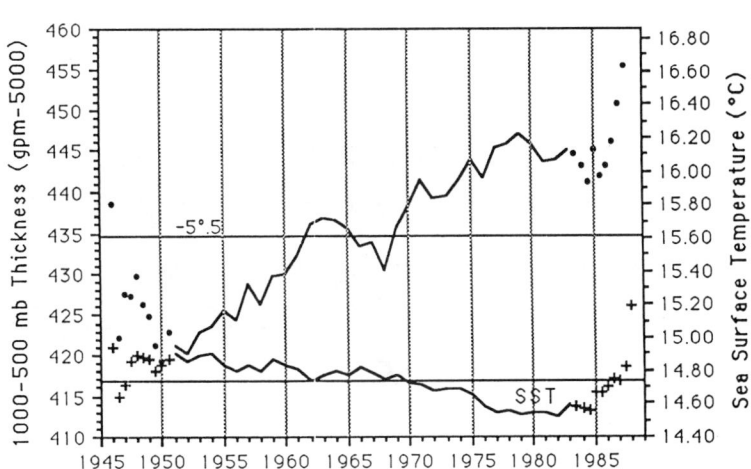

WINTER - EAST MEDITERRANEAN

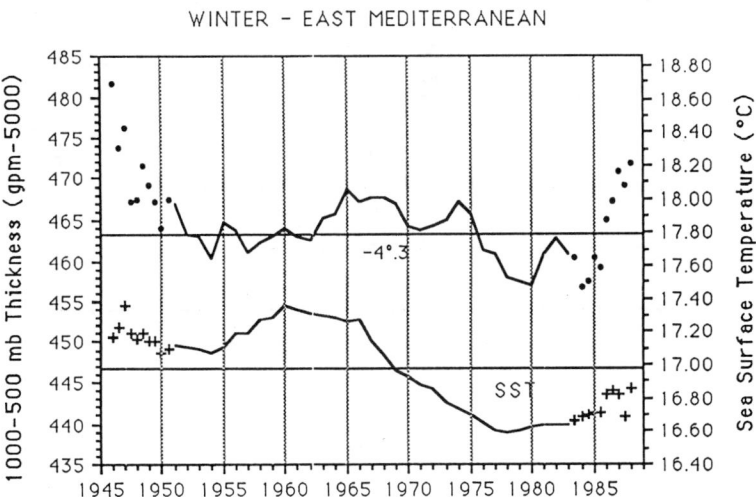

Figure 2e

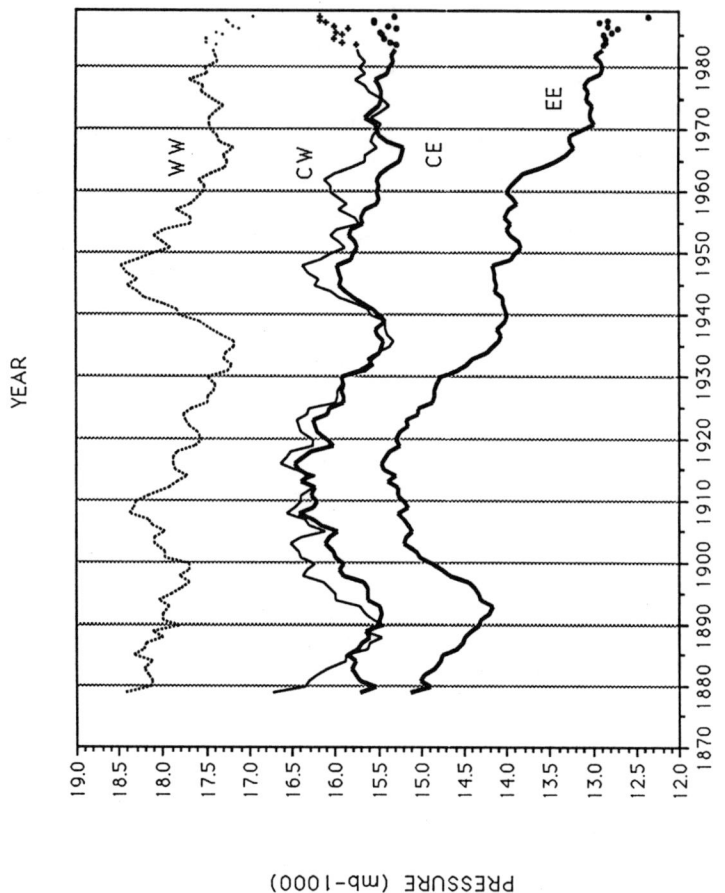

YEAR

PRESSURE (mb-1000)

Figure 3a

SPRING

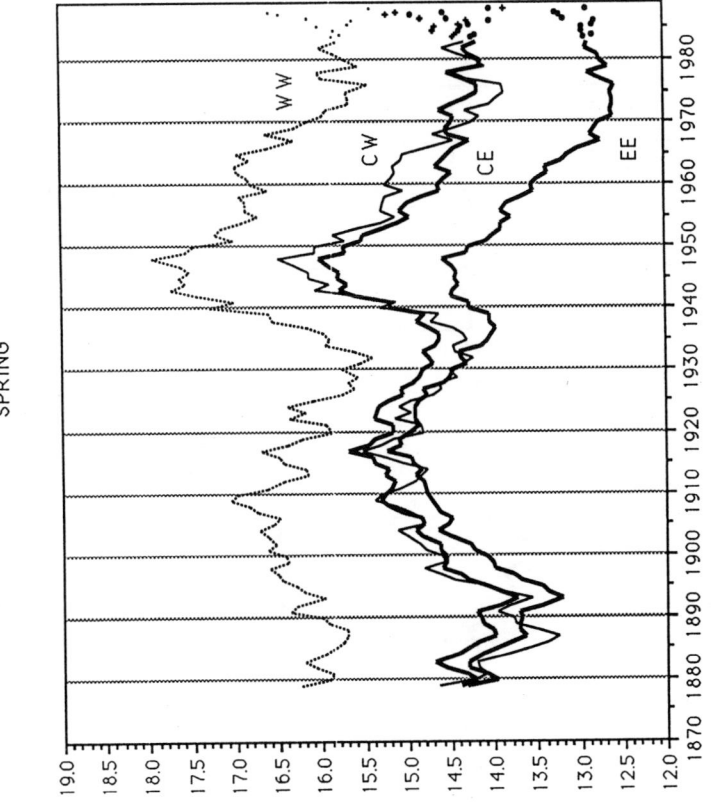

PRESSURE (mb-1000)

Figure 3b

289

SUMMER

PRESSURE (mb-1000)

Figure 3c

FALL

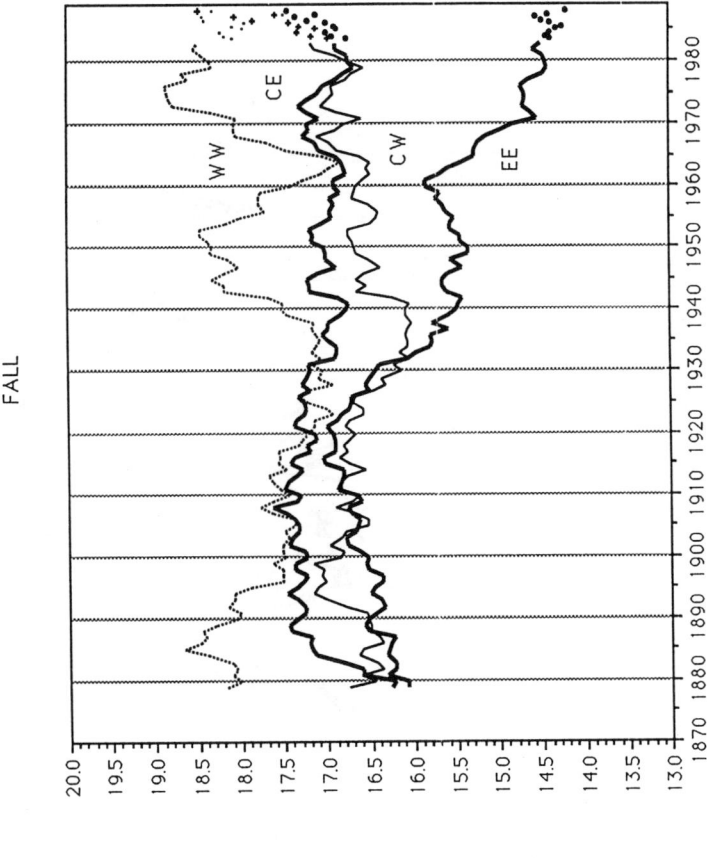

PRESSURE (mb-1000)

Figure 3d

291

WINTER

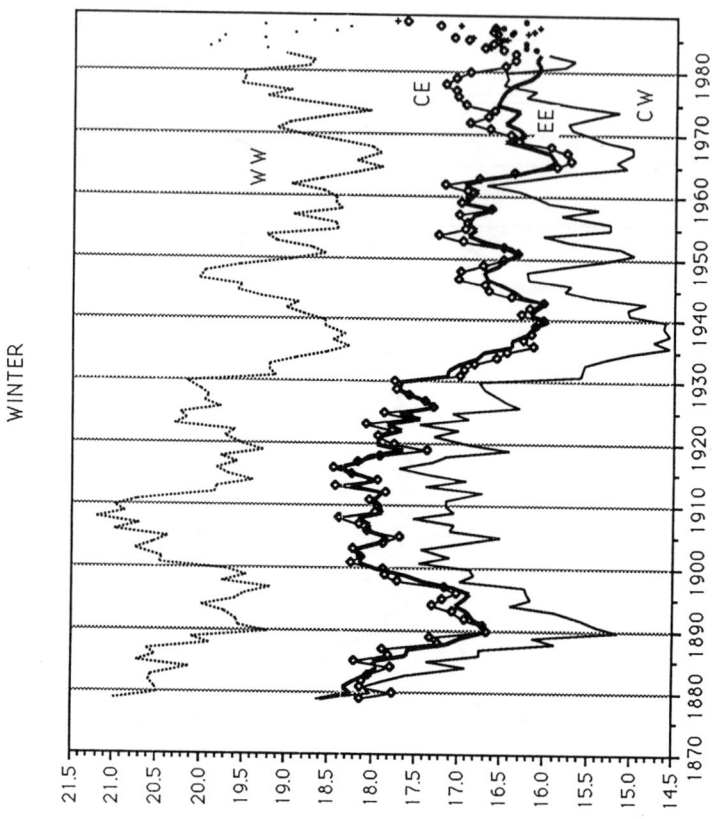

PRESSURE (mb-1000)

Figure 3e

292

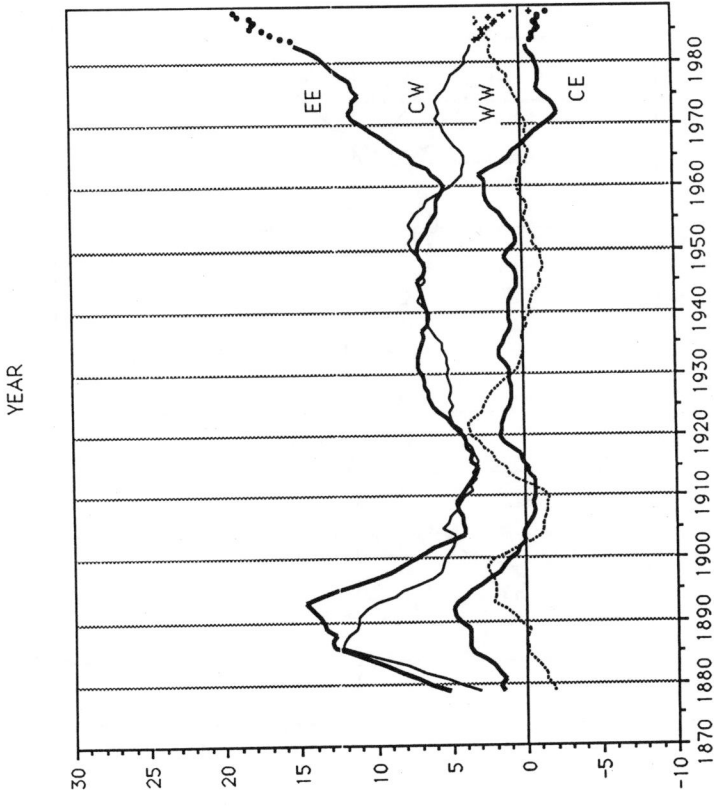

YEAR

GEOSTR. REL. VORTICITY ($s^{-1} \times 10^6$)

Figure 4a

SPRING

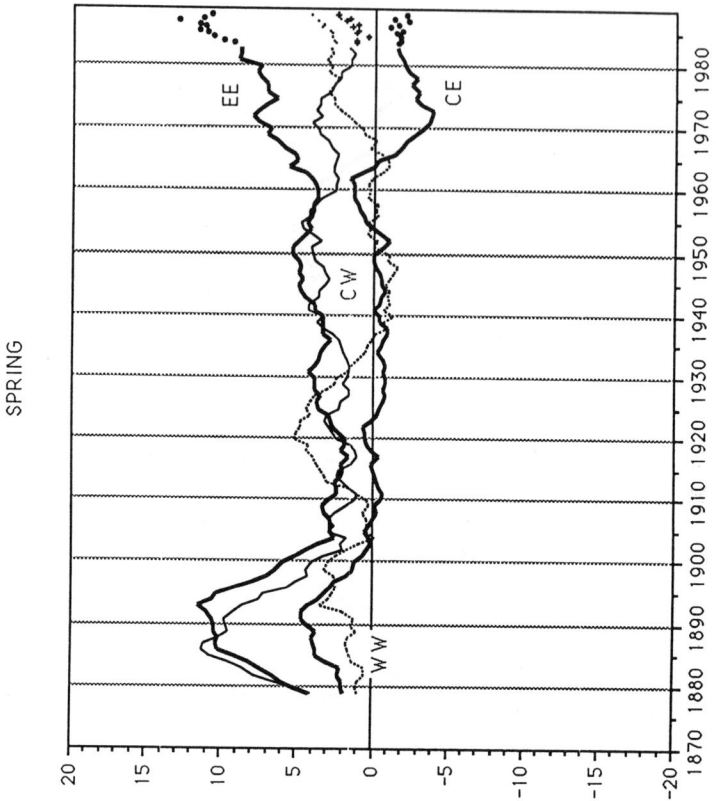

GEOSTR. REL. VORTICITY (s⁻¹ ×10⁶)

Figure 4b

294

SUMMER

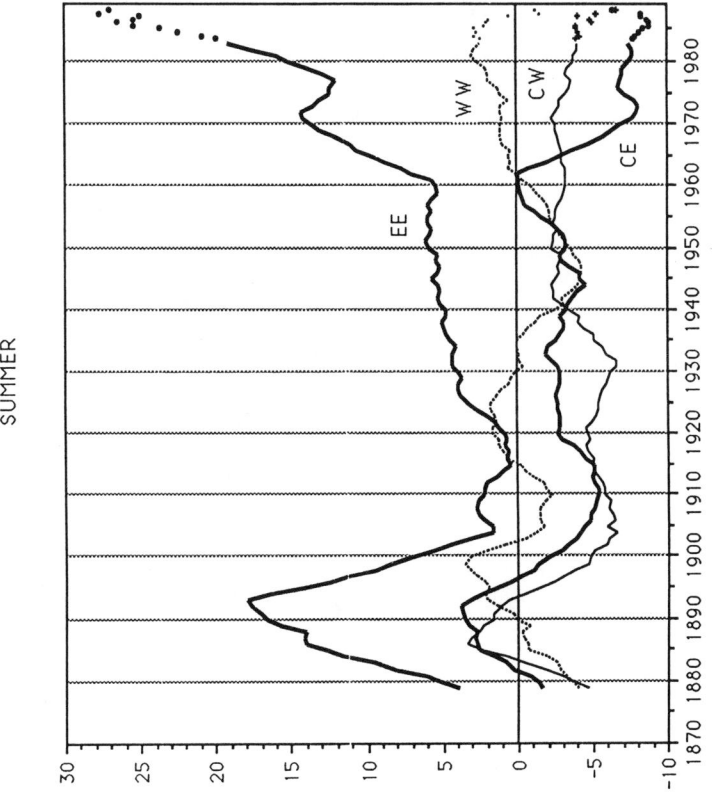

GEOSTR. REL. VORTICITY ($s^{-1} \times 10^6$)

Figure 4c

295

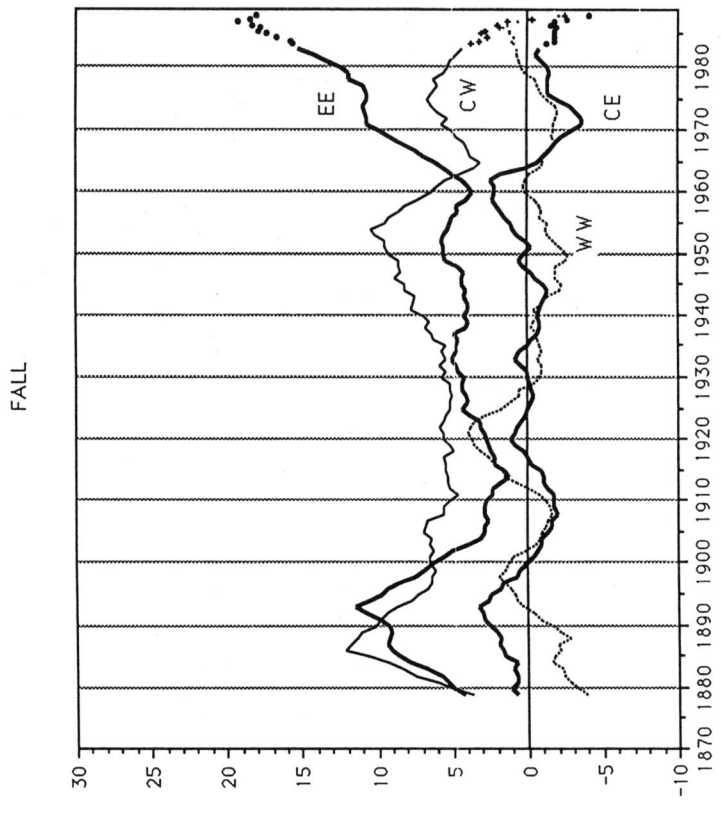

Figure 4d

WINTER

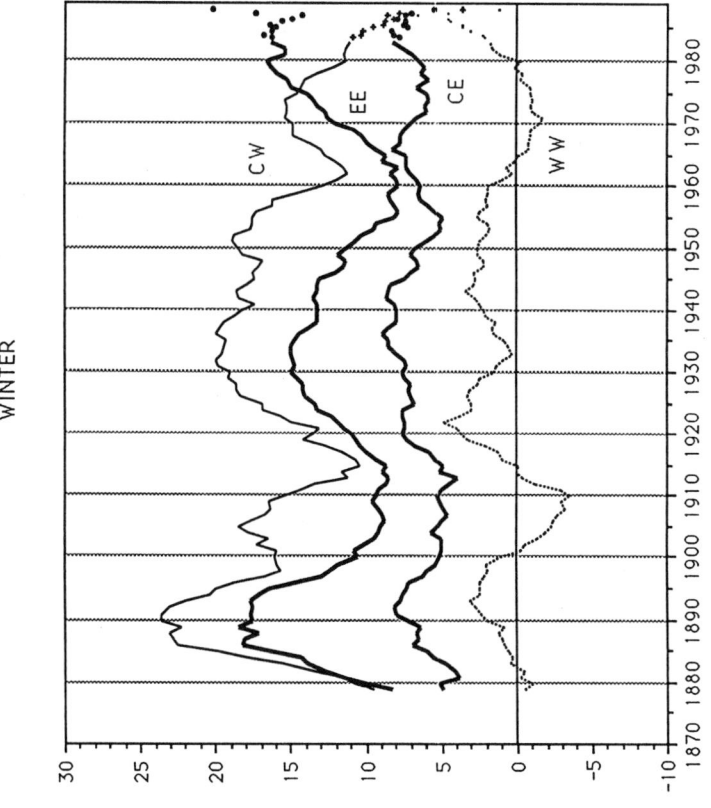

GEOSTR. REL. VORTICITY ($s^{-1} \times 10^6$)

Figure 4e

LANDSCAPE ECOLOGICAL IMPACT OF CLIMATIC CHANGE
some preliminary findings of the LICC Project

Matthias M. Boer
Department of Physical Geography, University of Utrecht
Utrecht, The Netherlands

INTRODUCTION

It has become clear that the increasing concentrations of atmospheric carbon dioxide and other radiatively-active gases will lead to more heat being trapped near the surface of the earth and to a rise in the global mean surface air temperature. Global warming and sea level rise are increasingly being recognized as one of todays most serious long-term environmental problems. The global dimension of the greenhouse problem and the uncertain, but potentially hazardous implications for the environment are leading to calls for a large research effort to establish a scientific basis for preventive and adaptive measures.

THE LICC PROJECT

As a contribution to the UNEP-WMO-ICSU World Climate Impact Programme and within the framework of the International Geosphere-Biosphere Programme of ICSU, The Netherlands' Ministry of Housing, Physical Planning and the Environment initiated the LICC Project (a) to address the potential effects of a future climatic change on (semi-) natural terrestrial ecosystems and landscapes in Europe. In order to cope with the large range of possible impacts of climatic change the LICC Project was concentrated on six case study areas or environments: alpine regions, the Fennoscandian (boreal and subarctic) region, the Mediterranean Region, fluvial systems, wetlands, and coastal dunes. These case studies cover those regions and landscape systems within Europe that were stated to be most sensitive and vulnerable to climatic change during the preceding European Workshop on Interrelated Bioclimatic and Land Use Changes (Noordwijkerhout, The Netherlands, 1987). As a follow-up of the Noordwijkerhout Workshop the LICC Project aimed to inventarise existing scientific information on the climate sensitivity of key-biotic and -abiotic processes operating in the case study environments, to exchange ideas on methodologies and techniques used for climate impact assessment and to draft recommendations for future research and monitoring programmes. To meet these objectives a preparatory literature review (1) was carried out and an international group of scientists from various environmental disciplines was formed for each case study. Over 70 scientists from 18 European and several other countries, such as Canada and the U.S.A., participated in the case study groups and were prepared to contribute to one of the discussion reports. Their papers were compiled into six separate discussion reports (2,3,4,5,6), one for each case study, and were discussed extensively

---

a. The LICC Project (Landscape Ecological Impact of Climatic Change) was coordinated by the Physical Geography Departments of the Universities of Amsterdam and Utrecht, and by the Nature Conservation Department of the Agricultural University of Wageningen.

during the European Conference on Landscape Ecological Impact of Climatic Change, the final stage of the LICC Project.

THE LICC CONFERENCE AND WORKSHOPS

From December 3-7, 1989, The European Conference on Landscape Ecological Impact of Climatic Change was held in Lunteren, The Netherlands, attended by about 200 participants from 22 countries. During the first part of the conference, workshop sessions were held to discuss, with the six case study discussion reports and other papers as a basis, the potential impact of climatic change on the selected case study environments. During the second and plenary part of the conference, the preliminary conclusions and recommendations of the working groups were presented to the other participants and set in a broader context by key-note papers on: climate models and scenarios, landscape ecological impact of past climatic changes, abiotic and biotic aspects of climatic change, methodological aspects of assessing the landscape ecological impact of climatic change, socio-economic aspects of climatic change, and the policy implications of landscape ecological impacts of climatic change. In this short paper, some general findings from the case study workshops can only be outlined very briefly. The final conclusions and recommendations of the working groups as well as the full texts of the key-note papers are included in the LICC Conference Proceedings (7).

CLIMATE SCENARIOS FOR IMPACTS STUDIES

The currently available climate models indicate that global mean surface temperatures may rise by 3.0 - 5.5 °C as a result of a doubling of the concentration of atmospheric CO_2 and other greenhouse gases such as methane (CH_4), nitrous oxide (N_2O), and chlorofluorocarbons (CFCs). For the purpose of regional climate-impact-assessment studies the figures on the global scale are of little use. For an indication of the greenhouse impact on the climate of a certain part of the globe we must rely on scenarios.

Essentially there are two kinds of climate scenarios. Those based on numerical models and those based on regional and seasonal patterns of past warm climates. Eybergen and Van Huis (1) compared recent climate scenarios for their applicability in studies on the landscape ecological impact of climatic change. They came to the conclusion that so far no scenario can forecast satisfactory well how the regional European climates may change due to a doubling of greenhouse gases.

Based on these conclusions a so-called "what...., if....approach" was recommended for the LICC Case studies. Contributors to the LICC Case study discussion reports were therefore asked to focus on the climate sensitivity of key-biotic and abiotic processes operating in terrestrial landscapes and ecosystems, instead of basing their assessment on one particular climate scenario.

ASSESSING THE LANDSCAPE ECOLOGICAL IMPACT OF CLIMATIC CHANGE

Forman and Godron (8) define landscape ecology as the study of the structure, function and change of a heterogeneous land area composed of interacting ecosystems. Similar to ecology, landscape ecology focusses on the interaction of interdependent living and non-living compartments or subsystems, such as the soil, water, vegetation or animal communities (9) and recognizes the ecosystem as a basic unit. Landscape ecology, however, differs from ecology by "translating" information and knowledge obtained at basic levels of, for example, a plant or animal population or community, a snow patch or a dune slack ecosystem to the higher level of the entire landscape system. Although the outlines of landscape ecology as a distinct field with its own concepts and principles are only of a recent origin, the integration at the landscape level of various environmental disciplines, such as physical geography, pedology, hydrology and system ecology, has proven to provide a useful approach for studies of complex landscape systems. To assess the landscape ecological impact of climatic change the LICC project identified a number of key-questions:

* what are the most important climate sensitive processes?
* to which (set of) climate parameters are they related?
* how sensitive are these processes for changes in climate parameter values; what is the response time; are there any ecological thresholds involved?
* what is the impact of climate induced alterations of those processes on the structure, function and change of ecosystems and landscapes; which equilibriums or feedback mechanisms could come into play?
* what are the implications for land use and how could land use modify the impact of climatic change on landscape systems?

PRELIMINARY FINDINGS FROM THE LICC CASE STUDIES

While inventorizing existing knowledge relevant for the assessment of the impact of climatic change on landscape systems, a serious lack in fundamental ecological knowledge showed up dramatically in all case study groups. Nevertheless, it appeared to be possible to identify a large range of key-ecological processes, response mechanisms and hazards that might be affected or triggered by CO_2 increment and associated climatic change. Assessment of the resulting impact of multiple factors on ecosystems and landscapes of the case study environment were, however, complicated due to expected synergistic and counteracting effects, which merely are poorly understood. Largely unknown properties of ecosystems such as their plasticity, resilience or buffering capacity to greenhouse induced changes of the environment represents another serious scientific obstacle for the assessment of the climate impact. Rapid changes in atmospheric chemistry and climate, such as those envisaged for the next decades, were however generally thought to lead to ecosystem destabilization or degradation.

Below, some remarks will be made on a number of the aspects and difficulties involved in landscape ecological impact assessment.

Changes in water and sediment fluxes

Climatic change will alter sediment and water fluxes in various ways. First of all through changes in the amount, nature (snow/rain) and timing of the precipitation. Secondly, through changes in the evapotranspiration rates and the storage of water in the soil. Future evapotranspiration and soil water storage cannot yet be assessed due to the current state of climate scenarios and the large uncertainties about (at least partly interrelated) changes in the vegetation, the effect of elevated CO_2 levels on plant transpiration rates, and about infiltration characteristics of the soil. The altered soil moisture conditions, groundwater flow, runoff and hydrologic regime of streams, which will ultimately result from these changes, will affect sediment yield, the transport and deposition of sediments and nutrients.

Changes in the vegetation cover

The vegetation cover, representing the climatologically active surface of most terrestrial ecosystems, is known of course to control to a large degree the micro-climatic conditions prevailing at a certain site. Through regulation of the temperature, moisture and wind regime and through the supply of habitats for wildlife or of litter and organic matter to the soil and streams, the vegetation determines features such as the abundance of animal species, the activity of physical, chemical and biological processes in the soil compartment or the quantity and quality of water available for infiltration, plant growth, erosion or runoff. The potential changes in the structure and composition of the vegetation cover is therefore of outmost importance for the assessment of the greenhouse impact on ecosystems or landscapes. The two most important effects of climate change on vegetation concern the direct physiological effects of CO_2 enrichment on plants and the potential shifts in the distribution of plant species and vegetation types.

Species response

Potential shifts in the distribution of plant and animal species can be made by extrapolating current relationships between bioclimate and species distribution. These scenarios are, however, not very realistic since they do not incorporate that plant species as well as animal species differ in their ability to respond to environmental changes. Some species will be able to profit (the "winners"), such as those with high regeneration rates or with strong migration capacities, while others (the "loosers"), such as longlived trees or species with very special ecological demands, will decline or may even become (locally) extinct. As a general conclusion, the rate of climatic change, as forcast by current scenarios, was thought to be too much for most species to cope with.

Dispersal and migration in a fragmentated landscape

In the European landscapes where relicts of (semi-)natural environments mostly occur fragmentarily due to agriculture, urbanization or other types of land use, many species depend for their dispersal on stepping stones or corridors provided by small landscape elements such as forest patches, hedgerows or streams. (It is important to note that an ecological infrastructure for one (class of) species may be a labyrinth of barriers for others). In a landscape experiencing climate changes such connective landscape elements are likely to change too or could even disappear. Especially in highly urbanized areas like western Europe where the connecting landscape elements often are no more than what human land use has left of the natural environment, the ecological infrastructure may be affected considerably by climate induced changes in land use patterns and the subsequent disappearance of such vital landscape elements. Changes in ecological connectivity may cause potential trouble for some species, while others might benefit from the same changes. Especially rare and endangered species living in isolated reserves and protected areas, often having very special ecological demands, are thought to be particularly vulnerable to climatic changes, since migration to other suitable habitats will be more difficult for these species than it will be for common ones.

Modification of climate impacts by man

In large parts of Europe the impact of climatic change on the landscape could be modified significantly by simultaneous changes in the landscape caused by demographic or socio-economic developments. Synergistic interactions of climatic change and human activities (e.g. land use, pollution) may amplify detrimental effects, for example where ecosystems that suffer from acid deposition or other kinds of disturbances already have a decreased ability to sustain stress. In other cases it may be very difficult or even impossible to assess natural, anthropogenic and climate change effects separately. The signal of climatic change may then be lost in the noise of other developments in the landscape. From the above it becomes clear that to assess the landscape ecological impact of climatic change, at least for the European region, demographic, sociologic and economic factors should be taken into account. A pressing need for integrated studies that encompass both human behaviour (economic, social, phychological) and biophysical effects on landscapes was therefore found to be of great importance for the future.

REFERENCES

(1)   EYBERGEN,F.A. AND VAN HUIS,J.C. 1988. Climate modeling and the climate sensitivity of biotic and abiotic processes. Report on Working phase 1. Landscape ecological impact of climatic change. Agri. univ. Wageningen, Univ. of Utrecht. Univ. of Amsterdam. pp. 34.

(2)   RUPKE, J. AND BOER, M.M. (eds.) 1989. Landscape Ecological Impact of Climatic Change; Alpine regions. Discussion report prepared for the European Conference on Landscape Ecological Impact of Climatic Change, held at Lunteren, The Netherlands, 1989. Univ. of Amsterdam, Univ. of Utrecht, Agr. Univ. Wageningen. pp. 139.

(3)   KOSTER, E.A. AND BOER, M.M. (eds.) 1989. Landscape Ecological Impact of Climatic Change; The Fennoscandian part of the boreal and subarctic zone. Discussion report prepared for the European Conference on Landscape Ecological Impact of Climatic Change, held at Lunteren, The Netherlands, 1989. Univ. of Amsterdam, Univ. of Utrecht, Agr. Univ. Wageningen. pp. 151.

(4)   IMESON, A.C. AND DE GROOT, R.S.(eds.) 1989. Landscape Ecological Impact of Climatic Change; Mediterranean regions. Discussion report prepared for the European Conference on Landscape Ecological Impact of Climatic Change, held at Lunteren, The Netherlands, 1989. Univ. of Amsterdam, Univ. of Utrecht, Agr. Univ. Wageningen.

(5)   JONGMAN, R.H.G. AND BOER, M.M. (eds.) 1989. Landscape Ecological Impact of Climatic Change; Fluvial systems. Discussion report prepared for the European Conference on Landscape Ecological Impact of Climatic Change, held at Lunteren, The Netherlands, 1989. Univ. of Amsterdam, Univ. of Utrecht, Agr. Univ. Wageningen. pp. 191.

(6)   STORTENBEKER, C.W. AND DE GROOT, R.S. (eds.) 1989. Landscape Ecological Impact of Climatic Change; Wetlands. Discussion report prepared for the European Conference on Landscape Ecological Impact of Climatic Change, held at Lunteren, The Netherlands, 1989. Univ. of Amsterdam, Univ. of Utrecht, Agr. Univ. Wageningen.

(7)   BOER, M.M. AND DE GROOT, R.S. (eds.) (1990) Landscape ecological impact of climatic change. Proceedings of a European Conference held at Lunteren, The Netherlands, Dec. 3–7, 1989. IOS, Amsterdam, Springfield (VA), Tokyo.

(8)   FORMAN, R.T.T AND GODRON, M. 1986. Landscape ecology. John Wiley & Sons, New York. pp. 619.

(9)   TIVY, J AND O'HARE,G. 1981. Human impact on the ecosystem. Oliver and Boyd. Edinburgh– New York.

RELATIONSHIPS BETWEEN HOLOCENE VEGETATION AND THE SPREADING OUT OF CIVILIZATIONS AND LANGUAGES IN SOUTHWESTERN-ASIA

F.DAVID

Laboratoire de botanique historique et palynologie
Faculté des sciences et techniques,St Jerôme. Marseille.

Summary
The reconstitution of holocene environments comes up against the definition and the estimate of human impact, specially in the upper part of holocene. This effect varies in time and space, as is the case in mediterranean countries, where advanced civilizations appeared very early.
A pluridisciplinary approach is proposed to deal with this question. Special attention is paid to palynology, archaeology and linguistic. The contraints and the main points that should be developped in each discipline are underlined.

1.INTRODUCTION

When we study the holocene environments, we always come up against the occurrence of the human impact. Several studies show that human disturbance in the environments became really strong in the upper part of the holocene, so much so that man can be considered as a "geological factor" (Frenzel,1980). The effects of this disturbance vary with time and space (Reille et all 1980, Pons et Quezel 1985). Our objective is to estimate it in the mediterranean zone.

Southwestern-Asia, that is to say: Turkey, Syria, Iraq, Iran, was choosen for the following reasons:

The greatest part of the area is located in the mediterranean zone, wich is known for its ecological richness. Nevertheless literature covering this area is not so abundant as in northern countries. The richness of the flora can be partly explained by the existence of refuges for meso-thermophilous species during the glaciations (Triat-laval,1978, Quezel,1985, Pons et all,1990). The presence of these species in the mediterranean zone during glacial times permitted a particularly swift answer of vegetation to climatic changes.

In the same time, the early expansion of civilizations in the region is indicated by a great number of archeological sites, throughout the holocene and it seems that the expansion of trees and men in this area occured simultaneously during the holocene. In this region also, the most ancient texts actually known have been found.

And finally, we found the opportunity for laying down and working out the rule of a pluridisciplinary study, including human sciences. As a matter of fact, there are three complementary approaches to this problem : palynology, archaeology and linguistic (cf Fig.1).

2.PALYNOLOGICAL APPROACH

The final environmental picture provided by this discipline must make clear the part of the local, regional and more distant vegetation (Janssen 1986). For the holocene, this picture depends essentially on climate and human impact. Five major points must be underlined:

The geomorphological features of the sites such as size, (Berglund 1979), inflow stream (Tauber 1977), substratum (Pennington 1978), determine the specific mixture of pollen rain and they must be taken into account in the choice of the sites (Webb et all 1978) and in the final discussion. The diversity of the plant cover is so high, and depends on so many factors such as local climate, soil, competition, history of vegetation itself etc..., that we need various coring sites from different ecological levels to take this diversity into account (Jacobson et Bradshaw 1981)

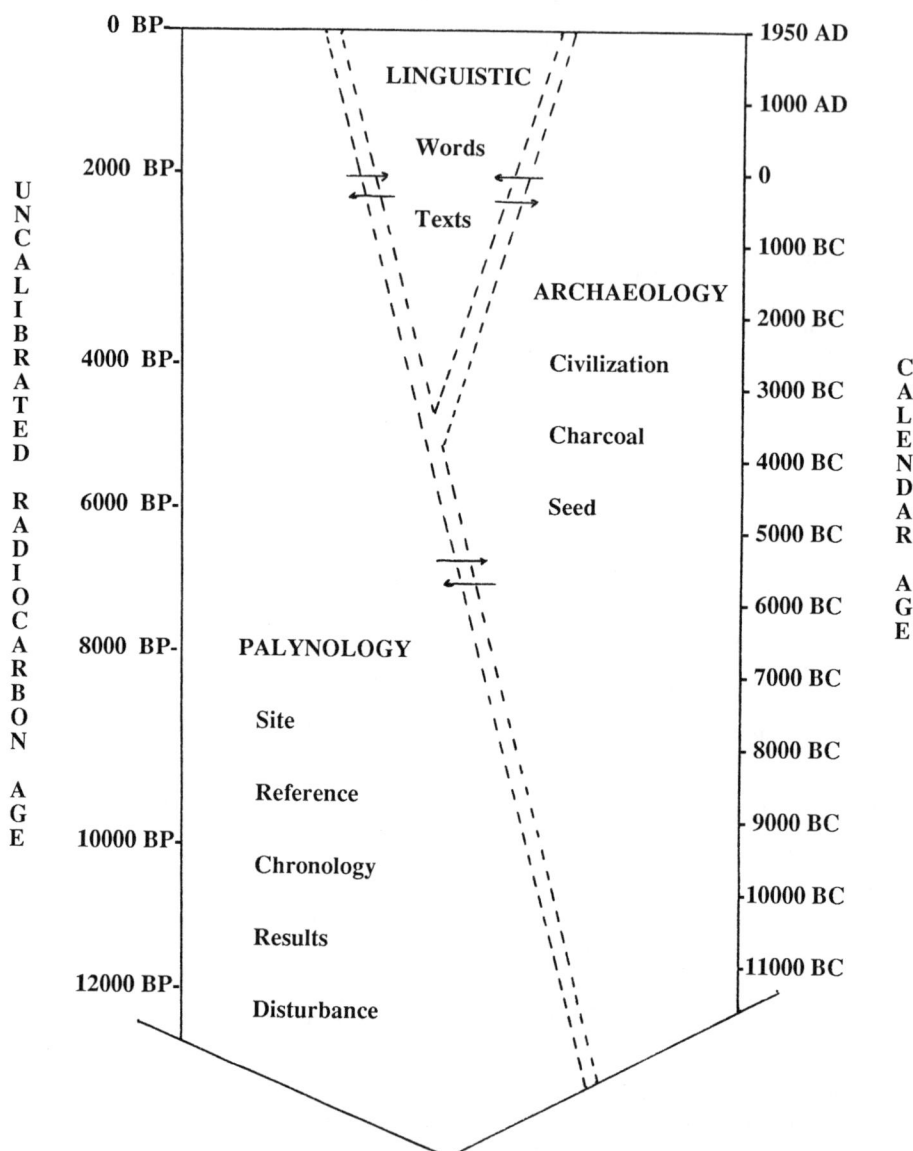

Fig 1: VEGETATION RECONSTRUCTION

Our final interpretation of the diagrams is based on the knowledge of the actual plant cover. It involves that our references follow the most recent works, defining the actual ecological levels (Chalabbi, 1980 , Quezel et all, 1980) and that our diagrams can always be reinterpreted in the light of these works (El moslimany, 1987).

If we aspire to give valuable results easily used for other disciplines, they are two main conditions to fulfil: firstly, a chronological scale permetting correlations not only within our own discipline (Janssen, 1980) but with the other disciplines. It becomes much more easy since the setting of the radiocarbon chronology and its continual perfecting (Pearson et all., 1986). Secondly, a standardized language corresponding with a strict methodology. It is particularly true, when we have to establish the boundaries of representation of the taxa or to describe their expansion (Triat-Laval, 1978).

For translating the vegetation changes in terms of climatic changes, we have to define the disturbance related to man activities. This disturbance has been bound to the definition of markers of the human activities. They are the anthropogenic species, wich have been defined and listed for northern Europe (Behre, 1981). We have to complete this work for mediterranean countries and this study is in progress (Saadi, 1990). But now, we have a complementary source of informations with the archaeological data. We can compare on maps, human settlements and coring sites. Such a work should lead us to make clear the difference between the real natural communities and the anthropogenic communities. In other respects, recent works show that the anthropogenic communities have developped from a stock of local species and can be different in each area. (Pons et all., 1990)

3.ARCHEOLOGICAL APPROACH

By this discipline, we have to define the main trends of human communities for each period and find informations about the populations structure, size, trade and technological advance. The most appropriate way for our purpose is to draw maps of the civilizations expansion, showing the different exchanges and relationships between them (Braidwood, 1975). These general data will lead us to further and finer investigations.

For example, the emergence of ceramic or metals technology increases the need for fuel and we can expect an increase in the vegetation clearance. This lead us to determine wich species were used for fuel, using special analysis such as charcoals-analysis. Nevertheless charcoal-analysis, as a general rule, inform us about woods and woods were used for many objects such as buildings, tools etc...(Taylor, 1981). We can determine many taxa at the species level and the most difficult will be to assign a determined use to each species but complementary data such as location within the site or ancient texts should be a help. A quantitative approach seems very difficult to establish (Thinon, 1979, Vernet, 1982), but a qualitative approach in terms of "presence, absence" will give satisfory results for a preliminary study.

In the archeological sites, one other interesting point is the analysis of macrorests such as seeds, including cultivated seeds and weeds. Men used plants for many activities such as medicine, perfumery, weaving, dyeing,etc...(Cerceau,.1985) but generally, crops weeds are the most frequent macrorests in these sites and they provide informations about crop-system, that is to say: season, height of cutting, surroundings and stockage (mixing or not of crops) (Van Zeist and Bakker-Heeres, 1988).

For our purpose, the comparison of lakes pollen diagrams with macrorests analysis in the archeological sites would complete the definition of the anthropogenic species and help to distinguish between crops weeds, grazing markers and local ecological species (Wazilikowa 1986). Actually, we can see that grazing in the mediterranean zone is one very efficient factor of deforestation but we cannot assume, without more specific investigations, that it was the same in the lower part of the holocene, where areas free of trees may have been large enough for culture without forest clearance. Finally, we should core in natural sites, at different distances of archeological sites to have a better idea of the human impact in pollen diagrams (Van Leeuwarden and Janssen 1985).

4.LINGUISTIC APPROACH

Writting is one of the evolution factors in the period between the time of the first agricultural communities and the classical societies. It appeared in the neolithic communities of Mesopotamia around 3500 B.C. for keeping the accounts of the crops and it spread out in mediterranean countries, during the third millenia, such as a tool useful to many languages. In outline, writting was used by two important empires, the sumero-accadians in Mesopotamia since third millenia and the hittites, in actual Turkey, since second millenia before hellenistic, roman and islamic empires.

But writting is a fixed form of the languages, wich can be boundlessly multiplied. It involves that we must never forget that language and writting are both a technical tool and a support to thought, even if the first aspect is mainly considered here. We ask questions that ancient men couldn't ask, owing to the fact that five millenia of experiments come between us. Our reasoning is not the same, however, we can call a scientific approach their methods of observation and classification of the phenomena. They lay down the relationships between these phenomena but they were unable to make clear the general rules in the abstract and they simply exhausted all the cases of a problem (Garelly,1972); it doesn't prevent our purpose to describe the environment and we have two main ways of working: the words and the texts.

Many tablets with lists of classified words such as trees, fruits, edible plants, etc... have been found in the archeological sites. These lists can be used to describe the environment as it was perceived by our ancesters. We come up against the fact that we observe a dead language with all the problems of conservation and selection of the data. Nevertheless, words are individually significant, if we assume that the most current words correspond with current objects. We can try to give a first shape to the maps of paleolanguages with a few words describing natural objects.

We can also use the texts. The subjects of the tablets are very varied. Following Oppenheim, they can be distributed in two groups: literature and practical books such as account books, administrative reports, letters, calendars etc...(Oppenheim,1968). Most of the time, we shall have to extract data of texts wich are not speculative but the source is so abundant that the problem is more a problem of choice and of time. One more time, it is the number of data and their cross checking, wich will give the weight value to the study. Finally, our knowledge of the surrounding should be a help in translating the ecological vocabulary. The feed back effect would be a picture of past environments from the actors themselves.

5.CONCLUSION

Each study has to be carried out with its own methodology, but it is more and more required to choose in agreement with the other disciplines the inquiring lines and the sites. The needs of the three disciplines converge to such an extent that it is surprising not to have more numerous common works.

The challenge is much more in the elaboration of a working agreement and in the will of the workers than in the elaboration of a light structure permitting investigations in several laboratories. In that case, it seems that the work that can be carried out in each laboratory will be more efficient than that effected in a large pluridisciplinary structure.

REFERENCES

BEHRE, K.E.1981. The interpretation of anthropogenic indicators in pollen diagrams. Pollen et spores 23(2):225-245.

BERGLUND, B.1979. Paleohydrological changes in the temperate zone in the last 15000 years.Subproject B.Lake and mire environments. International Geological Correlation Programme.Project 158.Lund

BEUG, H.-J.1975. Man as a factor in the vegetational history of the balkan peninsule. In "D.Jordaoc et all (eds) Bull.Geology,19: 101-110.

BRAIDWOOD, R.J.1975. Prehistoric men, 8th ed. Glewiew,Illinois: Scott, Foresman and Co.

CERCEAU, I.1985. Les ressources végétales et leur utilisation dans les régions égeennes au néolithique et à l'age du bronze. Documents archéologiques et documents épigraphiques. Thèse Université Paris I.

CHALABBI, M.1980. Analyse phytosociologique, phytoécologique, phytosociologique, phytoécologique, dendrometrique et dendroclimatologique des forêts de *Q. cerris subsp.pseudocerris* et contribution à l'étude taxonomique du genre *Quercus* en Syrie. Thèse es sciences. Faculté des sciences et techniques St Jerôme. Marseille.

EL MOSLIMANY, Ann P.1987. The late pleistocene climate of the lake Zeribar region (Kurdistan,Western Iran) deduced from the ecology and pollen production of non arboreal vegetation. Vegetation,72: 131-139.

FRENZEL, B.1979. L'homme comme facteur géologique en Europe. Bull.Ass.Fr.Qu. 4: 191-199.

GARELLY, P.1972. L'assyriologie. P.U.F.

JACOBSON, G.L.et BRADSHAW, R.H.W.1981. The selection of sites for paleovegetational studies. Quaternary Research 16.80-96.

JANSSEN, C.R.1980. Some remarks on facts and interpretation in quaternary palyno-stratigraphy. Bull. Ass. Fr. Qu.4:171-176.

JANSSEN, C.R.1986. The use of pollen indicators and of the contrast between regional and local pollen values in the assessment of the human impact on vegetation. In "Anthropogenic indicators in pollen diagram".

LEEUWAARDEN VAN, W. and JANSSEN C. R., 1985. A preliminary palynological study of peat deposits near an oppidum in the lower Tagus valley, Portugal.

OPPENHEIM, A.L.1968. Ancient Mesopotamia. Chicago.

PENNINGTON, W.1979. The origin of pollen in lake sediments:An enclosed lake compared with one receiving inflow stream. New Phytol.83-189-213.

PEARSON, G.W.et all 1986. High precision *C measurement of irish oaks to show the natural *C variations from AD 1840-5210 BC.Radiocarbon,v.28,N°2B,sp.issue.

PONS, A et QUEZEL, P.1985. The history of flora and vegetation and past and present human disturbance in the mediterranean region.In "Plant conservation in the mediterranean region"

PONS, A.,COUTEAUX ,M.,DE BEAULIEU, J.L.and REILLE, M.1990 .Plant invasions in Southern Europe from the paleocological point of view.In Biological invasions in Europe and the mediterranean basin.F.di Castri ed.

QUEZEL, P.1985. Définition of the mediterranean region and the origin of its flora.In "Plant conservation in the mediterranean area, Geobotany 7, Dr.W.Junk pub.pp.9-25.

QUEZEL, P.et BARBERO, M.1980. Carte de la végétation potentielle de la région méditerranéenne. Feuille 1. Ed. du C.N.R.S.

REILLE, M.,TRIAT-LAVAL, H.,VERNET, J.L.1980. Les témoignages des structures actuelles de végétation méditerranéenne durant le passé contemporain de l'action de l'homme. In "la mise en place, l'évolution et la caractérisation de la flore et de la végétation". Naturalia monspel. N° hors série.

SAADI, F.1990. Relationships between pollen rain and disturbed vegetation in Morocco. Anthropogenic pollen indicators. In "Human impact and abrupt climatic changes". SFAX 20-25 III 90.

TAUBER, H.1977. Investigation of aerial pollen transport in a forested area. Dansk Botanisk Arkiv 32.

THINON, M.1979. La pédoanthracologie: une nouvelle méthode d'analyse phyto-chronologique depuis le néolithique. C .R. Acad. Sc. Paris,287,D,1203-1206.

TRIAT-LAVAL, H.1978. Contribution pollenanalytique à l'histoire tardi et post glaciaire de la végétation de la basse vallée du Rhône. Thèse doctorat Aix-Marseille III.

VAN ZEIST,W.1987. Some reflections on prehistoric field weeds. Palaeoecology of Africa vol.18, pp.405-427.

VERNET, J.L.1982. L'analyse anthracologique,une méthode d'étude. Le courrier du C.N.R.S. sept.85: 22-32.

WASILIKOWA, K.1986. Plant macrofossils preserved in prehistoric settlements compared with anthropogenic indicators in pollen diagrams. In "Anthropogenic indicators in pollen diagrams".BEHRE K.E. ed.

THE PHENOMENON OF CYCLOGENESIS OVER THE AEGEAN SEA

E. A. FLOCAS
Department of Meteorology and Climatology
University of Thessaloniki, Greece

Summary

Cyclogenesis over the Aegean Sea is considered to be a very rare phenomenon, but highly related to the weather in Greece. For this reason,
a period of 18 years (1963-1980) of grid data information will be a-
nalyzed, in order to identify the cases of cyclogenesis over the area
of interest. This "Climatology" will result to certain categories,ac-
cording to their development criteria. The data used for this study is
grid points of geopotential height and temperature, being recorded at
the significant levels for the above mentioned period.

1. INTRODUCTION.
Previous study on the field of cyclogenesis in the area of Greece have
indicated that this phenomenon is mostly due to the cold outbreak   conditions and the strong temperature contrast between the cold land and   the
warm sea, and usually takes place during all year around except summer (Radinovic 1965, Tipaldi et al. 1980, and Michaelides 1987).
The main objective of the study is to be able to examine the synopti-
cal, dynamical and thermodynamical characteristics of the phenomenon of cyclogenesis over the Greek area, and particularly over the Aegean Sea. However, in order to meet this objective, it is firstly required to establish
a kind of easy to be used methodology, so that to be able identify the cases of cyclogenesis in the area of interest. This methodology will be the
focused area of interest in this particular study.

2. DATA USED.
Grid point values of geopotential height and temperature consist the
data used for this study. The grid point data information density is   2x2
degrees, and they cover the time period 1963-1980, inclusive. The measurements are refered to the significant levels of: 1000, 850, 700, 500, 300 ,
200 and 100 mb, and correspond at 00:00 GMT and 12:00 GMT. In addition to
these grid point data, the meteorological maps "European Meteorological
Bulletin" have been used for the same time period.

3. METHODOLOGY.
Since the phenomenon of cyclogenesis is somehow very high associated
with the geopotential height falls, temperature drops, positive vorticity,
and positive vorticity advection, these parameters, and particularly the
former two, are examined and used as predictors.
The main area of interest lies in northern Aegean Sea, which according
to the grid point data information can be identified from a grid point ha-
ving coordinates: $\phi=37.10^{\circ}$ and $\lambda=24.93^{\circ}$. For this particular point, being
used as a pilot to the research, the mean annual distribution of the geopotential height and the temperature at the level of 500 mb was calculated

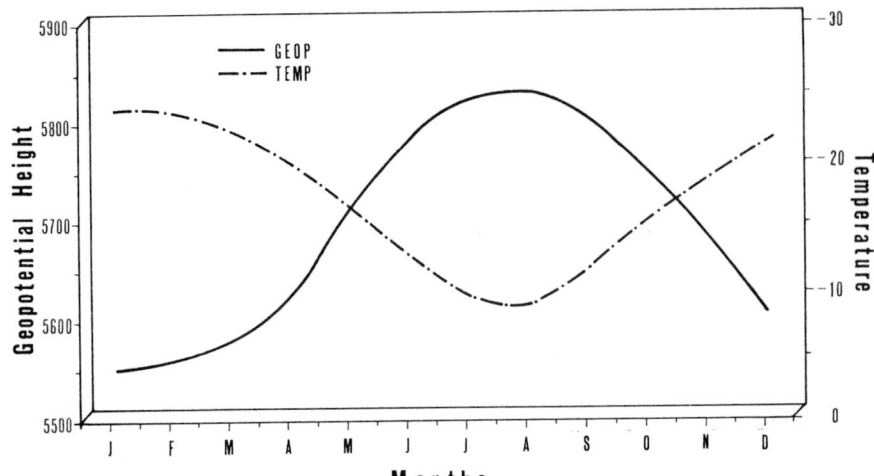

Fig. 1. Mean annual distribution of the geopotential height and temperature at the level of 500 mb, during the time period 1963-1980.

for the 18-year period (1963-1980) and is depicted in Fig. 1. These normals were furtherly used in an indicative way for the critical values adopted to the applied criteria.

The geopotential height and the temperature differences every 12 hours were calculated at that particular point situated over north Aegean Sea and at the level of 500 mb. Fig. 2(a, b) depicts the number of potential cases of cyclogenesis as a function of the adopted criteria, for each one of the two parameters. Fig. 2c indicates similarly the number of potential cases of cyclogenesis, but now as a function of several combinations of the previously adopted criteria. The examined data information correspond to the twice a day measurements of the 18-year period. From the presented figures become evident that, as the critical values of the above stated criteria become more severe, the frequency of occurrences of the potential cases of cyclogenesis decays exponentially.

Applying the finite difference approximation method, the geostrophic relative vorticity was calculated, through the horizontal Laplacian of the geopotential height, for the same grid point. This parameter was also used as an extra criterion to identify the potential cases of cyclogenesis.

In addition to the above stated criteria derived from the grid point data information, the surface and 500 mb meteorological charts were examined in order to identify the cases of cyclogenesis as those were indicated by the Greek National Meteorological Service. Fig. 2d depicts the number of potential cases of cyclogenesis as a function of the adopted criteria, incorporating: geopotential height differences, temperature differences, geostrophic relative vorticity, several combinations of the above stated parameters, and finally, the meteorological charts' observations.

4. CONCLUSIONS

The geopotential height differences, the temperature differences, the geostrophic relative vorticity, and finally the meteorological charts' observations were used to form a methodology to identify the potential cases of cyclogenesis over the northern Aegean Sea. This methodology has the advantage of being quite simple, realistic and easy to be used, since it re-

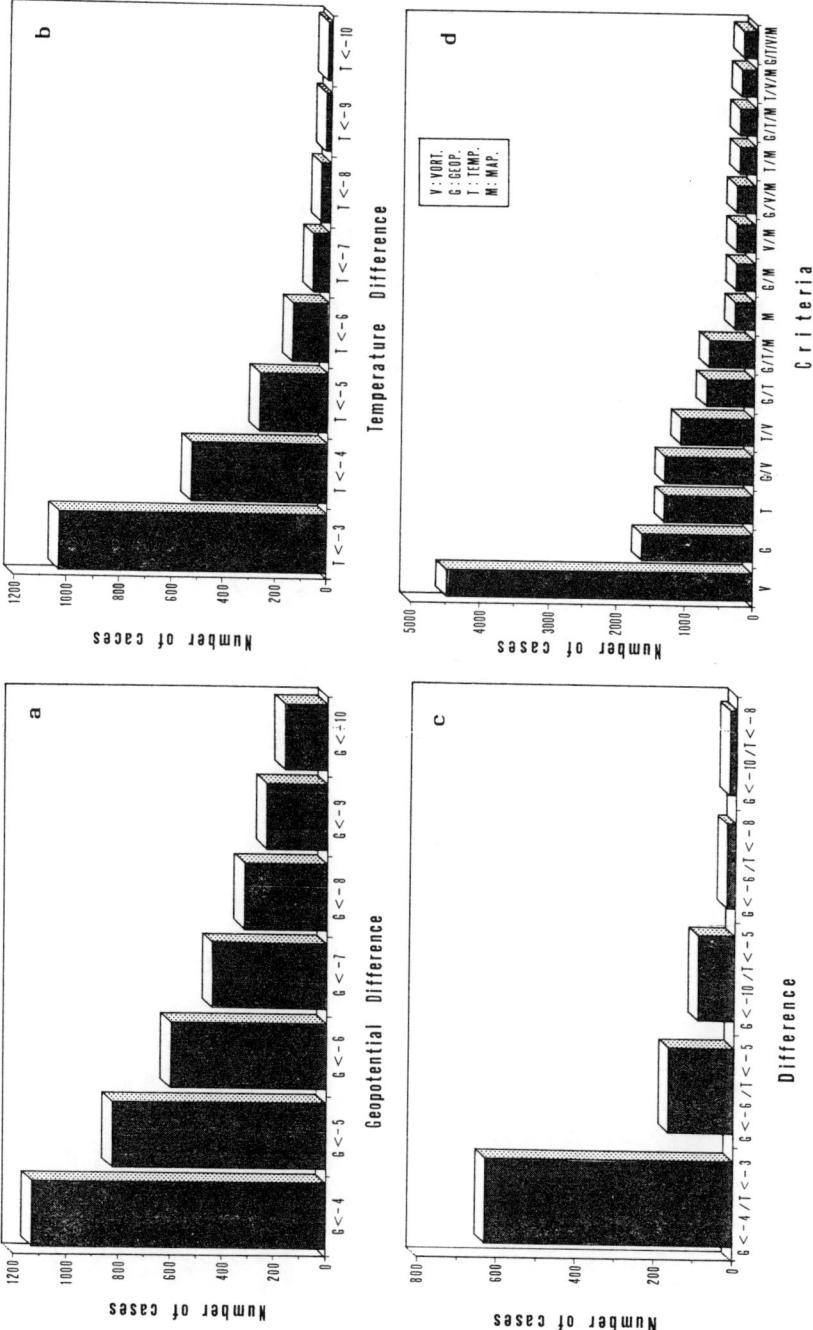

Fig. 2. The number of potential cases of cyclogenesis as a function of: (a) geopotential height differences, (b) temperature difference, (c) combinations of (a) and (b) figures, and (d) geopotential height and temperature differences, geostrophic relative vorticity, and meteorological charts' observations.

lies upon almost primitive data information.

REFERENCES

(1) MICHAELIDES, C.S., (1987): Limited Area Energetics of Genoa Cycloge-
     nesis. Mon. Wea. Rev., 115, 13-27 pp.

(2) RADINOVIC, D., (1965) : On Forecasting of Cyclogenesis in West Medi-
     terranean and Other Areas Bounded by Mountain Ranges by Baroclinic
     Model. Archiv. Met. Geoph. Bioklim. A, 14, 279-299 pp.

(3) TIBALDI, S., A.BUZZI and P.MALGUZZI, (1980) : Orographically Induced
     Cyclogenesis : Analysis of Numerical Experiments. Mon. Wea. Rev.,
     108(9), 1302-1314 pp.

SOME ASPECTS ABOUT THE CLIMATE/VEGETATION RELATIONSHIPS IN THE IBERIAN CENTRAL SYSTEM

R. GAVILAN

Departamento de Biología Vegetal II, Complutense University, E-28040 Madrid

Summary

The Iberian Central System (Spain-Portugal) has the suitable conditions to make a phytoclimatic study. We have realized as example of these bioclimatic procedures, a study about linear regression of *It* (thermicity index of Rivas-Martínez) on altitude. The results are given in the present paper.

1. INTRODUCTION

Climate constitutes one of the most important ecological factors for plant and vegetation distribution. In the XIX century we already knew that the vegetation change with altitude and latitude (Humboldt & Bonpland 1805). Climate/vegetation relationships can be focused at two levels:

a. as explanatory physio-ecological relationships.

b. as descriptive empirical relationships between macroclimatic parameters, or combination of them, and plant taxa, or vegetation unities.

The second approach, which we have taken, is useful for large areas where only macroclimatic (meteorological) data are available. Plants and their communities can be utilized with predictive aims since they are excellent bioindicators of climatic conditions.

One of the problems discussed in the phytogeographical literature consists of the attempt to delimit various vegetational regions or the areas of distribution of individual species or vegetation units in terms of climatic parameters or indices. There are numerous measures used to characterize climate, starting out from simple annual or monthly means of temperatures and amounts of precipitations and ending in complex indices that combining temperature, moisture, etc.

However these indices have two main functions: a descriptive function, because they can establish altitudinal limits or explain geographical patterns in the variation of the vegetation; and a predictive function, through the vegetation that lives in a territory, we can know what type of climate exists in it without the need of simple climatic parameters such as rainfall, temperature, etc.

2. GEOGRAPHIC LOCATION

We have chosen the Iberian Central System (Spain-Portugal) because of the suitable conditions of its patterns of climatic variation:

1. Altitudinal gradient: the range of altitudes reaching to 2592 m in Sierra de Gredos, permit us to study altitudinal variations of climate parameters and characterisation of four bioclimatic belts (thermotypes) according to Rivas-Martínez's thermicity index:

$$It = T + M + m = T + 2tm$$

where It: thermicity index

      T: mean annual temperature

      M: mean maximum temperature of the coldest month

      m: mean minimum temperature of the coldest month

      tm: mean temperature of the coldest month.

These bioclimatic belts are:

Mesomediterranean belt: *It* 210 to 350
Supramediterranean belt: *It* 210 to 70
Oromediterranean belt: *It* 70 to -10
Cryoromediterranean belt: *It* -10 to -100

These bioclimatic belts, that can be subdivided into horizons, show good fitness with the altitudinal changes of the vegetation (Rivas-Martínez 1982, 1985, 1987; Rivas-Martínez & al. 1986, 1987, 1990).

2. Continentality gradient: this mountain range extends from the Atlantic coasts to the inside of Iberian Peninsula, with an increase of continentality from west to east.

3. Asymmetry between N/S slopes caused by exposure factors, as well as altitudinal differences between the northern and the southern basins determinig atmospheric dynamics.

3. VEGETATION

The response of the vegetation to these climatic patterns is well marked and recognised through more than 20 distinct vegetation series (Rivas-Martínez 1987). Excluding edapho-hygrophylous vegetation series and other relic or poorly represented series, the natural potential vegetation of Iberian Central System comprises 3 types of evergreen-oak (*Quercus rotundifolia*) forests, 5 types of decidous oak (*Quercus pyrenaica*) forests, 4 shrubby communities or pine-woods that represent the oromediterranean natural potential vegetation and 3 cryoromediterranean psycroxerophilous pastures (Rivas-Martínz & al. 1987). The serial or substituting communities of the meso-supramediterranean forests comprise around 10 brush-shrubby communities, 8 *Cistus*-dominated scrubby communities, 3 *Erica*-dominated scrubby communities and more than 10 types of dry perennial grasslands. All these woody or perennial communities (syntaxa), that involve more than 200 diagnostic taxa, furnish bases enough to characterise the variation of climate and to establish bioclimatic correlations.

The aim of our study is the relationships between vegetation and climate, so on the one hand discriminant analyses of floristic and vegetational data and bioclimatic parameters and indices will be performed, and on the other hand altitudinal and geographical patterns of climatic variation and the corresponding variations of plant communities will try to be correlated.

4. ALTITUDINAL GRADIENT OF It IN SIERRA DE GUADARRAMA

As example of these procedures, we include here results of a preliminary survey about linear regression of *It* (and their components: *T* and *tm*) on altitude in Sierra de Guadarrama, an eastern mountain range of the Iberian Central System) (table 1). This procedure illustrates the relationships between altitude and It, that is the index used for delimiting the bioclimatic belts, and it will serve us later to establish the *It* intervals of different plant communities and species in the territory.

Elementary climatic data were supplied by the National Institute of Meteorology (I.N.M.). We have chosen a set of weather stations, that had at least 15 years of observations, in three well defined areas of this sierra, that are:

1) Northward slope: northern valleys of Eresma, Moros, Cega and Cerezuelo rivers.

2) Paular valley: an interior valley eastward oriented.

3) Southward slope: southern valleys of Guadarrama, Manzanares and Jarama rivers.

The vegetation unities involved in territorial characterisation of bioclimatic belts are refered in Rivas-Martínez & al. (op. cit. 1986) and Rivas-Martínez & al. (1990).

4.1 RESULTS

Linear regressions of *It*, *T* and *tm* on altitude showed highly significant probability values (> 99 %) for all their parameters (intercept, slope and regression coefficient), with correlation coefficients generally greater than 90 % (table 1). We have also tried several models of non-linear regression, but they improved worse correlations.

1) *Northward slope*

Regression of T on altitude presents better correlation coefficient than It, that seems affected by the moderate coefficient of *tm* regression. Although its high R-squared value, the linearity of *It* does not hide a marked deviation from linear model at low altitudes of the northern plateau, between 700-1000 m. In fact, though the regression line predicts mesomediterranean values of *It* at 800 m, real mesomediterranean values are not reached above 650-700 m. Taking into account this, the slope of the line regression above 1000 m can be shallowly corrected (slope: -0.169; intercept: 361.42).

The insufficient correlation It/altitude in the northern plateau may be due to the scarce altitudinal range combined with the effects of different exposures and topographical situations (hills or depressions), that influence specially the tm.

2) *Paular Valley*

Linear regressions in this valley reach the highest coefficients, probably because of its uniform topography and situation of stations. Correlation of T is also the greatest, but the one of *It* is very similar. The slope of *It* regression line is one *It* unit/100 m higher than in the northward slope regression, but also 0.56 *It* units/100 m lower than in the same regression corrected above 1000 m as it was yet explained. Therefore, these two regression lines reveal rather similarity. *It* is interesting to remark that this similarity, concordant with the vegetation patterns observed in both slopes, is not proportional with respect to the regressions of T and tm, which show divergent trends: the altitudinal gradient of T is more accentuated in the northward slope, while the altitudinal gradient of *tm* is more marked in Paular Valley.

3) *Southward slope*

The regression lines of southern valleys present the highest slope factors. Here the regression coefficient of *tm* is greater than the ones of *It* and T. Despite linearity of the thermic gradients, it can be detected some deviations from the linear model, stronger in T than in *tm* regression. At low altitudes in the southern plateau (600-900 m) the decreasing is more slow (8-10 *It* units/100 m) with moderately significant values. Between 900-1100 m the *It* seems to descend more quickly, with a ratio around twice the mean slope of regression. Finally, above 1100 m the *It* decreasing stabilises with a slope factor around 15 *It* units/100 m, hardly similar to the one of northern valleys. Though there are a little number of meteorological stations within these altitudinal intervals, insufficient to establish the accurate magnitude of such inflections, these are very significant, because reveal the assymetry of the intervals of bioclimatic belts and improve adjustments among the regressions performed on different slopes.

Thus, the strong thermic gradient of southward slope explains the relative constriction of the supramediterranean altitudinal interval and the descent of supra/oromediterranean limit observed in this slope. Moderation of the slope factor above 1200 m improves predictions for high altitudes concordant with the predictions of regressions performed in the northern valleys.

4) *Whole regression*

Whole regression including data of three zones shows correlation coefficients slightly lower; the one of *It* is now the highest. Comparisons between predictions infered from these regressions are showed in Table 2.

Summarizing, the *It* regression on altitude shows a high linearity in areas with well developed altitudinal gradients. Deviations from linearity are significant in topographically transitional areas and perhaps in areas with scarce altitudinal variation; these deviations are sometimes due to deviations of tm. Such deviations went considered in the adjustement of predictions ("corrected values" columns of table 2), specially to improve good fitting between different slope regressions.

5. REFERENCES

RIVAS-MARTINEZ, S. (1982). Étages bioclimatiques, secteurs chorologiques et séries de végétation de l'Espagne méditerranénne. Ecol. Medit. (Marseille) 8: 275-288. Marseille

RIVAS-MARTINEZ, ·S. (1985). Nuevo índice de termicidad para la Región Mediterránea. In: A. BLANCO DE PABLOS (ed.). Avances sobre la investigación en Bioclimatología: 377-380. Publ. C.S.I.C. Univ. de Salamanca.

RIVAS-MARTINEZ, S. (1987). Mapa de series de vegetación de España 1:400000 y Memoria. Publ. I.C.O.N.A., Serie Técnica. Madrid.

RIVAS-MARTINEZ, S., FERNANDEZ-GONZALEZ, F. and SANCHEZ-MATA, D. (1986) Datos sobre la vegetación del Sistema Central y Sierra Nevada. Opusc. Bot. Pharm. Complutensis 3: 3-136.

RIVAS-MARTINEZ, S., FERNANDEZ-GONZALEZ, F. and SANCHEZ-MATA, D. (1987). El Sistema Central: de la Sierra de Ayllón a Serra da Estrela. In: M. PEINADO & S. RIVAS-MARTINEZ (eds.), La vegetación de España: 419-451. Serv. Publ. Univ. Alcalá de Henares (col. Aula Abierta, 3). 544 pp. Alcalá de Henares. Madrid.

RIVAS-MARTINEZ, S., FERNANDEZ-GONZALEZ, F., SANCHEZ-MATA, D. and PIZARRO, J. (1990). Vegetación de la Sierra de Guadarrama. Itinera Geobotanica 4: 3-132. Secr. Publ. Univ. León.

TUHKANEN, S. (1980) Climatic parameters and indices in plant geography. Acta Phytogeogr. Suec. 67, 110 pp. Uppsala.

TABLE 1

	Regression	Coefficient	R»(%)	Slope	Intercept
	It/altitude	-0.968625	93.92	-0.153244	336.929
Northward slope	T/altitude	-0.983957	96.82	-0.0568956	170.615
	tm/altitude	-0.952856	90.79	-0.0481744	83.157
	It/altitude	-0.996308	99.26	-0.163328	344.129
Paular Valley	T/altitude	-0.997345	97.47	-0.0550923	164.161
	tm/altitude	-0.991628	98.33	-0.0541381	90.0288
	It/altitude	-0.977292	95.51	-0.184703	379.666
Southward slope	T/altitude	-0.969855	94.06	-0.0687695	187.816
	tm/altitude	-0.980798	96.20	-0.057967	95.9251
	It/altitude	-0.959573	92.08	-0.16893	357.775
Whole	T/altitude	-0.954204	91.05	-0.0634567	178.832
	tm/altitude	-0.954659	91.14	-0.0527349	89.4752

Values obtained by simple regression analysis

TABLE 2
Predictions of altitudinal limits of bioclimatic horizons based on regressions

It values/bioclimatic horizons	Northward slope Total	Corrected	Paular valley	Southward slope Total	Corrected	Whole
Cryoromediterranean -10	2260	2190	2170	2110	2210	2180
Upper oromediterranean 30	2000	1950	1920	1890	1940	1950
Lower oromediterranean 70	1740	1720	1680	1680	1660	1700
Upper supramediterranean 120	1400	1430	1370	1410	1350	1400
Middle supramediterranean 165	1120	1160	1100	1160	1150	1140
Lower supramediterranean 210	(820)	- -	(820)	900		880
Upper mesomediterranean						

OZONE DESTRUCTION ON POLAR STRATOSPHERIC AEROSOLS.

P. DI GIROLAMO
Physics Department, University "La Sapienza", 00185, Roma

Summary

The University of Rome is involved in a cooperative project devoted to the study of aerosols and cloud particles over Antarctica as well as their role in the aeronomy of high latitude regions. The nature and characteristics of stratospheric aerosols and clouds (PSCs), which represent a key element in the heterogeneous chemistry leading to the stratospheric ozone depletion, have been investigated by the use of an optical radar system.

1. INTRODUCTION

It is widely accepted that the stratospheric ozone layer is severely affected by a chain of reactions where halons molecules act as catalysts. Halons can control the abundance and distribution of stratospheric ozone.

High energy fluxes of UV radiation reaching the lower stratosphere can destroy CFCs molecules, i.e. the major anthropogenic source gases for stratospheric chlorine (1). After transport to the stratosphere, CFCs molecules are decomposed into active chlorine species (Cl and ClO) thus activating the catalytic removal of ozone.

Several reactions interfere with the catalytic cycle of chlorine. Active chlorine can combine with nitrose oxide or methane to form stable products behaving as inactive chlorine reservoirs ($ClONO_2$ and HCl).

Inactive chlorine species, almost stable in a gaseous enviroment, can react on the surface of stratospheric aerosols releasing active chlorine. The heterogeneous processes taking place on stratospheric aerosols are poorly known; an adequate assessment of their impact on stratospheric ozone cannot then be provided.

The key heterogeneous chemical reactions occurring in the polar stratosphere have been proposed to be (2):

$$ClONO_2 + HCl \longrightarrow Cl_2 + HNO_3$$

$$ClONO_2 + H_2O \longrightarrow HOCl + HNO_3$$

$HOCl$ and Cl_2 are both short-lived species in the spring sunlight, being rapidly converted through photolytical reactions into ClO_x radicals. These reactions do also behave as a final step in the conversion of reactive nitrogen (NO_x) to HNO_3. That is a necessary condition for chlorine activation because NO_2 induces rapid passivation of active chlorine.

Laboratory studies have been performed (3,4) to determine the time constants of heterogeneous processes, even if carried out under conditions quite different from those expected for the polar stratosphere. The rate at which reservoir molecules may react with a particle surface is dependent upon several parameters: the molecular mixing ratio, the collision frequency and the sticking coefficient. A great deal of further laboratory work is needed to definitively establish the reactive properties of PSCs particles.

The formation and nature of polar stratospheric aerosols and clouds have been studied by several authors (3,5,6,7,9). Nitric acid and water co-condensation have been proposed to occur in the winter polar stratosphere, leading to solid aerosol particles with a composition close to that of nitric acid thihydrate (NAT). NAT particles condensate at temperatures around 200 K, that is at temperatures a few degrees above the water ice point (3), in accordance with satellite observations that calculate the average relative humidity to be 58% near the PSC center (5). HNO_3 condensation can occur at temperature as warm as 205 K.

From a microphysical point of view, HNO_3 and H_2O are supposed to condensate onto supercooled sulphuric acidic aerosols nuclei (3). The initial condensation of nitric acid would then occur as a supercooled ternary solution of $H_2O/H_2SO_4/HNO_3$ (6). Pueshel et al. (7) suggest that most of the sulfate particles may be activated into nitric acid haze particles.

Water ice crystals also form in the polar stratosphere at lower temperatures. They show larger dimensions compared to NAT particles and do probably grow around a preexisting NAT core.

Direct evidence for the existence of nitric acid particles in the antarctic winter stratosphere has been found by Pueshel et al. (7), who performed total odd - nitrogen measurements; anyhow up to now no direct measurement of the composition of the stratospheric HNO_3/H_2O condensates have been carried out.

2. OPTICAL RADAR MEASUREMENTS

The optical radar, or lidar, is an active remote sensing device which has found wide applications in atmospheric studies. In its basic elastic backscattering configuration, the lidar can be used for detecting the presence of aerosols and determining their physical properties in a height region extending from the boundary layer to the upper stratosphere. The block diagram of a typical lidar system is reported in fig.1 (South Pole Lidar, Antarctica, operated by University of Rome(9)).

Remote measurements of stratospheric aerosols and clouds have been performed by the University of Rome in winter and early spring since 1988 (8,9,10). Typical lidar echoes from stratospheric heights are reported in fig.2. It was obtained in June 29, 1988. The figure also shows the temperature profile for the same day as well as the theoretical profile for the molecular echoes.

Aerosol lidar measurements are generally computed in terms of backscattering ratio, i.e. the ratio between the total detected signal and the computed molecular contribution. The proposed lidar echoes lead to backscattering ratios ranging from 3 to 5.

Lidar observations of polar stratospheric aerosols have also been reported by several other authors (11,12). Data reported by Kent et al. (11) show aerosol layers with backscattering ratios of less than 10.

Poole and McCormick (12) identified stratospheric aerosol layers over Greenland with backscattering ratios of the order of 5. All these observations are consistent with nitric acid clouds composed of small particles.

Water ice clouds composed of ice crystals show larger backscattering ratios. That is the case (backscattering ratios as large as 30-50) for several aerosol stratifications observed by Kent et al. (11) as well as, more recently, by Fiocco et al. (9,10). Such backscattering ratios implies the presence of a particle mass that can only be supplied by condensed water.

The optical depht of these ice stratifications has also been determined from lidar backscattering data. Fiocco et al. (10) computed values around 0.1. Similar computations have been reported by Kent et al. (11) who reported optical depths ranging from 0.006 to 0.11, combined with depolarizations of 0.3-0.5. Such observations suggest that ice particles concentrations are around 1 cm-3 and that particles have sizes of several microns.

REFERENCES

(1)   MOLINA, M.J. and ROLAND, F.S. (1974). Stratospheric sink for chlorofluoromethanes: chlorine-atom catalyzed destruction of ozone, Nature, 249, 810.

(2)   SOLOMON, S., GARCIA, R.R., ROWLAND, F.S., and WUEBBLES, D.J. (1986). On the depletion of Antarctic ozone, Nature, 321, 755.

(3)   TOON, O.B. et al. (1986). Condensation of HNO_3 and HCl in the winter polar stratosphere, Geophysical Research Letters, 13,1284.

(4)   ROSSI , M.J., MALHORTA, R., and GOLDEN, D.M. (1987). Heterogeneous chemical reaction of chlorine nitrate and water on sulfuric acid surfaces at room temperature, Geophysical Research Letters, 14, 127.

(5)   AUSTIN, J., REMSBERG, E.E., JONES, R.L., and TUCK, A.F. (1986). Polar stratospheric clouds inferred from satellite data, Geophysical Research Letters, 13, 1256.

(6)   TURCO, R.P., TOON, O.B. and HAMILL, P. (1989). Heterogeneous Physicochemistry of the Polar Ozone Hole, Journal of Geophysical Research, 94, 16493.

(7)   PUSHEL, R.F. et al. (1988). Condensed nitrate, sulfate, and chlorine in Antarctic stratospheric aerosols, Journal of Geophysical Research, 94.

(8)   FIOCCO, G., FUA', D., DE LUISI, J., DI GIROLAMO, P., CACCIANI, M. and DI SARRA, A. (1989). Optical radar observations of stratospheric clouds at South Pole during wiinter 88: Preliminary results, SIF: Conference Proceedings, 20.

(9)   FIOCCO, G., FUA', D., CACCIANI, M., DI GIROLAMO, P. and DE LUISI, J. (1990). On the temperature dependence of polar stratospheric clouds, Geophysical Research Letters (submitted).

(10) CACCIANI, M., DI GIROLAMO, P., FIOCCO, G., FUA', D. (1990). Identification of polar stratospheric clouds types from lidar measurements, SIF: Conference Proceeding.

(11) KENT, G.S., POOLE, L.R. and MCCORMICK, M.P. (1986). Characteristics of Arctic polar stratospheric clouds as measured by lidar, Journal of Atmospheric Sciences, 43, 2149.

(12) POOLE, L.R. and MCCORMICK, M.P. (1988). Polar stratospheric clouds and the Antarctic ozone hole, Journal of Geophysical Research, 93, 8423.

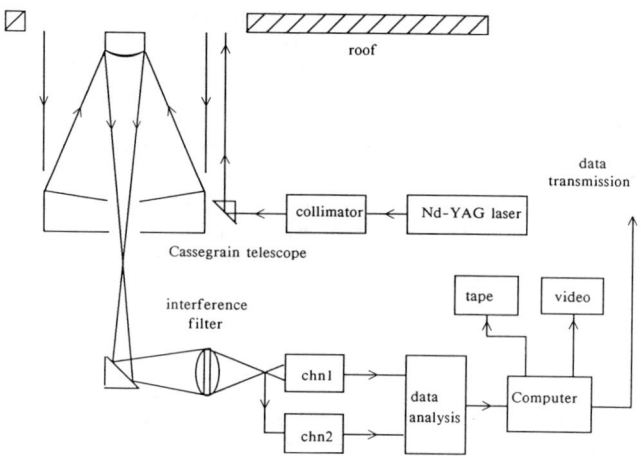

Fig.1. Block diagram of the South Pole Lidar

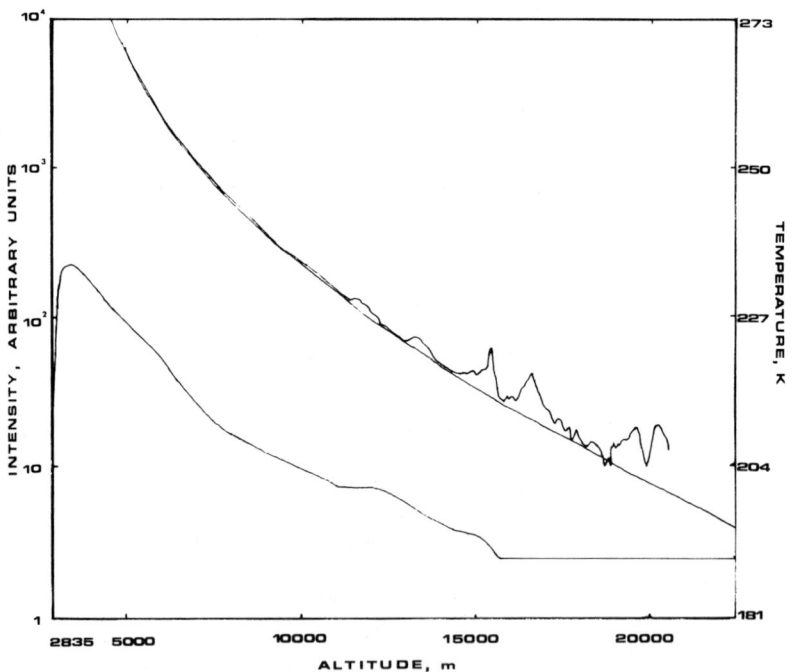

Fig.2. Lidar echoes for June 29, 1988.

HOLOCENE SEDIMENT STRATIGRAPHY
AND THE GENERATION OF OBSERVED DATA

An Example from Mounts Bay, Cornwall, South-West England.

M.G. Healy
Geography Department
University College Cork
Ireland

ABSTRACT

Much emphasis is currently placed on the necessity for model simulations capable of representing the dynamic interactions of those global components which affect climate and global change. The major difficulty with the modelling approach is the essential lack of observed and reproducable data on which model assumptions can be based and model outputs verified. This paper offers discussion of this problem and outlines how the development of additional techniques for data collection can contribute to a fuller understanding of how environmental change is manifested at a local and regional scale. The use of thermoluminescence as a possible dating mechanism receives special mention.

INTRODUCTION

Approaches to an examination of the overall question of the nature and rate of climate and global change require a diversity to match the complexity of the problem in hand. In the context of large-scale modelling of the behaviour of any number of individual interactive phenomena, or any combination of a series of such phenomena, the lines of investigation are necessarily cast wide, with emphasis placed primarily on the dissection of particular relationships so that their discreet and combined behaviour can be mathematically, statistically or otherwise reproduced. This approach has found favour particularly among scientists who seek to represent the dynamic atmospheric, oceanographic and terrestrial forcing which combine to produce variations in environmental conditions as computer simulations of changing input and throughput variables in a global climate regime (e.g.Barker and Davies,1989; Verstraete and Dickenson,1986; Yamazaki,1989). The most acute difficulty inherent in such an approach is that of the limited bank of observational data which can be drawn upon either to qualify the various parameters on which assumptions are based or to act as a validating framework against which the accuracy of findings may be established.

An alternative route to understanding the form and mechanisms of change may be pursued from the basis of observed data collection, where indicators of climatic change can be used directly to piece together the nature of environmental change over time, particularly for studies concerned with problems at the local to regional scale. The purpose of this paper is to show how considerable information relevant to the question of changing environmental conditions can be acquired through the maximisation of the utility of available techniques in field studies, thus expanding the necessary observational data bank against which hypothetical scenarios may be examined.

One of the blocks upon which an overview of climate and global change can be based is that of changing sea level and its relationship to long-term and large-scale climatic variation (e.g.Peltier,1988) As in the broader context outlined above, a number of alternative routes to the examination of the phenomenon of sea-level change and its variation in space and time are available for application (Shennan and Tooley,1987). For the purposes of this paper some opportunities available in the area of sedimentary geology are outlined, with particular reference to stratigraphy as a tool in the study of sea-level change, and similarly by extension in climatic change. The quantity of reproducible detail of changing environmental conditions is substantial and may clarify not alone the dynamic relationships at the land-sea interface zone, but also highlight the complexity of the nature of change and the problems inherent in establishing hypothetical assumptions where validating observational data is scarce.

ACCUMULATION OF STRATIGRAPHY

The accumulation of coastal deposits may be seen as a function of the land-sea interface relationship. For any particular site the nature of this relationship requires detailed examination before it becomes clear whether vertical interplay demonstrated in intercalated sediment layers can be attributed to isostatic or eustatic variations (Strief,1978). The incompletely understood nature of the effective relationship between these coastal evolution determinants has given rise to the concept of 'relative sea-level'(Tooley,1987) and the associated 'sea-level tendency'(Shennan,1986) theory in the field of sea-level studies. To determine how the isostatic-eustatic relationship functions at a particular site several lines of investigation may be followed, ranging from tectonic and seismic measurements through tidal observational data to historical accounts of extreme events recorded at coastal sites such as seiches or storms(Smith et al 1985) Where a range of interplaying factors combine to affect the evolutionary development of an individual coastal site evidence for the physical significance of such factors is often available in the sedimentary structures which go to formulate the stratigraphic sequence deposited at that site, assuming that contemporary or subsequent erosional or other processes allow the sequence to be preserved (Owens,1981). The data bank which the stratigraphic column represents when suitably preserved is often rich in the diversity of its potential information yield. The extraction of the full range of data requires the application of a wide range of analytical measures, each specific to the particular type of information required for specific research purposes.

In order to piece together the pattern of the land-sea level relationship for the purposes of sea surface studies within a framework of a postulated absolute rise in sea-level through the Holocene period several analyses can be applied to a retrieved stratigraphic column for various, though integrated, ends. Evidence for sea-level variation, whether positive or negative, can be sought in the analysis of many indicators of environmental change(Devoy,1987). Based on a sediment core from Mounts Bay, Cornwall, the following discussion presents some of the possibilities for analysis which are available and how the maximisation of the observable data can prove an interpretative aid in erecting hypotheses on which modelling assumptions can be based and model outputs can be refined and validated in the context of Holocene timescales. The extension of this methodological approach to several sedimentary monoliths at a particular site and the examination of a dense spatial network of sites over a geographical unit, as has been the approach in south-west England, provides a useful means of generating a bank of measured and observed data which can contribute positively to the understanding of the behaviour of dynamic phenomena under global, regional or local forcings.

CORNWALL : THE GEOGRAPHICAL SETTING

Cornwall county is broadly triangular in shape, extending south-west from the Devon border. The apex of the triangle ends in a two fingered protrusion, one the Penwith peninsula, the other the Lizard peninsula, with Mounts Bay indented between them. The river Fal dominates the south-facing coastal district to the north-east of the Lizard Point, beyond which the Cornwall coast extends in an east-north-east direction to Plymouth. The northern coast extends from St. Ives Bay, at the heel of the Penwith peninsula, north-eastward through Bude to its northern limit. The major feature of this northern coast is the estuary of the river Camel at Padstow.

*** See page 335 ***

Fig. 1   Mounts Bay, Cornwall, South-West England

MOUNTS BAY : THE CURRENT PHYSICAL LANDSCAPE

The terrestrial depositional area Mounts Bay is a flat back-beach coastal indentation to the east and north of the town of Penzance. The bay runs broadly parallel to the strip of roadway which joins Penzance to Long Rock and Marazion. Part of the bay has been reclaimed for industrial use, with the landward margins employed for agricultural purposes.

The Mounts Bay back-beach area is isolated from the present day beach by a natural barrier which supports the roadway from Long Rock to Marazion in addition to the BR Rail Line for part of its course. It is apparent that the barrier, topped by poorly formed sand dunes, is composed of coarse sand and gravel. The area behind the barrier and extending considerably inland is very wet and dominated by an extensive cover of phragmites and Salix scrub. The underlying geology at this site consists primarily of slate of the Mylor series to the east fronted by a deposit of 'alluvium' on head towards the foreshore area(Goode&Taylor,1988) The red River, which drains the hills north of Marazion, exits to the sea at the eastern end of the sand and gravel barrier, its course traversing the Mounts Bay back beach area. British Geological Survey maps describe this sedimentary basin as consisting of 'alluvium'.

A NOTE ON MINING IN CORNWALL

An active tin mining industry has existed in Cornwall since the seventeenth century, with a concentration around the area of Penzance which led to its establishment as a coinage town in 1663. The coastal areas of the west were favourable for lode mining, and a number of wheel pits were located on some of the smaller rivers. The majority of the large old mines are now abandoned but they have left a significant impact on the local landscape. The existence of mine waste or 'tailings' is a feature of many of the stratigraphic accumulations observable in Cornwall, and the precise impact of the additional sediment load which they represent, both on land and offshore cannot yet be estimated. The tendency of mine operators to release tailings into nearby streams and thus into river floodplains and estuaries has caused a disturbance to stratigraphic formations which may present as spurious sequences. Detailed examinations of the nature and behaviour of mine waste as depositional sediment (see (Coard et al, 1981) forms part of the overall study of coastal dynamics and sea-level change in Cornwall. Beyond recognising its significance as an influential variable in stratigraphic studies this aspect of investigation does not concern this discussion.

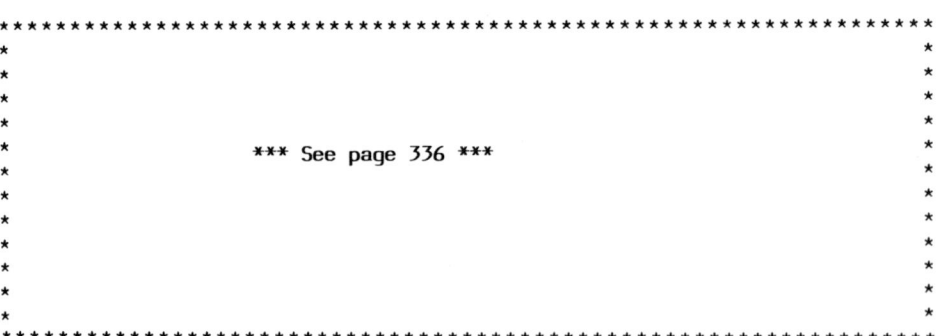

*** See page 336 ***

Fig. 2   Mounts Bay, Borehole 4 (BH4).

CORE STRATIGRAPHY : MOUNTS BAY

As part of a large scale investigation of changing relative sea-levels in south-west England several sites along the Cornish coast are in the process of being examined for evidence of change in the Holocene period. Mounts Bay represents but one of these sites and its use as the basis for this discussion is significant primarily because of the clarity of the stratigraphic sequence which it preserves. Within Mounts Bay an extensive programme of borehole extraction has been carried out with a view not only to examine stratigraphy in isolation but to carry out a range of analyses on the extracted sediments in an effort to present a comprehensive picture of the nature of change at this site through Holocene time and to offer an explanation of such change in the context of changing sea-levels and coastal evolutionary mechanics as dynamic processes as well as respondents to larger scale forcings.

The following is a description of the sediments retrieved from borehole number four (BH4) in Mounts Bay, as represented in fig.XX. Measurements shown are in centimeters.

0000-0.05	Vegetation (Gramineae) and root mat.
0.05-0.27	Coarse sand/fine gravel, rich in quartz.
0.27-0.35	Water table at base, sediments as above.
0.35-0.50	Fine gravel in coarse sand matrix.
0.50-0.55	Decreasing particle size to coarse sand only.
0.55-0170	Coarse /medium coarse sand.
0171-0191	Blackish medium coarse sand.
@0192	Sand contact with peat.
0192-0205	Peat, evidence of disturbance.
0205-0215	Peat, coarse sand incorporated.
0215-0225	Peat sequence contains coarse sand split-:
	0215-0220 Peat with silt and sand particles
	@0220 Sand split 3cm vertical depth
	0222-0225 Peat with sand fraction
0225-0241	Peaty deposit becomes more minerogenic downward, until at 0241 it is pure coarse sand.
0241-0253	Coarse sand with humic staining.
0253-0315	Coarse sand with significant organic content, crudely layered.
0315-0389	Medium coarse sand.
@0389	Sand continues. First appearance of shells, mostly broken fragments of Cardium, Littorina and Hydrobia.
0389-0431	Medium coarse sand with fine sand fraction. Shells continue.
0431-0487	Medium coarse sand with large gravel/small cobble size particles. Shells continue.
0487-0513	Coarse sand matrix with larger beach cobbles.
0513-0600	Coarse sand with shell fragments.
0600-0619	Coarse sand with isolated large particles of slate.
0619-0639	Coarse sand with very varied shell content-: Hydrobia, Cardium. Gastropods. Large quartz particles (>2cm axis).

```
0639-0645   Medium coarse sand with varied shells.
0645-0684   Coarse sand with large quartz particles (3-4cm axis).
            Rich in varied shells.
0684-0735   Medium coarse sand with shells and organic mottling.
0735-0806   Medium/fine sand. Contains chunks of wood.
0806-0860   Silty clay with high peaty organic content, medium
            coarse sand with woody fragments at base.
0860-0877   Medium sand with slight organic content.
0877-0897   Organics increase in medium sand dominated matrix.
            Shells reappear. Well preserved phragmites remains.
  @0923     Fine sand with gravel in matrix
  @0933     Gravel size ratio increases in sand matrix.
  @0945     Reduced gravel.
  @0998     Peat (with sand contamination).
1000-1017   Hiatus.
1017-1045   Peat dominates, clay fraction within the peat.
  @1045     Gravel particles appear in the peat.
1046-1062   Peat with clay fraction.
1062-1065   Weathered bedrock of slate.
```

INTERPRETATION OF STRATIGRAPHY

Traditional uses of stratigraphic logs have concentrated on the identification of biostratigraphic units within overall sequences(Eugster and Kelts,1983). These units could then be analysed in a systematic and thorough way for the purpose of the extraction of materials bearing microfossil, macrofossil and other types of indicators of environmental change, as well as material suitable for dating purposes. As the primary traditional dating tool has been that of 14/C, or radiocarbon dating, emphasis has naturally concentrated on the organic rich segments of stratigraphic sequences(Tooley and Switzur,1988). Though a range of other radiometric dating techniques, particularly using isotope decay ratios (e.g. Pb/210) has long been recognised, it is only in the relatively recent past that a significant interest and technological effort has been made to diversify the tools of sediment dating. Therefore in the absence of 'suitable' material likely to yield dated information, a tendency to divert attention away from unpromising sediment sequences has prevailed.

With increasing interest in the question of global warming and its effect on a dynamic water balance, linked strongly to ocean water behaviour with changing gross inputs and throughputs (Pirazzoli,1987), and the inevitable questions this poses about the nature of sea-level reaction to such change, the coastal zone has become increasingly important as an area which may store a record of such change in depositional environments. Because the essential nature of coastal sedimentary materials incorporates both organic and minerogenic materials, as well as complex combinations of the two, and because of the urgency in the search for reliable environmental data, it has become necessary to look more closely at the entire range of possibilities which may be available in deposited stratigraphic sediments. The purpose of the remainder of this paper is to briefly look at one such possibility, that of thermoluminescence as an investigative technique.

As is evident from the data presented here in Fig.2 and in the description of the stratigraphic sequence above, the material retrieved from BH4 in Mounts Bay is significantly variable in its nature. Perhaps the most striking feature of the deposited sequence is the substantial accumulations of minerogenic material in evidence. As described in the stratigraphic log, and confirmed in detailed particle size analysis of mineral sequences from the core, deposits of gravels, sands, silts and clays are significantly variable as discreet units within vertical sections. Such variability suggests differential depositional conditions through time(Troels-Smith,1955) either within the accumulation catchment or in the agency or agents which control deposition at this site. While such a suggestion may be borne out by a theoretical evaluation of the many controls on any site of deposition, it is only when such variation within minerogenic sequences can be attributed an inter-unit temporal relationship in a strict stratigraphic sense that the internal minerogenic record can achieve its potential as an additional databank, storing a record of environmental change through periods where other sources are absent or incapable encapsulating similar information. The central significance of developing a technique capable of providing a dating mechanism and thus a temporal framework for intra-unit and inter-unit minerogenic deposits within the context of stratigraphy is central to an advance in expanding the sedimentary data yield in this area.

Thermoluminescence as an investigative dating technique has a relatively long history of theoretical application, though the actual instances of its practical application have been largely restricted to the dating of pottery remains for archaeological and other purposes(Aitken,1985). More recent appreciation of the potential of this technique has led to the efforts at development which now look encouraging for application in mineral sedimentological studies. The background to this dating mechanism can be briefly outlined, though a detailed description of the technique would require an in depth discussion which is outside the range of our present purpose (for a detailed account see Aitken, 1985). The emission of light as a function of stored energy, usually heat, is common to many materials, emitted light being measured on the 'glow curve'. Contributions to the glow curve may come from both 'blackbody' radiation and irradiation from radioisotopes contained within the surrounding sediment matrix. Based on the glow curve characteristics a geochronological framework may be established for mineral sediments. A simple expression of this is as follows :

$$X = f(T,P,D) \qquad (A1, A2, A3)$$

where X is the measured glow curve corrected for blackbody radiation, T is the length of time of exposure to radiation, P is the parent material i.e. the capability of the material to act as a dosimeter, D is the average dosage of radiation (rads) per unit time, and assumptions A1 : the glow curve at the time of deposition is that of blackbody radiation only, A2 : homogeneity of the sample characteristics and A3 : no post depositional alteration of the acquired thermoluminescence has occurred (Aitken, 1985; Goudie, 1981).

While it is necessary to emphasise that thermoluminescence as a dating tool is as yet at a trial stage of development, the potential it may offer in providing additional observational data which cannot at this point be fully incorporated in interpretations of sedimentary geology is extremely significant. A mechanism whereby the stratigraphic variability of minerogenic deposits could be ascribed a temporal framework, thus clarifying sedimentation rates and ratios at periods of high energy activity at the coastal interface, such as in times of storms and surges, would considerably ease the interpretation of stratigraphic phenomena within, for example, the contexts of extreme events versus gradual coastal change.

CONCLUSION

In the investigation of the nature of climate and global change generally through the record of sea surface variability as stored in sediments and the possibilities for the interpretation of long term and short term environmental variation, an additional data source such as that offered by the development and application of techniques such as that described here provides a basis on which the erection of more theoretically sound parameters and assumptions for hypothetical models can be established.

References

Aitken, M.J.   (1985) Thermoluminescence Dating.
                                               Academic Press.
Barker, H.D. and Davies, J.A. (1989) 'Surface albedo estimates
                                               from Nimbus-7 ERB data and a
                                               two-stream approximation of
                                               the radiative transfer
                                               equation'.
                                               J. Cli. 2(5) 409-418
Devoy, R.J. (ed)   (1987) Sea Surface Studies.
Kauffman, E.G and Hazel,J.E.   (1977) Concepts and Methods of
                                               Biostratigraphy.
                                               Dowdon, Hutchinson and Ross,
                                               Pennsylvania.
Goode, A.J.and Taylor, R.T.   (1988) Geology of the country
                                               around Penzance.
                                               Memoir for 1:50 000
                                               geological sheets 351 and
                                               358 (England and Wales)
                                               British Geological Survey.
Simola, H., Coard, M. A. and
O' Sullivan, P.E.   (1981) 'Annual Laminations in the
                                               Sediments of Loe Pool,
                                               Cornwall'. Nature Vol 290
                                               p238.
Owens, R.   (1981) 'Holocene Sedimentation in
                                               the North-western North Sea'
                                               in: Holocene Marine
                                               Sedimentation in the North
                                               Sea Basin.

		Spec. Publs. Int. Ass. Sediment. Geol. Blackwell Scientific Press.
Peltier, W.R.	(1988)	Lithospheric thickness, Antatric deglaciation history, and ocean basin discretization effects in a global model of Postglacial sea level change: A summary of some sources of nonuniqueness. Quat. Res. 29 93-112.
Pirazzoli, P.A., Grant,D.R. and Woodworth, P.	(1987)	Trends of Relative Sea-level Change: Past, Present and Future XII Int. INQUA Congress Spec. Sess. 18. Global Change. (Reprint).
Shennan, I.	(1986)	Flandrian Sea-level Changes in the Fenland II: Tendancies of Sea-level Movement, Altitudinal Changes and Local and Regional Factors. J. of Quat. Sci. 1 155-179
Shennan, I. and Tooley, M.J.	(1987)	Conspectus of Fundamental and strategic research on sea-level changes. in:Tooley and Shennan (1985)
Strief, H.	(1978)	A New Method for the Representation of Sedimentary Sequences in Coastal Regions. Proc. of the 16th Coastal Engineering Conference. ASCE/Hamburg, W.Germany 1978 1245-1256
Smith, D.E., Cullingford, R.A. and Haggart, B.A.	(1985)	'A major coastal flood during the Holocene in Eastern Scotland' Eiszeitalter und Gegenwart 35 109-118.
Tooley, M.J.	(1987)	Sea-level Studies in: Tooley and Shennan
	(1987)	Sea-level Changes 1-24
Tooley, M.J. and Switzur, R.	(1988)	'Water level changes and sedimentation during the Flandrian age in the Romney Marsh area'. Romney Marsh: Evolution, Occupation, Reclamation. Eddison, J. and Green, C. Ox Uni Comm for Arch, Monograph 24. 53-71

Troels-Smith, J. (1955) Characterisation of unconsolidated sediments. Geol. Sur. of Denmark, Ser IV 3

Venstraete, M.M. and Dickinson R.E. (1986) 'Modeling surface processes in atmospheric general circulation models'. Geophysicae 4 (4) 357-364

Wilson, A.C. and Taylor, A.J. (1976) Stratigraphy and Sedimentation in West Cornwall. Trans. Roy. Soc. of Cornwall. Vol XX (4) 247-259

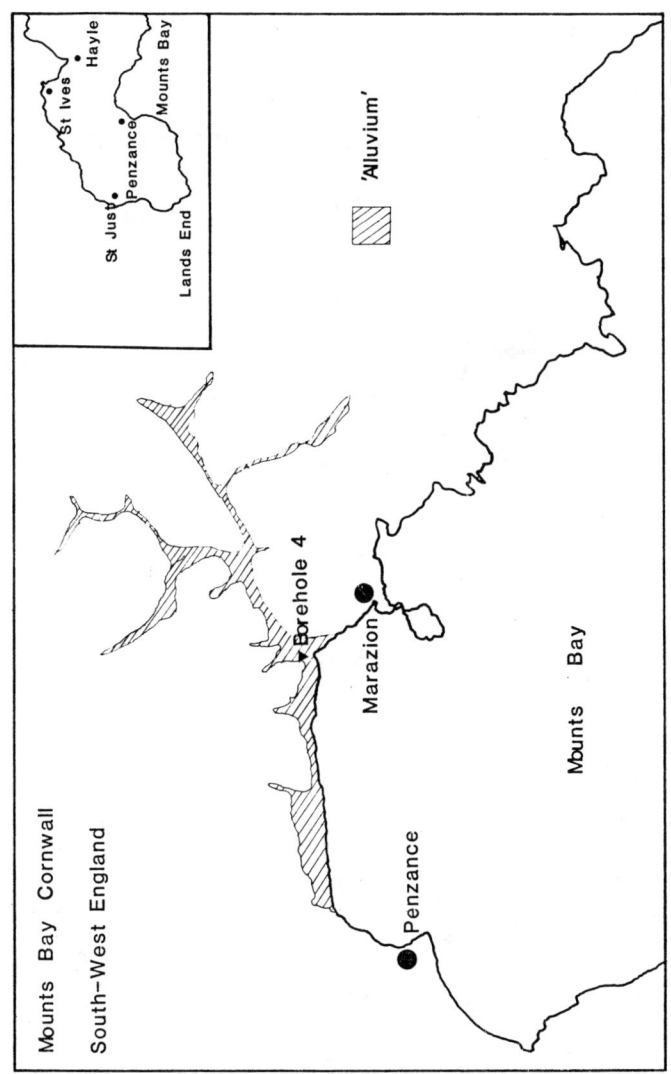

Figure 1

Figure 2 : MOUNTS BOREHOLE 2

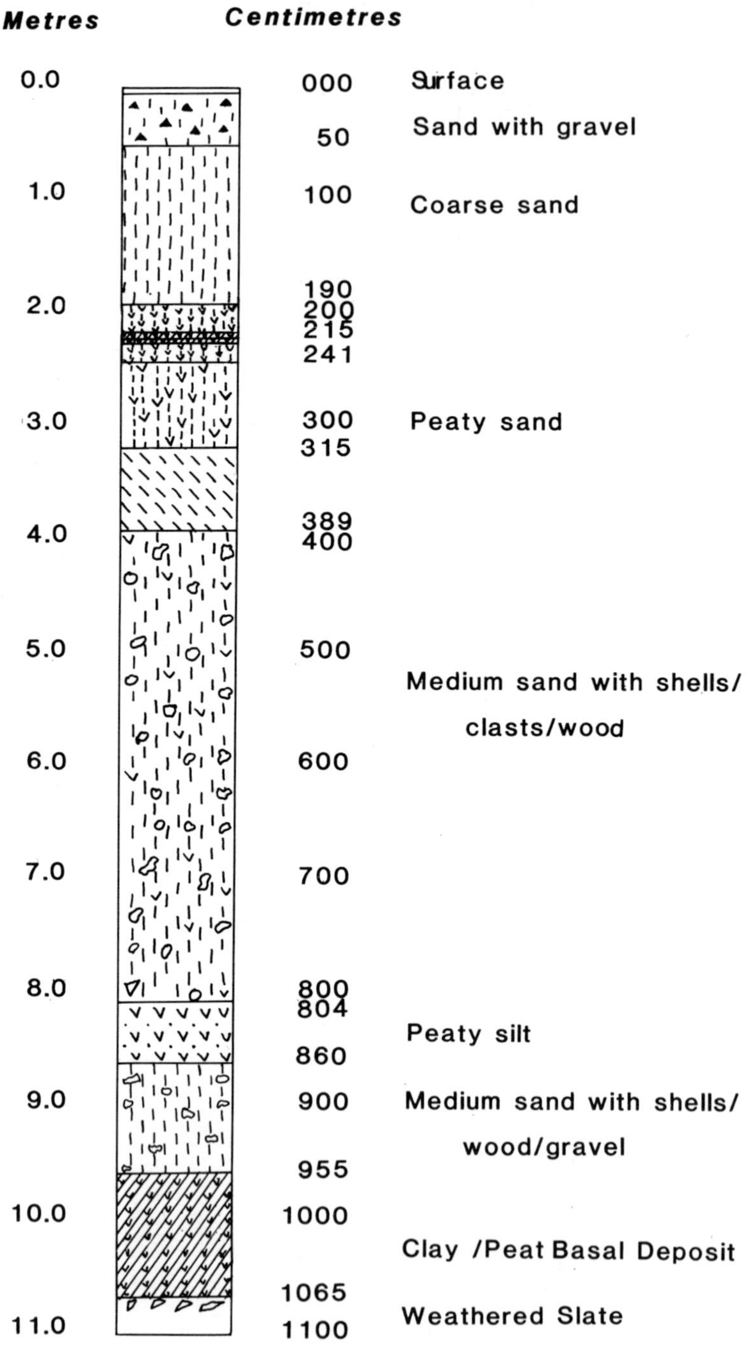

Metres		Centimetres	
0.0		000	Surface
		50	Sand with gravel
1.0		100	Coarse sand
2.0		190 200 215 241	
3.0		300 315	Peaty sand
4.0		389 400	
5.0		500	Medium sand with shells/ clasts/wood
6.0		600	
7.0		700	
8.0		800 804	
		860	Peaty silt
9.0		900	Medium sand with shells/ wood/gravel
		955	
10.0		1000	Clay /Peat Basal Deposit
		1065	
11.0		1100	Weathered Slate

VARIABILITY OF THE OCEAN CARBON CYCLE –
A MODELLING APPROACH TOWARDS AN EXPLANATION
OF THE GLACIAL pCO$_2$ REDUCTION IN THE ATMOSPHERE

C. Heinze

Max-Planck-Institut für Meteorologie
Bundesstraße 55
D-2000 Hamburg 13
Federal Republic of Germany
(Phone: (D)–(0)40/41173-319)

Abstract

Hypotheses for an explanation of the glacial pCO$_2$ reduction in the atmosphere were tested with a high resolution carbon cycle model and a simple linear response model. A decrease of ocean ventilation is suggested as the primary cause for the pCO$_2$ decrease.

Results of sensitivity studies with the Hamburg Carbon Cycle Model are presented focusing on changes in biogeochemical and physical parameters (nutrient cycling, composition of biogenic matter, solubility of CO$_2$, ocean ventilation, advective pattern of ocean circulation) on the background of the 80 ppm (ppm = parts per million of volume) reduction of atmospheric CO$_2$ partial pressure during the last glaciation. Evidence for this reduction of atmospheric pCO$_2$ from the interglacial (and pre-industrial) value of 280 ppm to the glacial level of 200 ppm comes from the analysis of ice core samples from Greenland and Antarctica (e.g. Vostok ice core, east Antarctica, Barnola et al., 1987).

The carbon cycle model is based on the velocity and thermohaline fields of the Hamburg Large Scale Geostrophic Ocean General Circulation Model (Maier-Reimer et al., 1990). It was used here in an annually averaged 11 layer and 3.5°×3.5° (horizontal resolution) version covering the whole globe (Maier-Reimer and Hasselmann, 1987; Bacastow and Maier-Reimer, 1990). The model simulates the inorganic carbon cycle and part of the organic carbon cycle of the ocean. In addition to the model described in Bacastow and Maier-Reimer (1990) CaCO$_3$ chemistry and interaction between ocean water and sediment pools for organic carbon and CaCO$_3$ are included (cf. Maier-Reimer and Bacastow, 1990).

As an explanation of the glacial CO$_2$ reduction several hypotheses have been suggested by different authors (see Berger and Keir, 1984, and Broecker and Peng,

1986, for a more comprehensive discussion). These hypotheses were tested with the carbon cycle model. A control run or reference run was defined with tracer distributions close to the pre-industrial situation (control run value for atmospheric CO_2 partial pressure: 278.5 ppm). Relative to this control run sensitivity experiments with the carbon cycle model were carried out. In each experiment one parameter of the carbon system was changed relative to the control run and the model was integrated again until the tracer distributions became stationary (20 CRAY-2s cpu hours per run). The following sensitivity runs were performed:

1. An increase of biological production at high latitudes (change in the parameterization for the growth conditions of biota).

2. A 30 % reduction of the Redfield ratio P:C (30 % increase of the nutrient utilization efficiency of marine biota).

3. A 30 % increase of the nutrient (PO_4, phosphate) inventory of the ocean.

4. An additional input of 1231 GtC POC (which corresponds to the same number of P-atoms as added in the PO_4 inventory experiment 3).

5. An additional input of 2462 GtC $CaCO_3$ (which corresponds to twice the amount of additional C input into the ocean as in the POC inventory experiment 4).

6. A 50 % decrease of the rain ratio $C_{CaCO3}:C_{organic}$ in newly formed biogenic particulate matter.

7. A 2 K reduction of sea surface temperature (i.e. a change of all temperature dependent chemical constants including the solubility coefficients for gaseous CO_2).

8. A 50 % reduction of ocean ventilation (50 % reduction of current velocities).

9. A change of the advective circulation pattern by use of a preliminary velocity field for the ocean with reconstructed ice age boundary conditions (Lautenschlager et al., 1989).

The resulting tracer distributions for atmospheric pCO_2, $\delta^{13}C$ and $CaCO_3$ saturation (calcite lysocline depth level) were compared to mean values for the last glaciation as derived from ice and sediment core analysis (for details see Heinze and Maier-Reimer, 1990). None of the experiments could match all of the observed tracer changes simultaneously (fig. 1).

Therefore combinations of parameter changes were tested. The model sensitivities provide a simple linear model for the paleo-climate tracer changes as a function of the carbon cycle parameter changes (relative to the control run or "pre-industrial" situation). The sensitivity for a change of the POC inventory here was replaced by

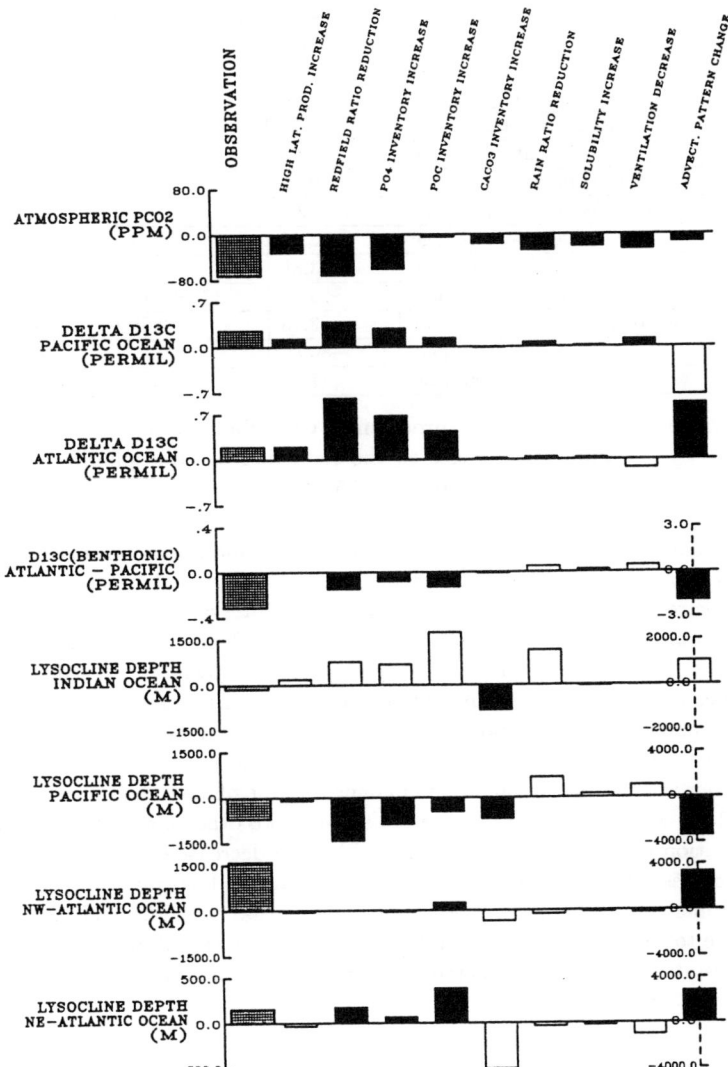

Figure 1: Results of the sensitivity studies. The light shaded bars (left column) denote the observed average tracer change for the 18–65 kyr BP interval relative to pre-industrial times. The dark shaded areas represent the modelled changes in tracer distributions resulting for the different sensitivity experiments that are in qualitative agreement with the observations.

Nr.	Scenario	all 9 parameters	"composition change"	"inventory change"
1	High latitude productivity change	−14.41	−	−
2	Redfield ratio change	2.76	−23.07	−
3	PO_4 inventory change	−4.05	−	−7.93
4	$\sum CO_2$ inventory change	−19.91	−	−19.63
5	$CaCO_3$ inventory change	−2.63	−	−5.68
6	Rain ratio change	3.69	14.61	−
7	Solubility change	−13.88	−	−
8	Ventilation change	−29.29	−59.13	−43.49
9	Change of the advective pattern	−3.13	−2.88	−3.53
	Total (model)	−80.86	−70.47	−80.27
	Observation	−72.5	−72.5	−72.5

Table 1: Result of fitting procedure. Contributions of the different parameter changes to the reduction of atmospheric pCO_2 (ppm) (averaged over 18–65 kyr BP).

that for a change of the $\sum CO_2$ inventory to separate additional inputs of carbon and phosphorus from one another. The simple linear response model was fitted to ice and sediment core data at subsequent time steps for the interval 0–120 kyr BP. Several parameters could not be resolved properly due to linear dependencies between different parameters. Nevertheless, the result suggests that the glacial pCO_2 reduction is primarily due to a decrease of ocean ventilation. Additionally the model result proposes a parallel increase of $CaCO_3$ production relative to organic matter production during glacial times that tends to a relative increase of atmospheric pCO_2. The resulting contributions of the different parameter changes to the pCO_2 reduction is summarized in tab. 1 for a fit of all nine parameters and for two other parameter combinations ("composition change", "inventory change"). The decrease of the $\sum CO_2$ inventory and the corresponding 20 ppm decrease in pCO_2 cannot be explained readily. It may be an artifact of the fitting procedure or indicate that an additional parameter did contribute to the CO_2 reduction that had not been accounted for in this study. The overall results should be considered with care due to the limitations of the models and observations.

REFERENCES:

BACASTOW, R. B. und E. MAIER-REIMER (1990) Circulation model of the oceanic carbon cycle. Climate Dynamics, 4, 95-125.

BARNOLA, J. M., D. RAYNAUD, Y. S. KOROTKEVICH, and C. LORIUS (1987) Vostok ice core provides 160,000-year record of atmospheric CO_2. Nature, 329, 408-414.

BERGER, W. H., and R. S. KEIR (1984) Glacial-Holocene changes in atmospheric CO_2 and the deep-sea record. In: *Climate processes and climate sensitivity.* J. E. Hansen and T. Takahashi, editors, AGU Geophysical Monographs, 29, pp. 337-351.

BROECKER, W. S., und T.-H. PENG (1986) Carbon cycle: 1985 - Glacial to interglacial changes in the operation of the global carbon cycle. Radiocarbon, 28, 309-327.

HEINZE, C., and E. MAIER-REIMER (1990) Glacial pCO_2 reduction by the World Ocean – Experiments with the Hamburg Carbon Cycle Model. Draft.

LAUTENSCHLAGER, M., U. MIKOLAJEWICZ, E. MAIER-REIMER, and K. HERT-ERICH (1989) Paleoclimate modelling. In: Research activities in atmospheric and oceanic modelling. G. J. Boer, ed., WMO/ICSU, CAS/JSC Working Group on Numerical Experimentation. Report No. 13, WMO/TD - No. 332, pp. 7.56-7.58.

MAIER-REIMER, E., and R. BACASTOW (1990) Modelling of geochemical tracers in the ocean. To be published in: Climate-Ocean Interaction, proceedings of NATO ARW at Oxford, October 1988, M. E. Schlesinger, ed..

MAIER-REIMER, E., and K. HASSELMANN (1987) Transport and storage of CO_2 in the ocean - an inorganic ocean-circulation carbon cycle model. Climate Dynamics, 2, 63-90.

MAIER-REIMER, E., K. HASSELMANN and U. MIKOLAJEWICZ (1990) On the sensitivity of the global ocean circulation to surface forcing. Draft.

A COMPARISON AMONG THE DIFFERENT TYPES OF CONFORMAL MAP PROJECTIONS

O. K. KAKALIAGOU
Department of Meteorology and Climatology
University of Thessaloniki, Greece

Summary

The different types of conformal map projections, which are mostly used in Meteorology, are presented and compared against each other. The main criterium used is the calculation of the scale factor as a function of geographical latitude. Since the studying area of interest lies at middle latitudes, the Lambert II conical projection is shown to be the more preferable than the other types of projections and particularly to the stereographic one, which is widely used. Hence, the Lambert II conical is adopted and incorporated to an appropriate objective analysis scheme.

1. INTRODUCTION.

It is known that the quasi-spherical or ellipsoidal surface of the earth can not be represented correctly in a plane. Therefore, there is not a single projection satisfying all purposes for which maps are usually required (Saucier, 1955). For the study of atmospheric motion and the associated phenomena, a true representation of the wind vector and the other dynamical and thermodynamical parameters would be very advantageous. Hence, the conformal map projection, according to which all angles are preserved (except possibly at the poles), is used exclusively.

The objective of this study is to compare the characteristics of the different types of the conformal map projections. Based upon the criteria used, in other words, the geographical location of the study area of interest (30° to 50°N and 10° to 30°E), the associated distortion, and the accurate calculations of dynamical parameters, the most adequate conformal map projection will be adopted and incorporated to an appropriate objective analysis scheme.

2. TYPES OF CONFORMAL MAP PROJECTIONS.

The different types of conformal maps which are mostly used in Meteorology are, the Mercator, the Stereographic and the Lambert conical projections. The mathematical expressions describing the coordinates and the scale factor for each one of the above stated types are presented on Table 1.

The Mercator projection is used for studies applied to equatorial regions. According to this, the spherical surface of the earth is projected on a cylinder, whose axis coincides with the axis of the earth (Godske et al., 1957). The circles of the geographical latitude are represented by horizontal lines, while the meridians are equidistant vertical lines.

The Stereographic projection is most suitable for studies applied to polar regions. It should be pointed out that this type can provide a continuous chart for the whole hemisphere, which is considered to be quite ad-

Table 1. Types of conformal map projections.

M E R C A T O R	S T E R E O G R A P H I C
$x = a\lambda\cos\varphi_0$ $y = a\cos\varphi_0 \log\tan(\frac{\pi}{4} + \frac{\varphi}{2})$ $\sigma = \dfrac{\cos\varphi_0}{\cos\varphi}$	$r = a(1+\cos\psi_0)\tan\frac{\psi}{2}$ $\vartheta = \lambda \quad , \quad \psi = \frac{\pi}{2} - \varphi$ $\sigma = \dfrac{1+\cos\psi_0}{1+\cos\psi}$
L A M B E R T I	L A M B E R T II
$r = a\tan\psi_0 \left[\dfrac{\tan\frac{\psi}{2}}{\tan\frac{\psi}{2}}\right]^n$ $\vartheta = n\lambda, \quad \psi = \frac{\pi}{2} - \varphi$ $n = \cos\psi_0$ $\sigma = \dfrac{\sin\psi_0}{\sin\psi} \cdot \left[\dfrac{\tan\frac{\psi}{2}}{\tan\frac{\psi_0}{2}}\right]^n$	$r = \dfrac{a}{n}\sin\psi_1 \left[\dfrac{\tan\frac{\psi}{2}}{\tan\frac{\psi_1}{2}}\right]^n$ $\vartheta = n\lambda \quad , \quad \psi = \frac{\pi}{2} - \varphi$ $n = \dfrac{\log\sin\psi_1 - \log\sin\psi_2}{\log\tan\frac{\psi_1}{2} - \log\tan\frac{\psi_2}{2}}$ $\sigma = \dfrac{\sin\psi_1}{\sin\psi} \cdot \left[\dfrac{\tan\frac{\psi}{2}}{\tan\frac{\psi_1}{2}}\right]^n$

vantageous. In this case, the spherical surface of the earth is projected on a plane being either tangent to the globe or intersecting at specified geographical latitude. However, the most commonly used Stereographic projection is the one at the 60°.

The Lambert conical projections are the most appropriate for studies applied to temperate regions. In this particular case, the spherical surface of the earth is projected onto a cone, whose axis coincides with the axis of the earth. When the cone is flattened out in a plane, the circles of longitude become radial straight lines and the parallels concentric circles (Godske et al., 1957). If this plane become tangential to the earth's surface, the one standard parallel conformal map projection is defined, hereafter denoted as Lambert I type. In the case of intersecting the earth's surface, the two constant parallels, or Lambert II, conformal map projection is defined. The most commonly used Lambert conical projection is the one with two constant parallels, at 30° and 60° (Saucier,1955 and Haltiner and Martin, 1957). It should be stated that the Mercator and the Stereographic projections are considered to be special cases of the Lambert conical projections (Saucier, 1955 and Haltiner and Williams,1980).

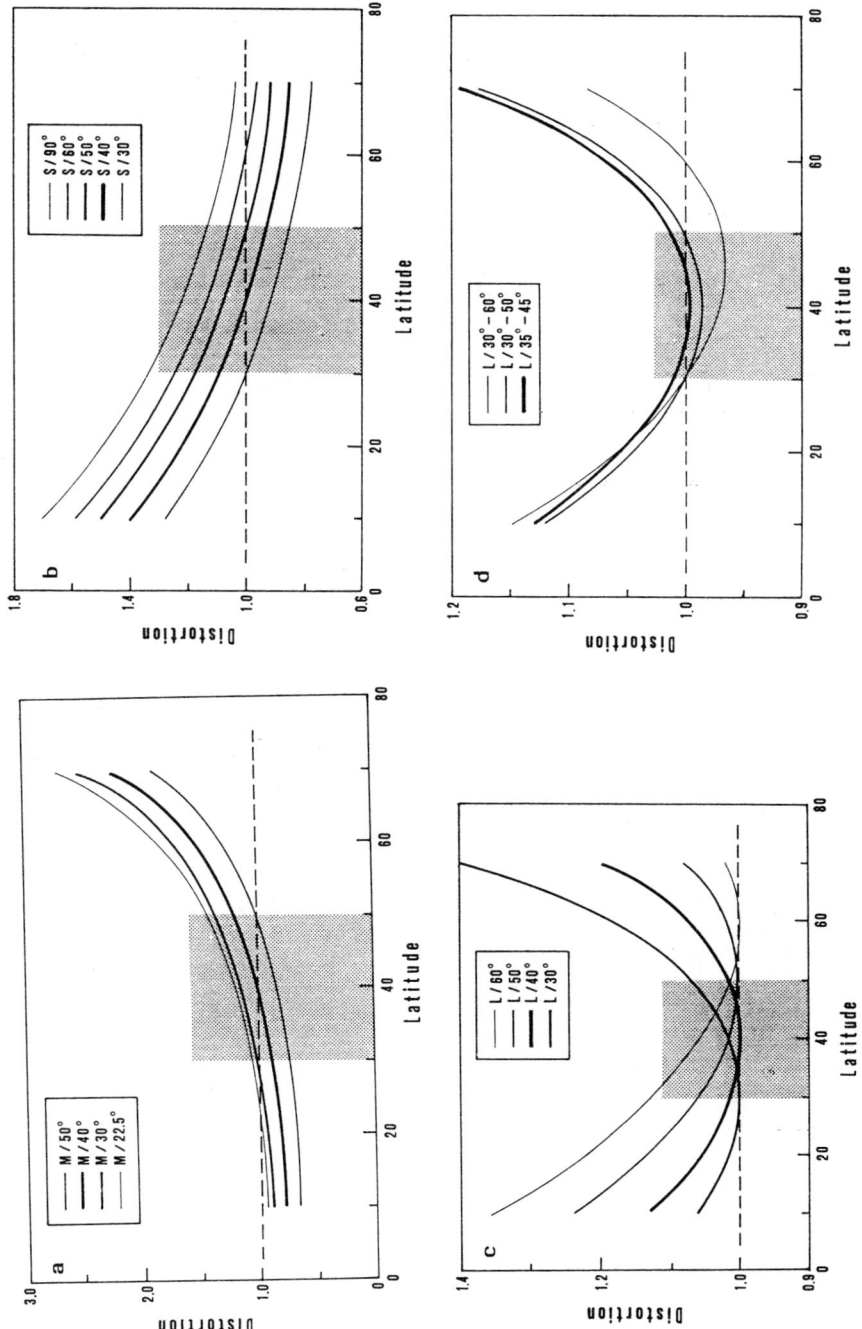

Fig. 1. The distortion as a function of the geographical latitude for the Mercator, Stereo-graphic, Lambert I and II conformal map projections.

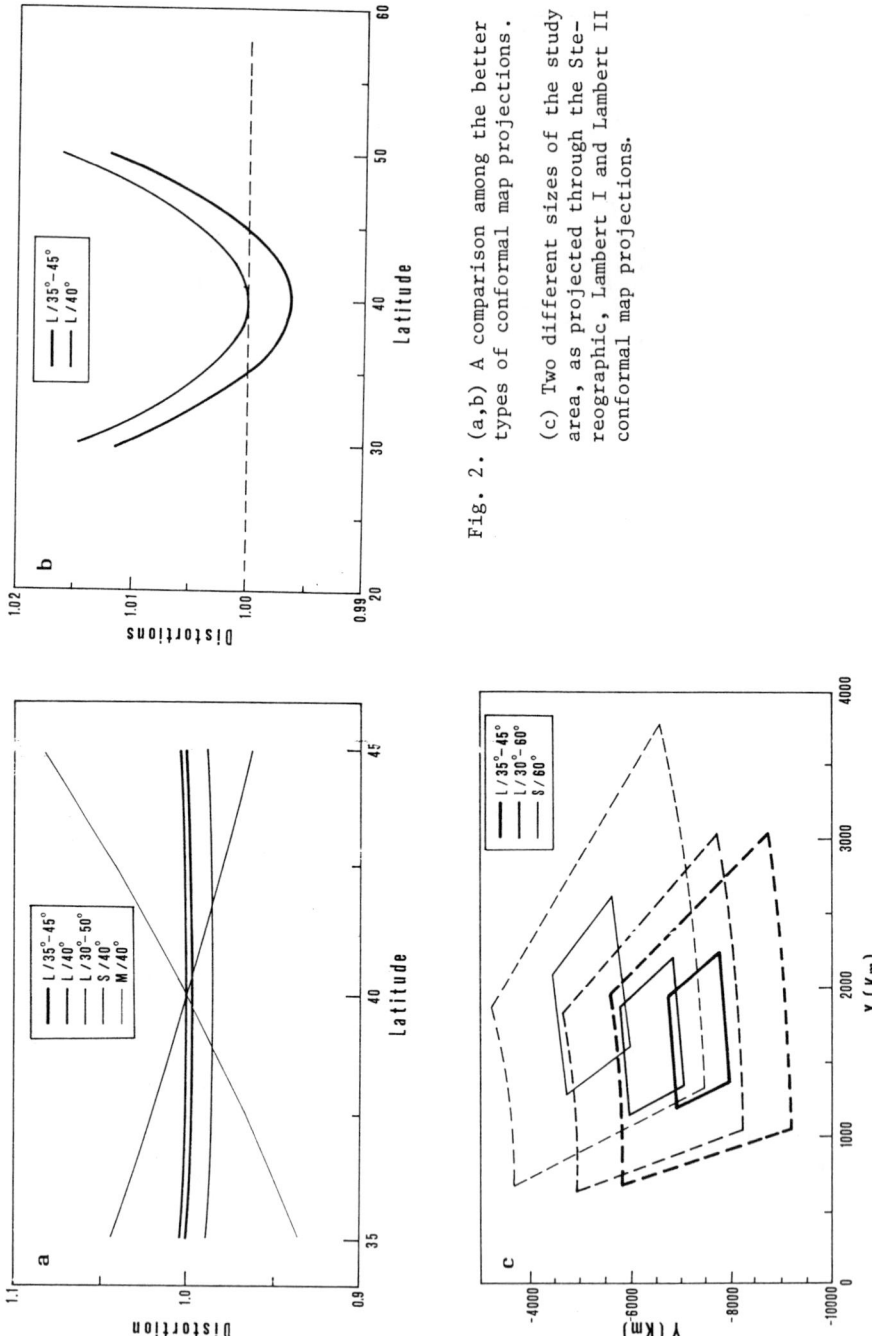

Fig. 2. (a,b) A comparison among the better types of conformal map projections.

(c) Two different sizes of the study area, as projected through the Stereographic, Lambert I and Lambert II conformal map projections.

As it is obvious, all the conformal map projection types have an in-
herited distortion, which is mathematically expressed on Table 1. In spite
of that, since the resulted conformal map projections will exclusively be
used in dynamic and synoptic Meteorological research, the minimum values
of the associated distortion, as a function of the geographical latitude
for the area of interest, has to be one of the major criteria for the correct choice of the most adequate type of conformal map projections.

3. COMPARISON OF THE TYPES OF CONFORMAL MAP PROJECTIONS.

The distortion, which is defined as the ratio of arc length of the
image surface to arc length of the earth, has been calculated for each one
of the above stated four types of conformal map projections. The formulae
used are those presented on Table 1 (Maling, 1973 and Saucier, 1955).

Figure 1 (a,b,c,d) depicts the distortion, as a function of the geographical latitude, for each one of the four types of conformal map projections (Mercator, Stereographic, Lambert I, and Lambert II). The shaded
areas correspond to the geographical latitude of the center of the primary
area of study. The dash lines depict the "true representation", in other
words, they are the lines where there is no distortion and hence the representations are perfect. For investigation purpose, all the types of projections are defined at several latitudes. The most appropriate and commonly used ones are, Mercator at 22.5^o, Stereographic at 60°, and Lambert II
at 30°-60°. Evaluating the calculated distortion for each one of the four
types for the area of interest (30°N to 50°N), the better representations
are encountered: for Mercator, Stereographic and Lambert I at 40°, and for
Lambert II at 35°-45°. The four better types of conformal map projections,
together with the Lambert II at 30°-50°, are shown in Fig. 2a, where a
close up look of the better five types of conformal map projections is provided. From Fig. 2a becomes evident that the Lambert I at 40° and the Lambert II at 35°-45° are the two better types of projections.

4. CONCLUSIONS.

A direct comparison of the two better types of conformal map projections is depicted in Fig. 2b. Taking into consideration the study area of
interest, which is centered around 40°N, the most adequate type of conformal map projection is concluded to be the Lambert II conical conformal map
projection defined at 35°-45°. This drawn conclusion is supported quite
well by the depiction indicated in Fig. 2c, where two different sizes of
the study area are projected, through the Stereographic at 60°, Lambert II
at 30°-60°, and Lambert II at 35°-45°. It becomes evident that the distortion encountered for Lambert II at 35°-45° is less, and hence, the Lambert
II conical conformal map projection, defined at 35°-45°, is the best type
of projection, which should be adopted and incorporated to an appropriate
objective analysis scheme.

REFERENCES

(1) GODSKE, C.L., BERGERON, T., BJERKNES, J. and BUNDGAARD, R.C. (1957) :
Dynamic Meteorology and Weather Forecasting, American Meteorological
Society. Boston, Massachusetts, 800 pp.

(2) HALTINER, G.J. and MARTIN, F.L. (1957):Dynamical and Physical Meteorology, Mc Graw-Hill Book Company, 470 pp.

(3) HALTINER, G.J., and R. T. WILLIAMS (1980): Numerical Prediction and
Dynamic Meteorology, John Wiley and Sons, 477 pp.

(4) MALING, D.H. (1973): Coordinate Systems and Map Projections, George Philip and Son LTD, 255 pp.

(5) SAUCIER, W.J. (1955): Principles of Meteorological Analysis, Univ. of Chicago Press, 438 pp.

LATE QUATERNARY PALAEOCEANOGRAPHY OF THE PERU MARGIN

S.C. KING
Department of Geology
Southampton University
Hampshire, U.K.

Summary

Late Quaternary stratigraphic assemblages of benthic and
planktonic foraminifer were recovered at O.D.P. site 680. Using
the ecology of the existant species a detailed interpretation
of late Quaternary environments is being built up. There is
seen a marked difference between glacial and interglacial
stages, with a decrease in upwelling and a more pronounced
oceanic influence.

Introduction

Site 680 is located on the Peruvian continental margin at 11°S , 78°W.
Its shallow depth (263 m) makes it sensitive to sealevel changes. It is
at the edge of present day upwelling cell, the effects of which are
reflected in the muddy organic carbon rich core. The core contains thin
laminations recording past El Niño events. From parallel isotopic work
the last interglacial, stage 5, has been provisionally identified.

Methods

Counts of at least 200 benthic individuals were made at 10 cm intervals
in the core. The planktonic to benthic ratio is based on a count of at
least 200 individuals.

Environmental Interpretation

Bolivinellina humilis is a benthic foraminifera indicative of low
oxygen environments (1). During stage 5 it is seen to decrease in
abundance in the total population by up to 40 %, while Nonionellas, a
genus associated with better oxygenated bottom conditions rapidly
increases.

This is coupled with a very rapid increase in the planktonic to benthic
ration too large simply to be due to sea level rise.

The central Pacific gyre is known to weaken in interglacial periods,
reflected by a decrease in the coastal upwelling. The northward
travelling Humbolt current appears to slacken in the interglacial
stage. This allows the increase of equatorial oceanic influences, shown
by the increase of tropical planktonic foraminiferia, e.g.
Neogloboquadrina dutertrei.

References

(1) Resig, J.M., Benthic foraminiferal stratigraphy and
paloenvironments off Peru, Ocean Drilling Program leg 112, in SUESS,
E., Von Huene et al 1989. Proc. ODP Sci. results, 112: College station,
TX (Ocean Drilling Program).

CO2 AND WATER VAPOUR EXCHANGES OF A MEDITERRANEAN MACCHIA CANOPY

R. Valentini
DI.S.A.F.RI. University of Tuscia Via S. Camillo de
Lellis 01100 Viterbo Italy

Summary
The role of terrestrial vegetation on the global
carbon cycle is becoming an important subject and a
growing need is felt for more precise and detailed
measurements of the response of terrestrial
ecosystems to the changing environment. Many
ecophysiological models developed in order to
assess the impact of climatic changes on vegetation
are based on physiological processes occurring at
cellular or leaf level. Scaling up these processes
from leaf or individual plant to the canopy level
is a complicate and rather difficult matter,
requiring accurate and reliable methods of results
testing. In this paper an application of the eddy
correlation technique to the determination of water
vapour and CO2 gaseous exchanges of a "macchia"
canopy is presented. Experimental data of the three
days of measurements are discussed in terms of
energy balance, carbon fluxes and carbon-water flux
ratio (CWFR).
Some considerations are derived from experimental
data about the possibility of using the eddy
correlation technique as a tool for testing
ecophysiological models.

1. Introduction

Fluxes of mass and energy from and to terrestrial
ecosystems play an important role in many biogeochemical
cycles and in recent years the gaseous exchanges at the
biosphere/atmosphere interface have gained a renewed and
increasing interest . Global climatic changes are to be
expected in the next century (Schneider 1989) and a
growing need is felt for more precise and detailed
measurements of the response of terrestrial ecosystem to
the changing environment.
Integrated measurements of CO_2 and water vapour exchanges
over vegetation are needed to explore the ecological
significance of scaling up from leaves to canopies and to
provide basic data to test ecophysiological models.
Although in the past two decades a certain number of
micrometeorological investigations of CO_2 and water
vapour exchanges have been carried out over forests, only
few studies have been focused on carbon flux
measurements. These studies have been carried out mainly

by flux-gradient methods (Jarvis et al. 1976) and for this reason they are inherently limited as it has been questioned in recent papers (Deanmead and Bradley 1987). Only recently, in consequence of the development of fast response analyzers (Othaki and Matsui 1982), the eddy correlation technique was employed to estimate CO2 fluxes over forests (Verma et al. 1986).
To our knowledge no measurements of CO_2 exchanges for an evergreen "macchia" Mediterranean ecosystem are reported in literature.
The aim of the present work is to provide an experimental test of the eddy correlation technique over a macchia stand in order to evaluate its performances for water vapour and carbon dioxide fluxes determination and its potential use for longer term studies on the ecophysiological significance of this particular Mediterranean formation.
Actually the "macchia" ecosystem represent a rather wide diffused natural vegetation community of Southern Europe and in Italy it represents almost the 20% of the forested area; because of its diffusion along the coastal line, it has an important ecological, recreational and protective function.

2. Materials and methods

The study was conducted at the ENEL (National Agency of Electric Energy) reservation near Montalto di Castro (Lat. 42°22' N, Long 11°32' E). The site is situated on the Tirrenian coast line on a flat plain limited at the shoreline by a sand dune. The vegetation cover is constituted by a Mediterranean "macchia" ecosystem. Principal woody species are Quercus ilex L., Juniperus oxycedrus var. macrocarpa S., Phyllirea latifolia L., Myrtus communis L. and Pistacia lentiscus L. The canopy height is rather uniform and on average is about 4 mt. Flux measurements were carried out with the eddy correlation instruments mounted at an height of 7 meters above the ground. The instruments included a three components sonic anemometer (Dobbie Instrument) equipped with a platinum fine wire temperature sensor , positioned about 10 cm away from the centre of the sonic head, and an infrared open path CO_2/H_2O analyzer (Othaki and Matsui 1982) (Analytical Corp.). The relative distance between the sonic anemometer and the open path head was about 30 cm.
Data acquisition was carried out by means of the anemometer processing unit equipped with an anti-aliasing filter centred at 5 Hz.
The fluxes (F) were obtained by the equation :

$$F = \overline{r'w'}$$

where r' and w' are the istantaneous fluctuations around the mean value for gas density and vertical wind speed respectively, and the overbar denotes the time average.

Fluxes were corrected to take into account the variations of density due to the simultaneous presence of a flux of heat and water vapour (Webb et al. 1980).
Soil heat flux (G) was estimated by means of the soil temperature flux-gradients relationship. Soil temperatures were measured at 5 and 30 cm depth in the soil and for the thermal conductivity a value of 0.0022 cal sec^{-1} cm^{-1} °C^{-1} was considered, which is a reasonable estimate for a sandy soil in the moisture conditions of the period of experiments.

3. Results and discussion

Diurnal patterns of the energy balances of the three study days are presented in fig.1. Fluxes directed toward the surface are considered positive and those directed upwards are negative.
Net radiation maxima range from 340 to 390 W/m2 for the three study days. These values are about one third lower than the summer ones, but however capable to drive evapotranspiration and the other energy exchanges.
Maxima usually occur between 12.00 and 13.00 hours while the two zero crossing points at about 8.30 and 16.00 hours. Sensible heat show a typical diurnal trends becoming positive usually at about 10.00 and 15.30, approximately one hour after and before the sign inversion of net radiation, respectively. Maximum values during the central hours of the day are 134, 178 and 135 W/m2 for 11th, 12th and 14th, while mean day-time values of H/Rn are 0.19, 0.38 and 0.26, respectively. On day 12th, starting from 13.00 until 15.00 the sensible heat flux is larger than latent heat.
Latent heat flux reach maximum values (233, 210 and 228 W/m2 on 11th,12th and 14th respectively) in coincidence with the maximum of net radiation.
The average values of the LE/Rn ratio for the three study days are 0.84, 0.59 and 0.79. Well watered agricultural crops have usually larger values of LE/Rn ratio, ranging from 0.8 to 1.8 (Kim et al. 1989.). Jarvis et al. (1976) reported for coniferous forests values similar to our data, while Rauner (1976) and Verma et al.(1986) on deciduous forests reported slightly higher values. On day 12th the reduction of the LE/Rn ratio and the increase of the H/Rn ratio is paralleled by an increase of Rn, which during this day reach its maximum values.
Carbon fluxes, expressed on a per ground area basis, for the three study days and the photosynthetic active photon flux density (PPFD) are presented in fig.2.
Carbon fluxes follow the PPFD trend and become positive (upward directed) when solar radiation approach zero.
Mean day-time values are -11.57 umoli/m^2 sec, -9.43 umoli/m^{2}sec and -7.91 umoli/m^{2}sec.
Maximum values of carbon flux, corresponding to about 400 W m^{-2}, on 11th, 12th and 14th are -22 umoli/m^{2}sec, -21.5 umoli/m^{2}sec and -14 umoli/m^{2}sec, respectively.

Jarvis et al (1976) reported carbon fluxes over coniferous forests ranging from -11 umoli/m^{2}sec to -29.5 umoli/m^{2}sec. Leuning and Aitwill (1978) found rates of CO2 exchanges over a mature eucalyptus forest, ranging from -2.3 umoli/m^{2}sec to -16 umoli/m^{2}sec, corresponding to a net radiation range from 300 W m^{-2} to 500 W m^{-2}. Measurements of net photosynthesis, carried out during the same days with a portable gas exchange system (ADC) show average values of -9 umoli/m^{2}sec, -7 umoli m^{-2}sec^{-1} and -6 umoli m^{-2}sec^{-1} for Pistacia lentiscus, Phyllirea angustifolia and Q. ilex, respectively. Considering a leaf area index for the macchia stand of about 2.5 m^2/m^2 (Ehleringer and Mooney 1983), we obtain for the three days mean day-time carbon fluxes, expressed on a per leaf area basis, of -4.6umoli m^{-2}sec^{-1}, -3.7 umoli m^{-2}sec^{-1} and -3.2 umoli m^{-2}sec^{-1}. The understimation of the eddy correlation values which amounts on average to 52% of measured net photosynthesis is due to the soil and biomass respiration, which is always present in the integrated flux over the canopy. The ratio of carbon flux and canopy net photosynthesis exceed 0.9 for a rubber plantation (Monteny et al. 1985), while Baldocchi et al. (1987) reported for a deciduous forest lower values between 0.5 and 0.8.
The respiration fluxes show a certain variability between the study days with maximum values reaching 7 umoli/m2sec, 3 umoli/m2sec and 4 umoli/m2sec. The ratio of the day-time mean respiration fluxes with the carbon ones range from 0.02 to 0.35 .
The carbon-water flux ratio (CWFR) of the stand during the three days of measurements is a function of the incident PPFD. For PPFD values greater than about 100 uE/m^{2}sec the CWFR mean value for all the study days is 9 mgCO2/gH2O. At PPFD values ranging from 0 to 400 the CWFR have a steep increase indicating a greater increase of the downward carbon flux than the water vapour one.

4. References

Baldocchi D.D., Verma S.B., Anderson D.E. 1987 - Canopy photosynthesis and water-use efficiency in a deciduous forest. J. Appl. Ecol. 24:251-260.

Deanmead O.T., Bradley E.F. 1987 - On scalar transport in Plant Canopies. Irr. Sci. 8:131-149.

Ehleringer J., Mooney H.A. 1983 - Productivity of Desert and Mediterranean climate plants in O.L. Lange, P.S. Nobel, C.B. Osmond and H. Ziegler, Encyclopedia of Plant Physiology Vol 12d pp. 205-226.

Jarvis P.G., Jones G.B., Landsberg J.J. 1976 - Coniferous forest, in J.L. Monteith (ed.) Vegetation and atmosphere, vol.2 pp.171-240. Academic Press, New York 439 pp.

Kim J., Verma S.B., Rosenberg N.J. 1989 - Energy balance and water use of cereal crops. Agric. For. Meteor. 48:135-147.

Leuning R., Aitwill P.M: 1978 - Mass Heat and Momentum exchange between a mature Eucalyptus forest and the atmosphere. Agric. Meteor. 19:215-241.

Monteney B.A., Barbier J.M., Bernos C.M. 1985 - Determination of the energy exchanges of a forest type culture : Hevea brasilianensis in Hutchinson B.A., Hicks B.B eds. The Forest-Atmosphere Interaction pp. 211-234. Reidel, Dordrecht.

Othaki E., Matsui T. 1982 - Infrared device for simultaneous measurement of fluctuations of atmospheric carbon dioxide and water vapour. Bound. Lay. Meteor. 24:109-119.

Rauner JU. L. 1976 - Decidous forest, in J.L. Monteith Vegetation and Atmosphere vol. 2 pp. 241-264. Academic Press New York 439 pp.

Schneider S.H. 1989 - The greenhouse effect : Science and Policy. Science 243:771-781.

Verma S.B., Baldocchi D.D., Anderson D.E., Matt D.R., Clement R.I. 1986. Eddy fluxes of CO_2, water vapour and sensible heat over a decidous forest. Bound. Lay. Meteor. 36:71-91.

Webb E.K., Pearman G.I., Leuning R. 1980 - Corrections of flux measurements for density effects due to heat and water vapour transfer. Q.J.R. Meteor. Soc. 106:85-100.

11 Nov 89

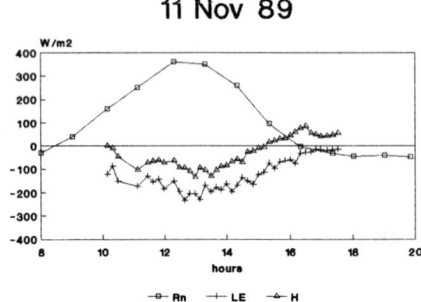

12 Nov 89

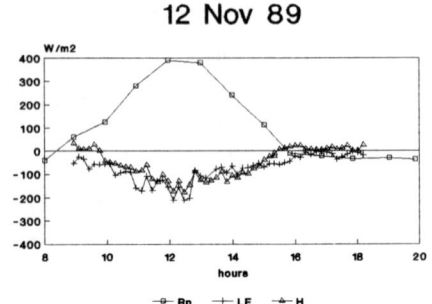

14 Nov 89

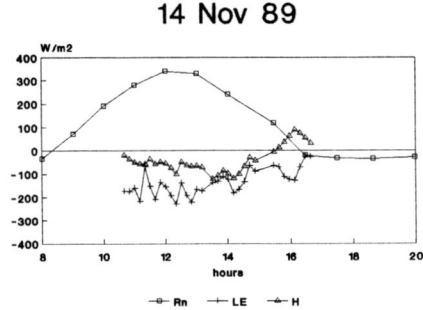

Fig.1 Energy balances of the three study days
Rn = net radiation LE = Latent Heat flux
C = Sensible Heat flux

11 NOV

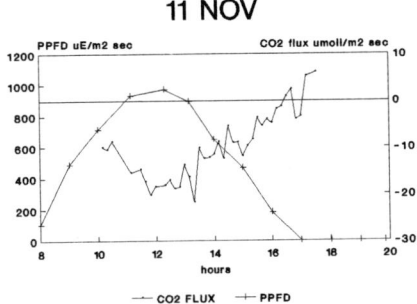

12 NOV

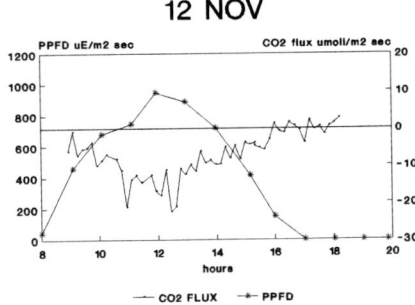

14 NOV

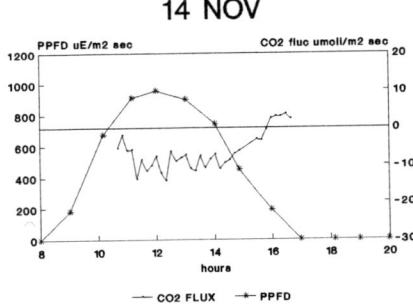

Fig.2 Carbon fluxes and incident radiation (PPFD) for the three study days

European Communities — Commission

EUR 13149 — Climate and global change
Proceedings of the European School of Clima-
tology and Natural Hazards course, held in
Arles/Rhône, France, from 4 to 12 April 1990

Edited by : *J. C. Duplessy, A. Pons, R. Fantechi*

Luxembourg: Office for Official Publications of the European
Communities

1991 — 357 pp. — fig., tab. — 16.2 x 22.9cm

Environment and quality of life series

ISBN 92-826-2779-9

Catalogue number: CD-NA-13149-EN-C

Price (excluding VAT) in Luxembourg: ECU 32.50

The European School of Climatology and Natural Hazards is a part of the training and education activities of Epoch (European programme on climatology and natural hazards).

It annually organizes courses open to graduating, graduate or postgraduate students in appropriate fields of climatology, natural hazards and closely related fields.

The courses are organized in cooperation with European institutions involved in the carrying out of the Community's R&D programmes on climatology and natural hazards, and are aimed at allowing students to attend formal lectures and to participate in informal discussions with leading research workers. The opportunity for demonstrations, case studies or presentation of posters is given to the students attending the courses.

The teachers are selected from among European scientists who are leading authorities in their respective fields.

The present volume contains the lessons delivered at the course held in Arles/Rhône, France, from 4 to 12 April 1990 on the subject 'Climate and global change', together with short presentations by the students of their own research activities and interests.